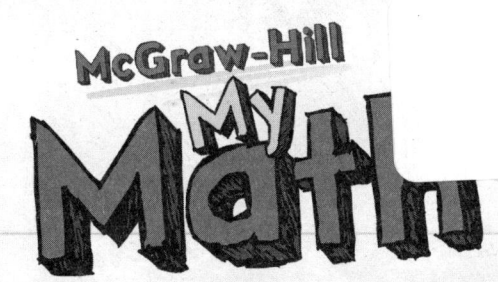

McGraw-Hill My Math

Interactive Guide
Teacher Edition

Grade 5

PROPERTY OF CKSD 401
9210 Silverdale Way NW
Silverdale, WA 98383

McGraw Hill Education

ConnectED.mcgraw-hill.com

Mc Graw Hill Education

Copyright © 2014 McGraw-Hill Education

All rights reserved. No part of this publication may be reproduced or distributed in any form or by any means, or stored in a database or retrieval system, without the prior written consent of McGraw-Hill Education, including, but not limited to, network storage or transmission, or broadcast for distance learning.

STEM McGraw-Hill is committed to providing instructional materials in Science, Technology, Engineering, and Mathematics (STEM) that give all students a solid foundation, one that prepares them for college and careers in the 21st century.

Send all inquiries to:
McGraw-Hill Education
8787 Orion Place
Columbus, OH 43240

Selections from:
ISBN: 978-0-02-130831-6 *(Grade 5 Student Edition)*
MHID: 0-02-130831-4 *(Grade 5 Student Edition)*
ISBN: 978-0-02-130237-6 *(Grade 5 Teacher Edition)*
MHID: 0-02-130237-5 *(Grade 5 Teacher Edition)*

Printed in the United States of America.

Visual Kinesthetic Vocabulary® is a registered trademark of Dinah-Might Adventures, LP.

1 2 3 4 5 6 7 8 9 ROV 20 19 18 17 16 15

Contents

How To Use This Book
Proficiency Level Descriptors ... vii
Strategies for EL Success ... viii
How to Use the Teacher Edition ... ix
How to Use the Student Edition ... xii
English/Spanish Cognates ... xvi
Template Masters ... xix

Chapter 1 Place Value
Mathematical Practice 1/Inquiry ... T1
Lesson 1 Place Value Through Millions ... T2
Lesson 2 Compare and Order Whole Numbers Through Millions ... T3
Lesson 3 Inquiry/Hands On: Model Fractions and Decimals ... T4
Lesson 4 Represent Decimals ... T5
Lesson 5 Inquiry/Hands On: Understand Place Value ... T6
Lesson 6 Place Value Through Thousandths ... T7
Lesson 7 Compare Decimals ... T8
Lesson 8 Order Whole Numbers and Decimals ... T9
Lesson 9 Problem-Solving Investigation: Use the Four Step Plan ... T10

Chapter 2 Multiply Whole Numbers
Mathematical Practice 3/Inquiry ... T11
Lesson 1 Prime Factorization ... T12
Lesson 2 Inquiry/Hands On: Prime Factorization Patterns ... T13
Lesson 3 Powers and Exponents ... T14
Lesson 4 Multiplication Patterns ... T15
Lesson 5 Problem-Solving Investigation: Make a Table ... T16
Lesson 6 Inquiry/Hands On: Use Partial Products and the Distributive Property ... T17
Lesson 7 The Distributive Property ... T18
Lesson 8 Estimate Products ... T19
Lesson 9 Multiply by One-Digit Numbers ... T20
Lesson 10 Multiply by Two-Digit Numbers ... T21

Chapter 3 Divide by a One-Digit Divisor
Mathematical Practice 8/Inquiry ... T22
Lesson 1 Relate Division to Multiplication ... T23
Lesson 2 Inquiry/Hands On: Division Models ... T24
Lesson 3 Two-Digit Dividends ... T25
Lesson 4 Division Patterns ... T26
Lesson 5 Estimate Quotients ... T27
Lesson 6 Inquiry/Hands On: Division Models with Greater Numbers ... T28
Lesson 7 Inquiry/Hands On: Distributive Property and Partial Quotients ... T29
Lesson 8 Divide Three-and Four-Digit Dividends ... T30
Lesson 9 Place the First Digit ... T31
Lesson 10 Quotients with Zeros ... T32
Lesson 11 Inquiry/Hands On: Use Models to Interpret the Remainder ... T33
Lesson 12 Interpret the Remainder ... T34
Lesson 13 Problem-Solving Investigation: Extra or Missing Information ... T35

Chapter 4 Divide by a Two-Digit Divisor

Mathematical Practice 8/Inquiry	T36
Lesson 1 Estimate Quotients	T37
Lesson 2 Inquiry/Hands On: Divide Using Base-Ten Blocks	T38
Lesson 3 Divide by a Two-Digit Divisor	T39
Lesson 4 Adjust Quotients	T40
Lesson 5 Divide Greater Numbers	T41
Lesson 6 Problem-Solving Investigation: Solve a Simpler Problem	T42

Chapter 5 Add and Subtract Decimals

Mathematical Practice 2/Inquiry	T43
Lesson 1 Round Decimals	T44
Lesson 2 Estimate Sums and Differences	T45
Lesson 3 Problem-Solving Investigation: Estimate or Exact Answer	T46
Lesson 4 Inquiry/Hands On: Add Decimals Using Base-Ten Blocks	T47
Lesson 5 Solve Hands On: Add Decimals Using Models	T48
Lesson 6 Add Decimals	T49
Lesson 7 Addition Properties	T50
Lesson 8 Inquiry/Hands On: Subtract Decimals Using Base-Ten Blocks	T51
Lesson 9 Inquiry/Hands On: Subtract Decimals Using Models	T52
Lesson 10 Subtract Decimals	T53

Chapter 6 Multiply and Divide Decimals

Mathematical Practice 2/Inquiry	T54
Lesson 1 Estimate Products of Whole Numbers and Decimals	T55
Lesson 2 Inquiry/Hands On: Use Models to Multiply	T56
Lesson 3 Multiply Decimals by Whole Numbers	T57
Lesson 4 Inquiry/Hands On: Use Models to Multiply Decimals	T58
Lesson 5 Multiply Decimals	T59
Lesson 6 Multiply Decimals by Powers of Ten	T60
Lesson 7 Problem-Solving Investigation: Look for a Pattern	T61
Lesson 8 Multiplication Properties	T62
Lesson 9 Estimate Quotients	T63
Lesson 10 Inquiry/Hands On: Divide Decimals	T64
Lesson 11 Divide Decimals by Whole Numbers	T65
Lesson 12 Inquiry/Hands On: Use Models to Divide Decimals	T66
Lesson 13 Divide Decimals	T67
Lesson 14 Divide Decimals by Powers of Ten	T68

Chapter 7 Expressions and Patterns

Mathematical Practice 4/Inquiry	T69
Lesson 1 Inquiry/Hands On: Numerical Expressions	T70
Lesson 2 Order of Operations	T71
Lesson 3 Write Numerical Expressions	T72
Lesson 4 Problem-Solving Investigation: Work Backward	T73
Lesson 5 Inquiry/Hands On: Generate Patterns	T74
Lesson 6 Patterns	T75
Lesson 7 Inquiry/Hands On: Map Locations	T76
Lesson 8 Ordered Pairs	T77
Lesson 9 Graph Patterns	T78

Chapter 8 Fractions and Decimals

Mathematical Practice 3/Inquiry	T79
Lesson 1 Fractions and Division	T80
Lesson 2 Greatest Common Factor	T81
Lesson 3 Simplest Form	T82
Lesson 4 Problem-Solving Investigation: Guess, Check, and Revise	T83
Lesson 5 Least Common Multiple	T84
Lesson 6 Compare Fractions	T85
Lesson 7 Inquiry/Hands On: Use Models to Write Fractions as Decimals	T86
Lesson 8 Write Fractions as Decimals	T87

Chapter 9 Add and Subtract Fractions

Mathematical Practice 5/Inquiry	T88
Lesson 1 Round Fractions	T89
Lesson 2 Add Like Fractions	T90
Lesson 3 Subtract Like Fractions	T91
Lesson 4 Inquiry/Hands On: Use Models to Subtract Unlike Fractions	T92
Lesson 5 Add Unlike Fractions	T93
Lesson 6 Inquiry/Hands On: Use Models to Subtract Unlike Fractions	T94
Lesson 7 Subtract Unlike Fractions	T95
Lesson 8 Problem-Solving Investigation: Determine Reasonable Answers	T96
Lesson 9: Estimate Sums and Differences	T97
Lesson 10 Inquiry/Hands On: Use Models to Add Mixed Numbers	T98
Lesson 11 Add Mixed Numbers	T99
Lesson 12 Subtract Mixed Numbers	T100
Lesson 13 Subtract with Renaming	T101

Chapter 10 Multiply and Divide Fractions

Mathematical Practice 5/Inquiry	T102
Lesson 1 Inquiry/Hands On: Part of a Number	T103
Lesson 2 Estimate Products of Fractions	T104
Lesson 3 Inquiry/Hands On: Model Fraction Multiplication	T105
Lesson 4 Multiply Whole Numbers and Fractions	T106
Lesson 5 Inquiry/Hands On: Use Models to Multiply Fractions	T107
Lesson 6 Multiply Fractions	T108
Lesson 7 Multiply Mixed Numbers	T109
Lesson 8 Inquiry/Hands On: Multiplication as Scaling	T110
Lesson 9 Inquiry/Hands On: Division with Unit Fractions	T111
Lesson 10 Divide Whole Numbers by Unit Fractions	T112
Lesson 11 Divide Unit Fractions by Whole Numbers	T113
Lesson 12 Problem-Solving Investigation: Draw a Diagram	T114

Chapter 11 Measurement

Mathematical Practice 6/Inquiry	T115
Lesson 1 Inquiry/Hands On: Measure with a Ruler	T116
Lesson 2 Convert Customary Units of Length	T117
Lesson 3 Problem-Solving Investigation: Use Logical Reasoning	T118
Lesson 4 Inquiry/Hands On: Estimate and Measure Weight	T119
Lesson 5 Convert Customary Units of Weight	T120
Lesson 6 Inquiry/Hands On: Estimate and Measure Capacity	T121
Lesson 7 Convert Customary Units of Capacity	T122
Lesson 8 Display Measurement Data on a Line Plot	T123
Lesson 9 Inquiry/Hands On: Metric Rulers	T124
Lesson 10 Convert Metric Units of Length	T125
Lesson 11 Inquiry/Hands On: Estimate and Measure Metric Mass	T126
Lesson 12 Convert Metric Units of Mass	T127
Lesson 13 Convert Metric Units of Capacity	T128

Chapter 12 Geometry

Mathematical Practice 7/Inquiry	T129
Lesson 1 Polygons	T130
Lesson 2 Inquiry/Hands On: Sides and Angles of Triangles	T131
Lesson 3 Classify Triangles	T132
Lesson 4 Inquiry/Hands On: Sides and Angles of Quadrilaterals	T133
Lesson 5 Classify Quadrilaterals	T134
Lesson 6 Inquiry/Hands On: Build Three-Dimensional Figures	T135
Lesson 7 Three-Dimensional Figures	T136
Lesson 8 Inquiry/Hands On: Use Models to Find Volume	T137
Lesson 9 Volume of Prisms	T138
Lesson 10 Inquiry/Hands On: Build Composite Figures	T139
Lesson 11 Volume of Composite Figures	T140
Lesson 12 Problem-Solving Investigation: Make a Model	T141

Visual Kinesthetic Vocabulary®	VKV1
VKV® Answer Appendix	VKV35

Proficiency Level Descriptors

	Interpretive (Input)		Productive (Output)	
	Listening	**Reading**	**Writing**	**Speaking**
An Emerging Level EL • New to this country; may have memorized some everyday phrases like, "Where is the bathroom", "My name is....", may also be in the "silent stage" where they listen to the language but are not comfortable speaking aloud • Struggles to understand simple conversations • Can follow simple classroom directions when overtly demonstrated by the instructor	• Listens actively yet struggles to understand simple conversations • Possibly understands "chunks" of language; may not be able to produce language verbally	• Reads familiar patterned text • Can transfer Spanish decoding somewhat easily to make basic reading in English seem somewhat fluent; comprehension is weak	• Writes labels and word lists, copies patterned sentences or sentence frames, one- or two-word responses	• Responds non-verbally by pointing, nodding, gesturing, drawing • May respond with yes/no, short phrases, or simple memorized sentences • Struggles with non-transferable pronunciations.
An Expanding Level EL • Is dependent on prior knowledge, visual cues, topic familiarity, and pretaught math-related vocabulary • Solves word problems with significant support • May procedurally solve problems with a limited understanding of the math concept.	• Has ability to understand and distinguish simple details and concepts of familiar/previous learned topics	• Recognizes obvious cognates • Pronounces most English words correctly, reading slowly and in short phrases • Still relies on visual cues and peer or teacher assistance	• Produces writing that consists of short, simple sentences loosely connected with limited use of cohesive devices • Uses undetailed descriptions with difficulty expressing abstract concepts	• Uses simple sentence structure and simple tenses • Prefers to speak in present tense.
A Bridging Level EL • May struggle with conditional structure of word problems • Participates in social conversations needing very little contextual support • Can mentor other ELs in collaborative activities.	• Usually understands longer, more elaborated directions, conversations, and discussions on familiar and some unfamiliar topics • May struggle with pronoun usage	• Reads with fluency, and is able to apply basic and higher-order comprehension skills when reading grade-appropriate text	• Is able to engage in writing assignments in content area instruction with scaffolded support • Has a grasp of basic verbs, tenses, grammar features, and sentence patterns	• Participates in most academic discussions on familiar topics, with some pauses to restate, repeat, or search for words and phrases to clarify meaning.

Strategies for EL Success

Surprisingly, content instruction is one of the most effective methods of acquiring fluency in a second language. When content is the learner's focus, the language used to perform the skill is not consciously considered. The learner is thinking about the situation, or how to solve the problem, not about the grammatical structure of their thoughts. Attempting skills in the target language forces the language into the subconscious mind, where useable language is stored. A dramatic increase in language integration occurs when multiple senses are involved, which causes heightened excitement, and a greater investment in the situation's outcome. Given this, a few strategies to employ during EL instruction that can make teaching easier and learning more efficient are listed below:

- Activate EL prior knowledge and cultural perspective
- Use manipulatives, realia, and hands-on activities
- Identify cognates
- Build a Word Wall
- Modeled talk
- Choral responses
- Echo reading
- Provide sentence frames for students to use
- Create classroom anchor charts
- Utilize translation tools (i.e. Glossary, eGlossary, online translation tools)
- Anticipate common language problems

Common Problems for English Learners

	Cantonese	Haitian Creole	Hmong	Korean	Spanish	Vietnamese
Phonics Transfers						
Pronouncing the /k/ as in cake	●		●	●		
Pronouncing the digraph /sh/		●	●		●	●
Hearing and reproducing the /r/, as in *rope*	●		●	●		●
/j/			●		●	
Hearing or reproducing the short /u/		●	●			
Grammar Transfers						
Adjectives often follow nouns		●	●		●	●
Adjectives and adverb forms are interchangeable		●	●			
Nouns have feminine or masculine gender					●	
There is no article or there is no difference between articles *the* and *a*		●	●			●
Shows comparative and superlative forms with separate words			●		●	
There are no phrasal verbs				●	●	

How to Use the Teacher Edition

The Interactive Guide provides scaffolding strategies and tips to strengthen the quality of mathematics instruction. The suggested strategies, activities, and tips provide additional language and concept support to accelerate English learners' acquisition of academic English.

English Learner Instructional Strategy

Each lesson – including Inquiry/Hands On and Problem-Solving Investigation – references an English Learner Instructional Strategy that can be utilized before or during regular class instruction. These strategies specifically support the Teacher Edition and scaffold the lesson for English learners (ELs).

Categories of the scaffolded support are:
- Vocabulary Support
- Language Structure Support
- Sensory Support
- Graphic Support
- Collaborative Support

The goal of the scaffolding strategies is to make each individual lesson more comprehensible for ELs by providing visual, contextual and linguistic support to foster students' understanding of basic communication in an academic context.

Lesson 4 Inquiry/Hands On: Sides and Angles of Quadrilaterals
English Learner Instructional Strategy

Vocabulary Support: Activate Prior Knowledge

Display anchor charts, word webs, KWL charts, or any other graphic organizers from previous lessons related to polygons, including examples of how to classify triangles by their attributes. Be sure to include the classroom cognate chart as well. Have students take turns coming up to the graphic organizers and sharing a piece of information about polygons with the other students. Ask students how their knowledge of classifying triangles might be applied to classifying other kinds of polygons, including quadrilaterals.

Display the following sentence frames to help students participate during the lesson:

____ congruent sides
____ congruent angles
____ parallel sides
Figures ____ have congruent ____.

Since ELs benefit from visual references to new vocabulary, many of the English Learner Instruction Strategies suggest putting vocabulary words on a Word Wall. Choose a location in your classroom for your Word Wall, and organize the words by chapter, by topic, by Common Core domain, or alphabetically.

How to Use the Teacher Edition *continued*

English Language Development Leveled Activities

These activities are tiered for Emerging, Expanding, and Bridging leveled ELs. Activity suggestions are specific to the content of the lesson. Some activities include instruction to support students with lesson-specific vocabulary they will need to understand the math content in English, while other activities teach the concept or skill using scaffolded approaches specific to ELs. The activities are intended for small group instruction, and can be directed by the instructor, an aide, or a peer mentor.

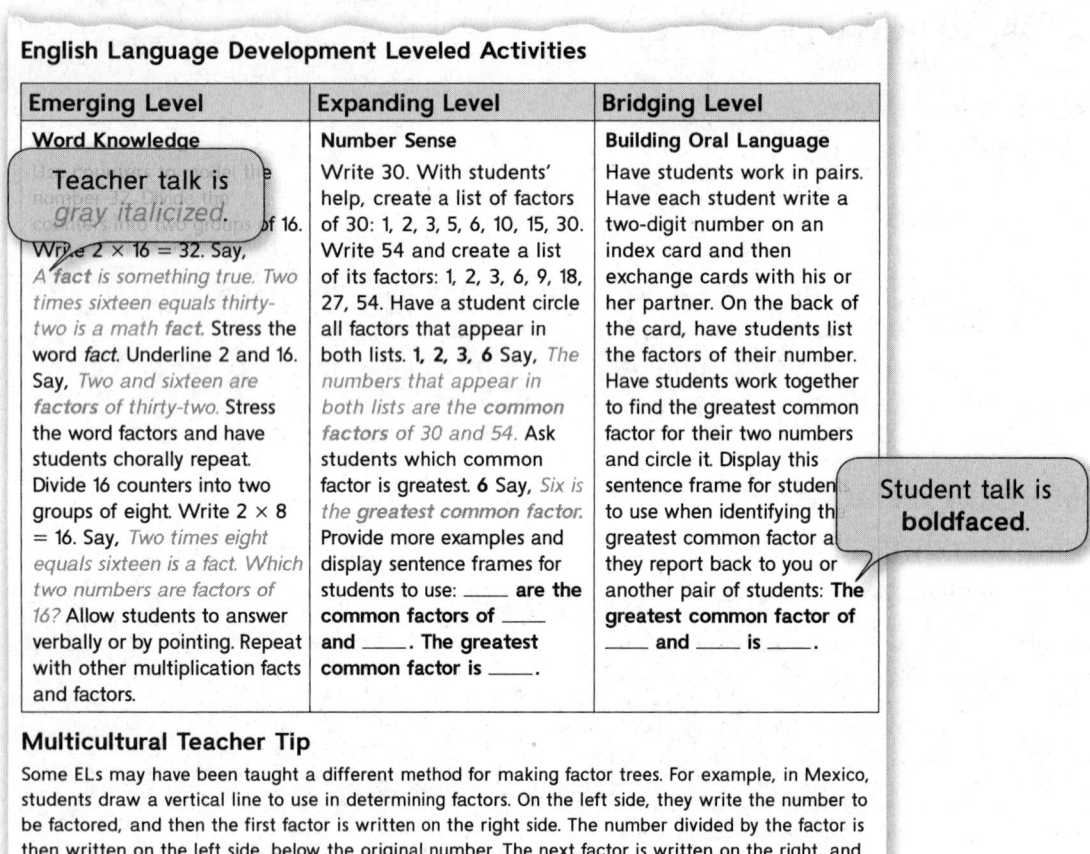

English Language Development Leveled Activities

Emerging Level	Expanding Level	Bridging Level
Word Knowledge *[Teacher talk is gray italicized.]* ...of 16. Write 2 × 16 = 32. Say, *A fact is something true. Two times sixteen equals thirty-two is a math fact.* Stress the word *fact*. Underline 2 and 16. Say, *Two and sixteen are factors of thirty-two.* Stress the word *factors* and have students chorally repeat. Divide 16 counters into two groups of eight. Write 2 × 8 = 16. Say, *Two times eight equals sixteen is a fact. Which two numbers are factors of 16?* Allow students to answer verbally or by pointing. Repeat with other multiplication facts and factors.	**Number Sense** Write 30. With students' help, create a list of factors of 30: 1, 2, 3, 5, 6, 10, 15, 30. Write 54 and create a list of its factors: 1, 2, 3, 6, 9, 18, 27, 54. Have a student circle all factors that appear in both lists. **1, 2, 3, 6** Say, *The numbers that appear in both lists are the common factors of 30 and 54.* Ask students which common factor is greatest. **6** Say, *Six is the greatest common factor.* Provide more examples and display sentence frames for students to use: **___ are the common factors of ___ and ___. The greatest common factor is ___.**	**Building Oral Language** Have students work in pairs. Have each student write a two-digit number on an index card and then exchange cards with his or her partner. On the back of the card, have students list the factors of their number. Have students work together to find the greatest common factor for their two numbers and circle it. Display this sentence frame for students to use when identifying the greatest common factor as they report back to you or another pair of students: **The greatest common factor of ___ and ___ is ___.** *[Student talk is boldfaced.]*

Multicultural Teacher Tip
Some ELs may have been taught a different method for making factor trees. For example, in Mexico, students draw a vertical line to use in determining factors. On the left side, they write the number to be factored, and then the first factor is written on the right side. The number divided by the factor is then written on the left side, below the original number. The next factor is written on the right, and the process continues until there are no more factors. In factoring 18, for example, the result would be *18, 9, 3, 1* listed on the left side, and the prime factors *2, 3, 3* listed on the right.

Multicultural Teacher Tip

These tips provide insight on academic and cultural differences you may encounter in your classroom. While math is the universal language, some ELs may have been shown different methods to find the answer based on their native country, while cultural customs may influence learning styles and behavior in the classroom.

What's the Math in this Chapter?

Mathematical Practice
The goals of the Mathematical Practice activity are to help students clarify the specific language of the Mathematical Practice and rewrite the Mathematical Practice in simplified language that students can **relate** to. Examples are discussed to make connections, showing students how they use this Mathematical Practice to solve math problems.

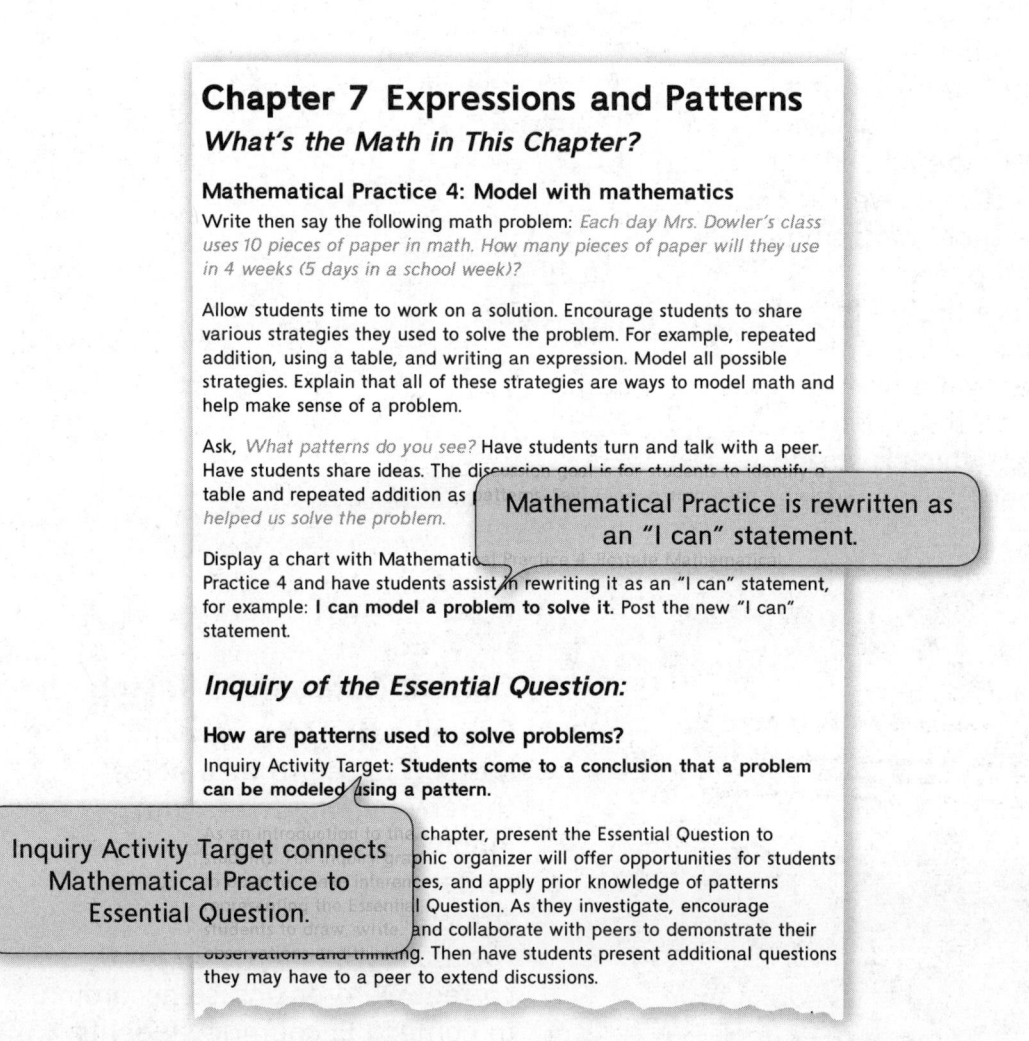

Chapter 7 Expressions and Patterns
What's the Math in This Chapter?

Mathematical Practice 4: Model with mathematics
Write then say the following math problem: *Each day Mrs. Dowler's class uses 10 pieces of paper in math. How many pieces of paper will they use in 4 weeks (5 days in a school week)?*

Allow students time to work on a solution. Encourage students to share various strategies they used to solve the problem. For example, repeated addition, using a table, and writing an expression. Model all possible strategies. Explain that all of these strategies are ways to model math and help make sense of a problem.

Ask, *What patterns do you see?* Have students turn and talk with a peer. Have students share ideas. The discussion goal is for students to identify a table and repeated addition as [...] helped us solve the problem.

Display a chart with Mathematical Practice 4 and have students assist in rewriting it as an "I can" statement, for example: **I can model a problem to solve it.** Post the new "I can" statement.

Inquiry of the Essential Question:

How are patterns used to solve problems?
Inquiry Activity Target: **Students come to a conclusion that a problem can be modeled using a pattern.**

As an introduction to the chapter, present the Essential Question to [...] phic organizer will offer opportunities for students [...] ces, and apply prior knowledge of patterns [...] Question. As they investigate, encourage [...] and collaborate with peers to demonstrate their [...]ng. Then have students present additional questions they may have to a peer to extend discussions.

(Callout: Mathematical Practice is rewritten as an "I can" statement.)

(Callout: Inquiry Activity Target connects Mathematical Practice to Essential Question.)

Inquiry of the Essential Question
As an introduction to the Chapter, the Inquiry of the Essential Question graphic organizer activity is designed to introduce the Essential Question. The activity offers opportunities for students to observe, make inferences, and apply prior knowledge of samples/models representing the Essential Question. Collaborative conversations drive students toward the Inquiry Activity Target which is to make a connection between the "Mathematical Practice of the chapter" and the "Essential Question of the chapter."

How to Use the Student Edition

Each student page provides EL support for vocabulary, note taking, and writing skills. These pages can be used before, during, or after each classroom lesson. A corresponding page with answers is found in the Teacher Edition.

Inquiry of the Essential Question

Students observe, make inferences, and apply prior knowledge of chapter specific samples/models representing the Essential Question of the chapter. Encourage students to have collaborative conversations as they share their ideas and questions with peers. As students inquire the math models, present specific questions that will drive students toward the Inquiry Activity Target which is stated on the Teacher Edition page.

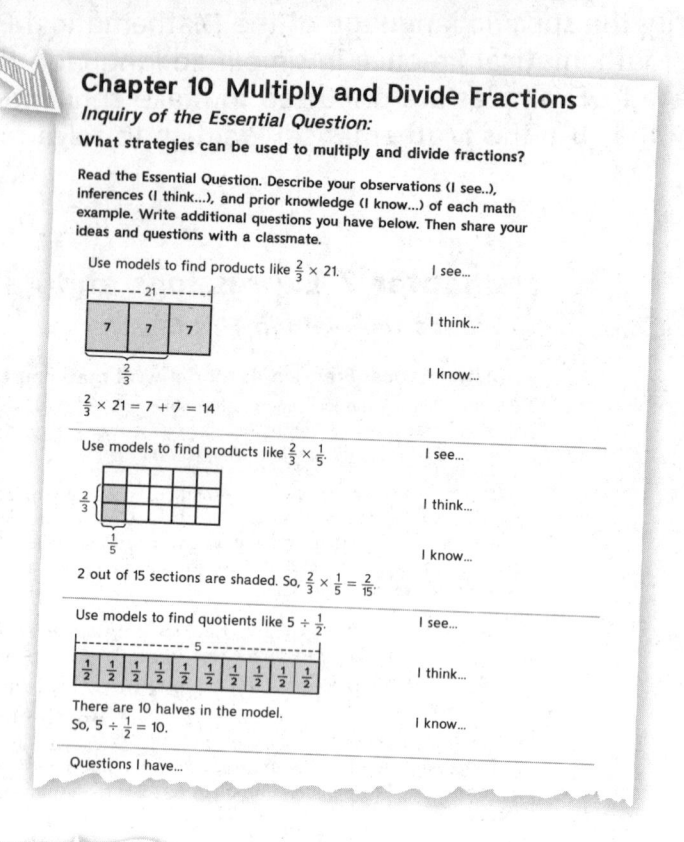

Cornell Notes/Note Taking

Cornell notes offer students a method to use to take notes, thereby helping them with language structure. Scaffolded sentence frames are provided for students to fill in important math vocabulary by identifying the correct word or phrase according to context. Encourage students to refer to their books to locate the words needed to complete the sentences. Each note taking graphic organizer will support students in answering the Building on the Essential Question.

xii

Vocabulary Cognates

Students define each vocabulary word or phrase and write a sentence using the term in context. Space is provided for Spanish speakers to write the definition in Spanish. A blank Vocabulary Word Boxes template is provided on page xx in this book for use with non-Spanish speaking ELs.

Guided Writing

Guided writing provides support to help ELs meet the stated Lesson Objective. Content specific questions are scaffolded to build language knowledge in order to answer the question. Give bridging level students the opportunity to mentor and assist emerging and expanding level students when answering the questions, if needed.

xiii

How to Use the Student Edition *continued*

Vocabulary Chart

Three-column charts concentrate on English/Spanish cognates. Students are given the word in English. Encourage students to use the Glossary to find the word in Spanish and the definition in English. As an extension, have students identify and highlight other cognates which may be in the definitions. A blank Vocabulary Chart template is provided on page xix in this book for use with non-Spanish speaking ELs.

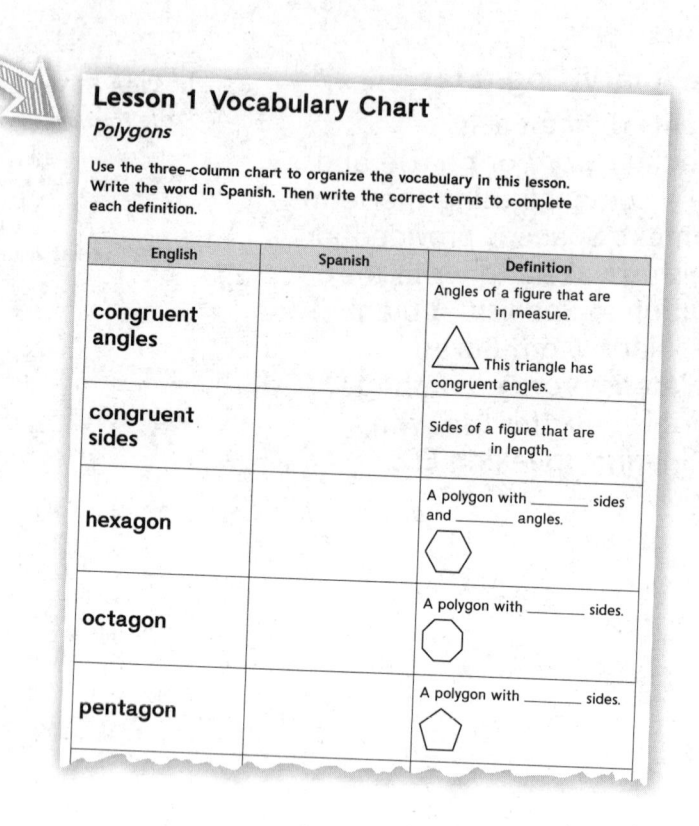

Concept Web

Concept webs are designed to show relationships between concepts and to make connections. As each concept web is unique in design, please read and clarify directions for students. Encourage students to look through the lesson pages to find examples or words they can use to complete the web.

Definition Map

The definition maps are designed to address a single vocabulary word, phrase, or concept. Students should use the Glossary to help define the word in the description box. Sentence frames are provided to scaffold characteristics from the lesson. Students can refer to the lesson examples and Glossary to assist them in completing the sentence frames as well as creating their own math examples. Make sure you review with students the tasks required.

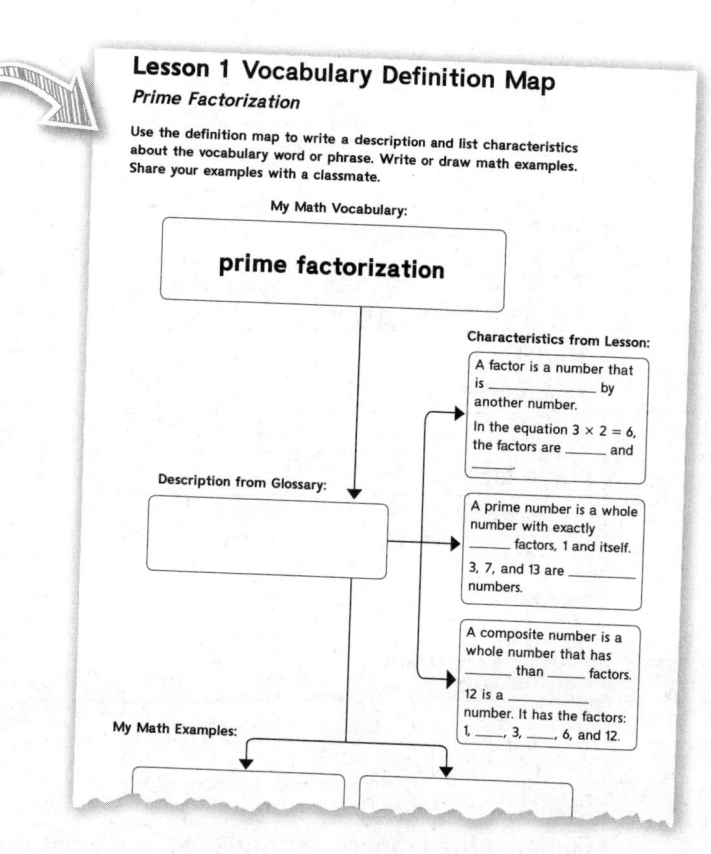

Problem-Solving Investigation

Each Problem-Solving Investigation page focuses on scaffolding Exercises 1 and 2 from the Apply the Strategy portion of the lesson in the book. The text for each exercise highlights signal words and phrases to help students decipher the key information in the problem. Visual images as well as tables and sentence frames are included to assist students with the problem-solving process. A blank Problem-Solving Investigation template is provided on page xxii in this book if students need additional assistance with other exercises.

xv

English/Spanish Cognates Used in Grade 5

Chapter	English	Spanish
1	decimal	decimal
	decimal point	punto decimal
	equivalent decimals	decimales equivalentes
	period	período
	standard form	forma estándar
2	base	base
	compatible numbers	números compatibles
	cubed	al cubo
	Distributive Property	propiedad distributiva
	estimate	estimación (noun) estimar (verb)
	exponent	exponente
	factor	factor
	prime factorization	factorización prima
	product	producto
3	dividend	dividendo
	divisor	divisor
	multiple	múltiplo
	quotient	cociente
	variable	variable
4	*No New Cognates*	
5	Associative Property (of Addition)	propiedad asociativa (de la suma)
	Commutative Property (of Addition)	propiedad conmutativa (de la suma)
	difference	diferencia
	Identity Property (of Addition)	propiedad identidad (de la suma)
	inverse operation	operaciones inversas
	place value	valor posicional
	sum	suma
6	Associative Property of Multiplication	propiedad asociativa de la multiplicación
	Commutative Property of Multiplication	propiedad conmutativa de la multiplicación
	Identity Property of Multiplication	propiedad de identidad de la multiplicación
7	coordinate plane	plano de coordenadas
	numerical expression	expresíon numérica
	order of operations	orden de las operaciones
	ordered pair	par ordenado
	origin	origen
	sequence	secuencia
	term	término
	x-coordinate	coordenada x
	y-coordinate	coordenada y

Chapter	English	Spanish
8	common factor	factor común
	common multiple	múltiplo común
	denominator	denominador
	equivalent fractions	fracciones equivalentes
	fraction	fracción
	(greatest) common factor (GCF)	(máximo) común divisor (M.C.D.)
	(least) common denominator (LCD)	(mínimo) común denominador (m.c.d.)
	(least) common multiple (LCM)	(mínimo) común múltiplo (m.c.m.)
	multiple	múltiplo
	numerator	numerador
	simplest form	forma simplificada
9	mixed number	número mixto
10	improper fraction	fracción impropia
	unit fraction	fracción unitaria
11	capacity	capacidad
	centimeter (cm)	centímetro (cm)
	convert	convertir
	fluid "liquid" ounce (fl oz)	onza líquida
	gallon (gal)	galón (gal)
	gram (g)	gramo (g)
	kilogram (kg)	kilogramo (kg)
	kilometer (km)	kilómetro (km)
	liter (L)	litro (L)
	mass	masa
	meter (m)	metro (m)
	metric system	sistema métrico (SI)
	mile (mi)	milla (mi)
	milligram (mg)	miligramo (mg)
	milliliter (mL)	mililitro (mL)
	millimeter (mm)	milímetro (mm)
	ounce (oz)	onza (oz)
	pint (pt)	pinta (pt)
	quart (qt)	cuarto (ct)
	yard (yd)	yarda (yd)
12	acute triangle	triángulo acutángulo
	attribute	atributo
	composite figure	figura compuesta
	congruent angles	ángulos congruentes
	congruent figures	figuras congruentes
	congruent (sides)	(lados) congruentes
	cubic unit	unidad cúbica

English/Spanish Cognates Used in Grade 5 *continued*

Chapter	English	Spanish
12	equilateral triangle	triángulo equilátero
	hexagon	hexágono
	isosceles triangle	triángulo isósceles
	obtuse triangle	triángulo obtusángulo
	octagon	octágono
	parallelogram	paralelogramo
	pentagon	pentágono
	polygon	polígono
	prism	prisma
	quadrilateral	cuadrilátero
	rectangle	rectángulo
	rectangular prism	prisma rectangular
	regular polygon	polígono regular
	rhombus	rombo
	right triangle	triángulo rectángulo
	scalene triangle	triángulo escaleno
	three-dimensional figure	figura tridimesional
	trapezoid	trapecio
	triangle	triángulo
	triangular prism	prisma triangular
	unit cube	cubo unitario
	vertex	vértice
	volume	volumen

NAME _____ DATE _____

Vocabulary Chart
Chapter _____, Lesson _____

Use the three-column chart to organize the vocabulary in this lesson. Write the word in your own language. Then write each definition.

English	Native Language	Definition

NAME _____ DATE _____

Vocabulary Word Boxes
Chapter _____, Lesson _____

Use the word boxes to define the math word in English and in your native language. Write a sentence using your math word.

Definition	

My math word sentence:

Definition	

My math word sentence:

NAME _____ DATE _____

Vocabulary Definition Map
Chapter _____, Lesson _____

Use the definition map to write a description and list characteristics about the vocabulary word or phrase. Write or draw math examples. Share your examples with a classmate.

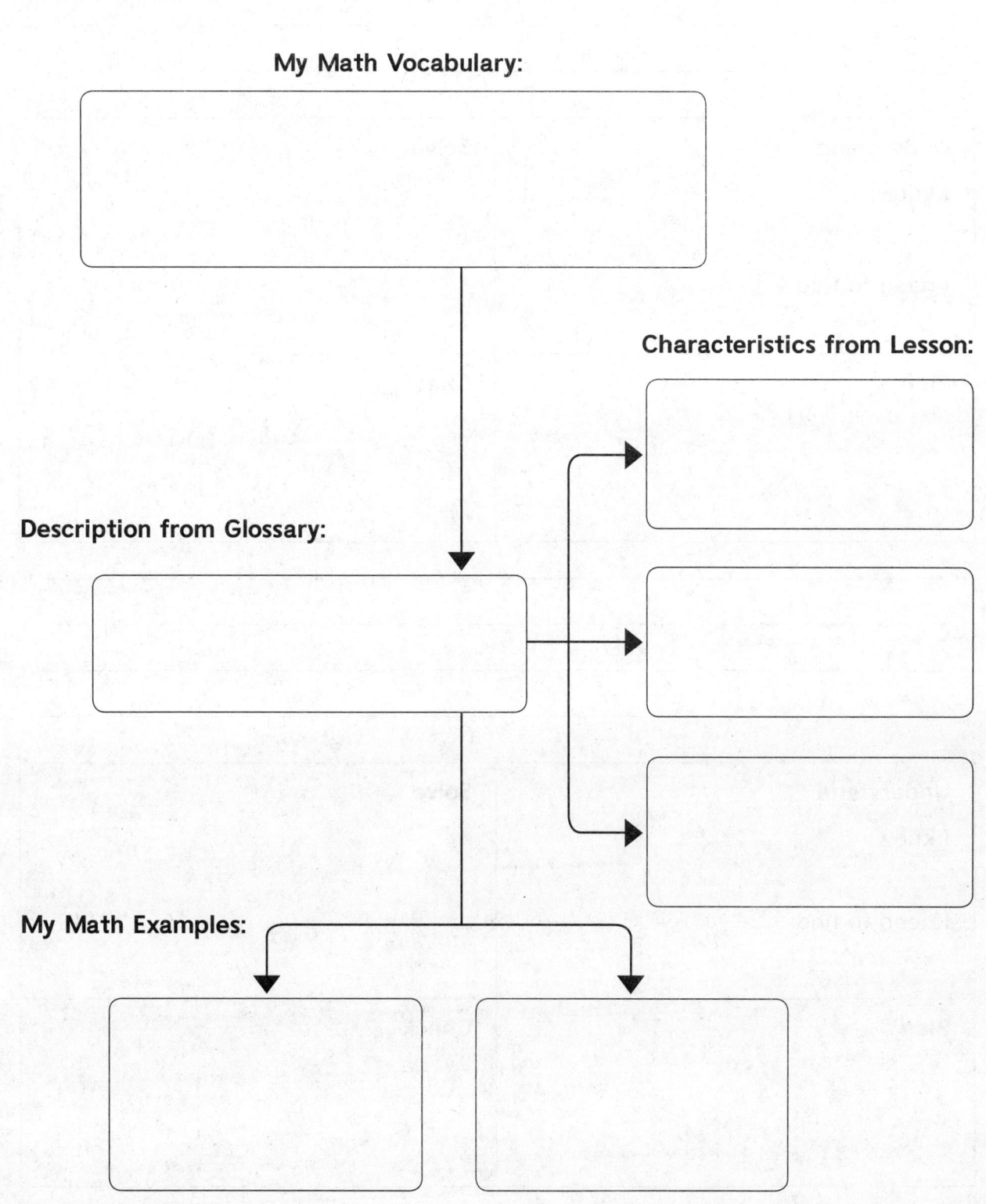

My Math Vocabulary:

Characteristics from Lesson:

Description from Glossary:

My Math Examples:

xxi

NAME _____ DATE _____

Problem-Solving Investigation

Chapter ____, Lesson ____

1. _____
   _____
   _____
   _____

Understand	Solve
I know:	
I need to find:	
Plan	**Check**

2. _____
   _____
   _____
   _____

Understand	Solve
I know:	
I need to find:	
Plan	**Check**

Chapter 1 Place Value

What's the Math in This Chapter?

Mathematical Practice 1: Make sense of problems and persevere in solving them

Write and read the following word problem. *Elijah has 375 songs on his digital audio player. His brother, William has 122 songs. Elijah and William decide to combine their songs. How many songs do they have in all?* Discuss what they know and what they need to find in the problem. Ask, *How do we solve?* **Add the numbers together.** On the board write 375 + 122 vertically but do not line the numbers up correctly. Put the 1 in 122 under the 7 in 375 instead of the 3. Keep solving the problem, even if students are protesting.

Say, *Okay, I solved the problem. Elijah and William have 3,872 songs. Wait! Does 3,872 songs make sense?* Allow students time to think about the way you solved the problem. When a student identifies that you didn't add correctly say, *Oh you are right. I understand what I did wrong. I didn't line up the numbers correctly.*

Draw a place-value chart and write the numbers correctly aligned in the chart. Model proper addition showing the sum to be 497. Discuss with students how using place value helps us to **make sense of problems** and add correctly.

Display a chart with Mathematical Practice 1. Restate Mathematical Practice 1 and have students assist in rewriting it as an "I can" statement, for example: **I can make a plan and continue trying until I solve a problem.** Post the new "I can" statement in the classroom.

Inquiry of the Essential Question:

How does the position of a digit in a number relate to its value?

Inquiry Activity Target: **Students come to a conclusion that they can use place value to solve problems.**

As an introduction to the chapter, present the Essential Question to students. The inquiry graphic organizer will offer opportunities for students to observe, make inferences, and apply prior knowledge of problem solving representing the Essential Question. As they investigate, encourage students to draw, write, and collaborate with peers to demonstrate their observations and thinking. Then have students present additional questions they may have to a peer to extend discussions.

Regroup students and restate Mathematical Practice 1 and the Essential Question. Pose questions to reflect on what has been learned to guide students in making connections between the Mathematical Practice and the Essential Question.

NAME ______________________________ DATE ______________

Chapter 1 Place Value

Inquiry of the Essential Question:

How does the position of a digit in a number relate to its value?

Read the Essential Question. Describe your observations (I see...), inferences (I think...), and prior knowledge (I know...) of each math example. Write additional questions you have below. Then share your ideas and questions with a classmate.

Thousands Period			Ones Period		
hundreds	tens	ones	hundreds	tens	ones
2	7	7,	3	8	9

I see ...

I think...

I know...

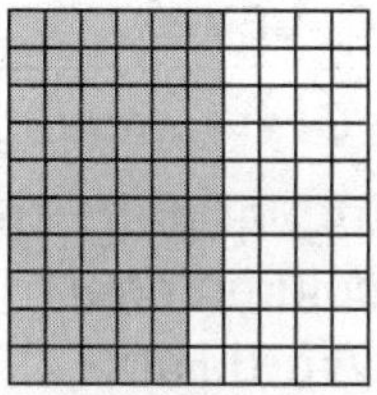

Fraction: $\frac{58}{100}$

Word form: fifty-eight hundredths

Decimal: 0.58

I see ...

I think...

I know...

Step 1 Understand the facts, and what needs to be found.
Step 2 Plan the strategy.
Step 3 Solve the problem.
Step 4 Check that the answer makes sense.

I see ...

I think...

I know...

Questions I have...

____

____

____

Grade 5 • Chapter 1 *Place Value* **1**

Lesson 1 Place Value Through Millions

English Learner Instructional Strategy

Graphic Support: Utilize Resources

Direct students to review the Glossary definition for *place value* in both English and Spanish. Using play money, discuss with students the corresponding values of a penny, dime, and dollar. Provide the following sentence frames to aid students in the discussion: **The value of a dime is ____ pennies. The value of a dollar is ____ dimes.** Display a place-value chart. Then model how money corresponds to the ones, tens and hundreds place on the chart.

During Talk Math, emphasize the similarity in how the value of the money increased to how value increases when moving left in the place-value chart.

English Language Development Leveled Activities

Emerging Level	Expanding Level	Bridging Level
Number Sense	**Recognize and Act It Out**	**Building Oral Language**
Write *standard form* and its Spanish cognate, *la forma estándaron* on a classroom cognate chart. Display a place-value chart. Write 352,654 in the chart. Point to each number and say, *Three is in the hundred-thousands place, five is in the ten-thousands place, two is in the thousands place . . . and so on.* Stress the /s/ in pla*c*e. Write another whole number to the millions place in the chart. Randomly point to a place value and ask, *Which number is in the ____ place?* Have students answer chorally.	Write *period* and the Spanish cognate, *el period* on a classroom cognate chart. Display a nine-digit number frame. Label each period and place value up to the millions place. Say one place value of a digit in the number 6,042,962. For example, say, *40,000.* Have a student volunteer write the digit in the correct place value and say, **Four is in the ten-thousands place.** Continue with the other place values, not in order, until the frame is filled. Provide the following sentence frame: ____ **is in the ____ place.**	Have bilingual pairs create a three-column chart labeled Word Form, Standard Form, and Expanded Form. Have one student name a seven-digit number. Have the other student write the number in words, in standard form, and in expanded form in the appropriate columns. Have pairs repeat the activity, switching roles each time. Afterward, discuss with students how to write the word form of each number in their native language.

Teacher Notes:

NAME _____ DATE _____

Lesson 1 Vocabulary Chart
Place Value Through Millions

Use the three-column chart to organize the vocabulary in this lesson. Write the word in Spanish. Then write the correct terms to complete each definition.

English	Spanish	Definition
period	período	Each group of __three__ digits on a place-value chart.
standard form	forma estándar	The usual or common way to write a number using __digits__.
expanded form	forma desarrollada	A way of writing a number as the __sum__ of the __values__ of its digits.
place	posición	The __position__ of a digit in a __number__.
place value	valor posicional	The value given to a __digit__ by its __position__ in a number.
place-value chart	table de valor posicional	A chart that shows the __value__ of the __digits__ in a number.

2   Grade 5 • Chapter 1 *Place Value*

Lesson 2 Compare and Order Whole Numbers Through Millions

English Learner Instructional Strategy

Language Structure Support: Tiered Questions

During the lesson, structure questions so students of different English proficiencies can answer in ways suitable to their level of understanding. For example, when comparing place values in two different numbers, you may ask, *Is eight greater than or less than 2?* Emerging students can answer with the short phrase **greater than**. For expanding students, provide a sentence frame so they can answer in a complete sentence, such as **____ is greater than/less than ____**. Encourage bridging students to provide more complex answers. For example, the student may explain what conclusion can be made from the comparison. **Eight is greater than two, so I know 9,085,216 is greater than 9,022,673.**

English Language Development Leveled Activities

Emerging Level	Expanding Level	Bridging Level
Number Sense	**Recognize and Act It Out**	**Academic Language**
Write the word *equal* and the Spanish cognate, *igual* on a cognate chart. Distribute counters. Ask one student to count his or her counters. Write the number. Repeat with another student. Write that number. Ask, *Which number is greater?* Give students a chance to answer by pointing. Then point to each number as you say, *____ is greater than ____.* Write the appropriate symbol (< or >) to compare the numbers. Repeat with other students' counters, using the less than and is equal to symbols as needed.	Have students name two whole numbers up to seven digits. Write the numbers in standard form. Ask, *Which number is greater?* Have students answer and name the place value that determines which number is greater. Provide the following sentence frame: **____ is greater than ____. I know because of the digits in the ____ place.** Repeat with three numbers. Ask, *Which number is greatest? Which number is least?* Alter the sentence frame to help students respond appropriately.	Create number cards with whole numbers up to seven digits on each card. Distribute them confidentially to each student in bilingual pairs. Explain that students will be guessing their partner's number by asking comparison questions about the number's value. For example, the student may ask, **Is the number less than ten thousand? Is the number greater than one million?** And so on until the number is guessed. Have students switch roles and repeat the activity.

Teacher Notes:

T3   Grade 5 • Chapter 1 *Place Value*

NAME _____ DATE _____

Lesson 2 Concept Web

Compare and Order Whole Numbers through Millions

Use the concept web to identify which symbol to use, greater than (>), less than (<), or equal to (=), for each example.

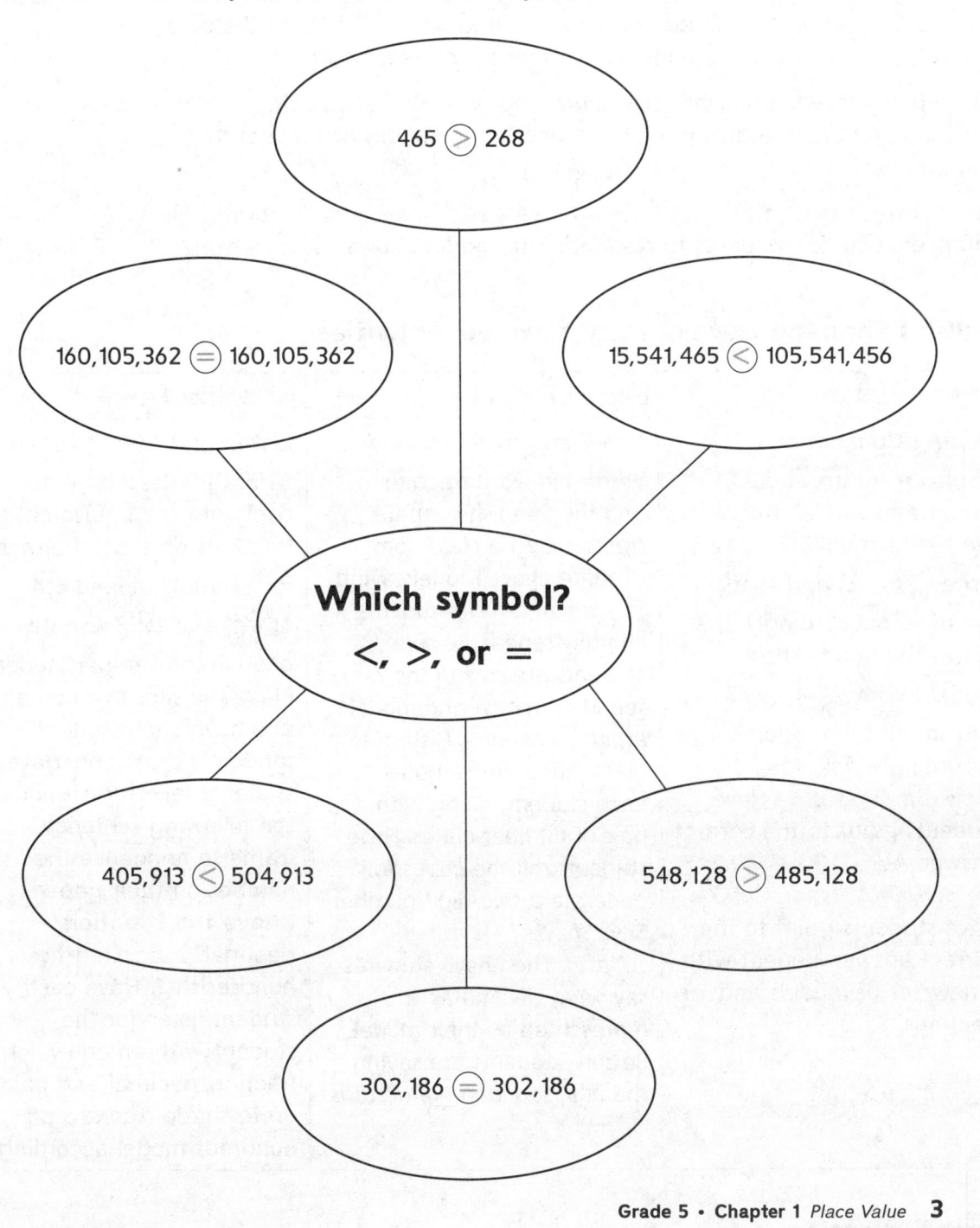

- 465 (>) 268
- 160,105,362 (=) 160,105,362
- 15,541,465 (<) 105,541,456

Which symbol?
<, >, or =

- 405,913 (<) 504,913
- 548,128 (>) 485,128
- 302,186 (=) 302,186

Grade 5 • Chapter 1 *Place Value*   **3**

Lesson 3 Inquiry/Hands On: Model Fractions and Decimals

English Learner Instructional Strategy

Sensory Support: Models

Before the lesson, create a set of index cards with either a tenths or hundredths written in words on each, for example *nine tenths, thirty-two hundredths, two tenths, sixteen hundredths*, and so on. Distribute the cards so that equal numbers of students have tenths and hundredths.

Pair students so each pair has one tenth and one hundredth card. Distribute one blank tenths model and one blank hundredths model to each pair. Say, *Shade the models to show your fractions.*

Then ask each student, *What decimal does your model show?* Provide this sentence frame for students to respond: **The model shows ___ [tenths/hundredths].**

English Language Development Leveled Activities

Emerging Level	Expanding Level	Bridging Level
Making Connections	**Non-Transferrable Sounds**	**Listen and Identify**
Display a tenths and a hundredths model. Above the tenths model, write $\frac{3}{10} = 0.3$ and shade the model accordingly. Above the hundredths model, write $\frac{28}{100} = 0.28$ and shade the model accordingly. Ask, *Which shows three tenths?* Have students point to the correct answer. Ask, *Which shows twenty-eight hundredths?* Have students point to the correct answer. Repeat with a new set of models and fractions.	Write the word *fraction* and the Spanish cognate, *fraccióne* on a classroom cognate chart. Model saying *fraction* and have students chorally repeat. Be sure students are saying the /sh/ sound correctly, and model again as needed. Distribute a blank hundredths model to each student, along with a pair of number cubes. Have students roll the cubes to generate a two-digit number. Say, *Shade that many squares.* Then have students say what decimal is represented by their model. Be sure students are saying the /th/ sound in *hundredths*.	Before the lesson, write decimals and equivalent fractions on slips of paper, for example $\frac{6}{10}$ and *0.6* or $\frac{52}{100}$ and *0.52*. Create enough for one per student. Place the slips in a container and have each student randomly draw one. Have students take turns using the following sentence frame to announce the number on their paper: **I have the [fraction/decimal] ___ [tenths/hundredths].** Have each student listen for the student with an equivalent fraction/decimal and pair up to shade a tenths or hundreds model accordingly.

Teacher Notes:

NAME _____ DATE _____

Lesson 3 Vocabulary Cognates

Inquiry/Hands On: Model Fractions and Decimals

Use the Glossary to define the math word in English and in Spanish in the word boxes. Write a sentence using your math word.

decimal	decimal
Definition A number that has a digit in the tenths place, hundredths place, and beyond.	**Definición** Número que tiene al menos un dígito en la posición de los décimos o centésimos, o en cualquier posición posterior.

My math word sentence:

Sample answer: The decimal form of $\frac{8}{10}$ is 0.8. The decimal form of one-quarter is 0.25.

decimal point	punto decimal
Definition A period separating the ones and the tenths in a decimal number.	**Definición** Punto que separas las unidades y los décimos en un número decimal.

My math word sentence:

Sample answer: The period between 0 and 8 in 0.8 and the period between 3 and 4 in $3.47 are decimal points.

Lesson 4 Represent Decimals

English Learner Instructional Strategy

Collaborative Support: Pass the Pen

During the Guided Practice part of the lesson, have three volunteers take turns writing each decimal in word form on sentence strips. Afterward, collect the strips and randomly distribute them to three other students. Each student will identify the shaded grid that corresponds to his or her sentence strip. Provide the following sentence frame for students to use when they identify the grid: **This grid shows _____.**

Be sure native Spanish speakers are correctly saying the /th/ digraph in tenths, hundredths, and thousandths.

English Language Development Leveled Activities

Emerging Level	Expanding Level	Bridging Level
Activate Prior Knowledge Write the word *decimal* and the Spanish cognate, *el decimal* on a classroom cognate chart. Write then say the sentence: *Count the money.* Using play money, have students count out $1.25 chorally with you. Write then say: *We counted $1.25.* Point to the period at the end of the sentence. Say, *The period shows where the sentence ends.* Point to the decimal point and say, *The decimal point shows where the ones place ends.*	**Making Connections** Say, *I will write the fraction one tenth.* Write $\frac{1}{10}$. Say, *Now I will write the decimal one tenth.* Write 0.10. Say, *I wrote one tenth as a fraction and as a decimal.* Repeat with $\frac{1}{100}$ and $\frac{1}{1,000}$. Display a labeled place-value chart (hundreds to thousandths). Write a fraction, such as $\frac{56}{100}$. Have student volunteers read the fraction aloud, write the decimal in the chart, and then say what they did. Provide a sentence frame: **I wrote the fraction _____ as a decimal.**	**Academic Language** Have bilingual pairs create a three-column chart labeled Fraction, Word Form, and Decimal. Have one student write a fraction with a denominator of 10, 100, or 1,000 in the first column. The other student will complete the chart by writing the word form and the decimal. Have pairs repeat the activity switching roles each time. Discuss with students how to write the word form in their native language.

Teacher Notes:

NAME _____   DATE _____

Lesson 4 Vocabulary Definition Map
Represent Decimals

Use the definition map to write a description and list characteristics about the vocabulary word or phrase. Write or draw math examples. Share your examples with a classmate.

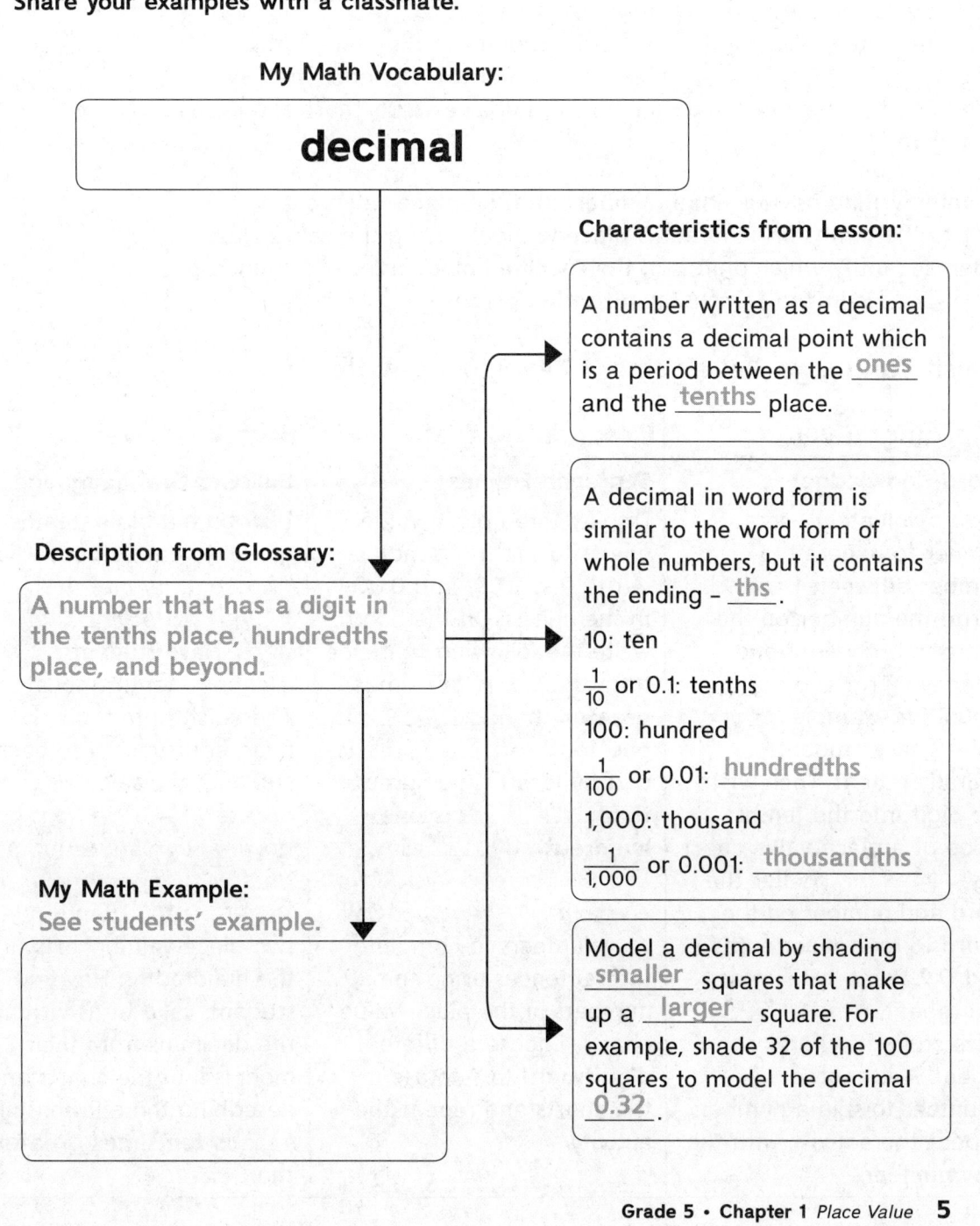

My Math Vocabulary:

decimal

Description from Glossary:
A number that has a digit in the tenths place, hundredths place, and beyond.

My Math Example:
See students' example.

Characteristics from Lesson:

A number written as a decimal contains a decimal point which is a period between the __ones__ and the __tenths__ place.

A decimal in word form is similar to the word form of whole numbers, but it contains the ending – __ths__.

10: ten
$\frac{1}{10}$ or 0.1: tenths
100: hundred
$\frac{1}{100}$ or 0.01: __hundredths__
1,000: thousand
$\frac{1}{1,000}$ or 0.001: __thousandths__

Model a decimal by shading __smaller__ squares that make up a __larger__ square. For example, shade 32 of the 100 squares to model the decimal __0.32__.

Grade 5 • Chapter 1 *Place Value*   5

Lesson 5 Inquiry/Hands On: Understand Place Value

English Learner Instructional Strategy

Vocabulary Support: Modeled Talk

During the Model the Math section of the lesson, begin by displaying a place-value chart with a decimal number to the hundredths place written in it. Point to the tenths place and say, *The digit ____ is in the tenths place.* Have students chorally repeat. Then point to the hundredths place and say, *The digit ____ is in the tenths place.* Have students chorally repeat. Write a new decimal number in the place-value chart and repeat the activity.

Continue writing new decimal numbers into the place-value chart, but point to the tenths or hundredths place without saying the value. Have students identify which digit is in the specified place using a sentence frame: ____ **is in the [tenths/hundredths] place.**

English Language Development Leveled Activities

Emerging Level	Expanding Level	Bridging Level
Word Knowledge	**Sentence Frames**	**Building Oral Language**
Have a volunteer use a spinner to generate a number between 1 and 9. Write the number on the board in both word and standard form and say it aloud, for example, *Nine and 9.* Have students chorally repeat. Then write the digit into the tenths place of a place-value chart. Say, *Nine tenths.* Alter the word and number on the board to read *nine tenths* and *0.9*. Point to them as you repeat, *Nine tenths.* Have students chorally repeat. Ask another volunteer to spin a number. Repeat the activity with the new number.	Display three place-value charts to the thousandths. Write 0.9, 0.09, and 0.009 in the charts. On the board, write the following sentence frames: ____ **is ten times greater than** ____. ____ **is one tenth of** ____. ____ **is one hundred times greater than** ____. ____ **is one hundredth of** ____. Say, *Use the sentence frames to compare the numbers.* Call on volunteers to complete the sentences using the numbers in the place-value charts. Choose a different digit from 1 to 9 to use in the charts and repeat the activity.	Distribute a blank tenths model to each student. Say, *Choose a number from 1 to 9, and shade that many parts.* Have students exchange tenths models. Then distribute a blank hundredths model to each student, and say, *Shade the hundredths model to show a decimal number ten times greater than the one in the tenths model.* Then display two place-value charts to the hundredths. Have students take turns writing the decimals from their models into the charts and describing the relationship: ____ **is ten times greater than** ____.

Teacher Notes:

T6 Grade 5 • Chapter 1 *Place Value*

NAME _____ DATE _____

Lesson 5 Guided Writing

Inquiry/Hands On: Understand Place Value

How do you use place value to understand decimals?

Use the exercises below to help you build on answering the Essential Question. Write the correct word or phrase on the lines provided.

1. Rewrite the question in your own words.
   See students' work.

2. What key words do you see in the question?
   place value, decimal

3. Place value is the value given to a __digit__ by its __position__ in a number.

4. The value of the digit 2 in the **ones** place is __2__ or __two__ ones.
   The value of the digit 2 in the **tens** place is __20__ or __two__ tens.

tens	ones
2	2

5. The value of the digit 2 in the **tens** place is __10__ times as much as the value of the digit 2 in the ones place. The value of the digit 2 in the **ones** place is $\frac{1}{10}$ times as much as the value of the digit 2 in the tens place.

6. The value of the digit 2 in the **tenths** place is __0.2__ or __two__ tenths.
   The value of the digit 2 in the **hundredths** place is __0.02__ or __two__ hundredths.

ones	tenths	hundredths
0	2	2

7. The value of the digit 2 in the **tenths** place is __10__ times as much as the value of the digit 2 in the hundredths place. The value of the digit 2 in the **hundredths** place is $\frac{1}{10}$ times as much as the value of the digit 2 in the tenths place.

8. How do you use place value to understand decimals?
   Sample answer: Each digit to the left of another digit has a value that is ten times as much as that same digit would have in the place to its right.

Lesson 6 Place Value Through Thousandths
English Learner Instructional Strategy

Vocabulary Support: Review Vocabulary

Before the lesson, write the following place-value names on separate index cards: *thousandths, hundredths, tenths, ones, tens, hundreds, thousands.* Also create an index card for *decimal.* Display an unlabeled place-value chart consisting of seven place values. Randomly distribute the index cards to students. Have students work together to correctly label the place-value chart using the index cards.

Afterward, discuss the order of each place in the chart. Provide the following sentence frames to aid students in the discussion:

____ is ____ place(s) to the left of ____.
____ is ____ place(s) to the right of ____.

English Language Development Leveled Activities

Emerging Level	Expanding Level	Bridging Level
Number Sense	**Show What You Know**	**Building Oral Language**
Display some play money, for example, $2.45. Model counting the money. Write the amount using a decimal point. Point to the decimal point and say, *A decimal point separates the whole dollars and the cents. Which side shows the whole dollars? Which side shows the cents?* Have students answer by pointing. Write a decimal, such as 6.418. Say, *A decimal point separates whole numbers from tenths, hundredths, and thousands. Which number is in the tenths place?* Have students answer by pointing. Repeat with other place values.	Display a labeled place-value chart with decimals through thousandths. Say the digit of one place value in 462.941. Have a student volunteer write the digit in the correct place on the chart. Continue naming the place values in random order until the chart is filled. Next have students take turns directing you to write a new number in the chart. Provide the following sentence frame: **Write ____ in the ____ place.** Afterward, write the decimal in expanded form and word form. Discuss the different forms.	Have bilingual pairs write decimals in standard form, word form, or expanded form. One student will say a decimal number through thousandths and then direct his or her partner to write the number in a specified form. For example, the student may say, **Write two and fifty-four hundredths in expanded form.** The other student will write the number as directed. Students will switch roles and repeat the activity until each student has written at least one example of each form.

Multicultural Teacher Tip

Students from Latin American countries may write their numbers in slightly different forms than their American peers. In particular, ones and sevens can be easily confused. Latin American ones are written with a short horizontal line at top, and at first glance, they may appear to be American-style sevens. To clearly distinguish a seven from a one, a Latin American seven will include a cross-hatch at the middle of the upright. Other differences to note: Eights and fours are drawn from the bottom up. As a result, fours at times appear as nines. Nines may also be drawn with a curved descender, making them look like lowercase "g"s.

NAME _____ DATE _____

Lesson 6 Vocabulary Chart

Place Value Through Thousandths

Use the concept web to identify the place of each digit in the decimal. Write in word form.

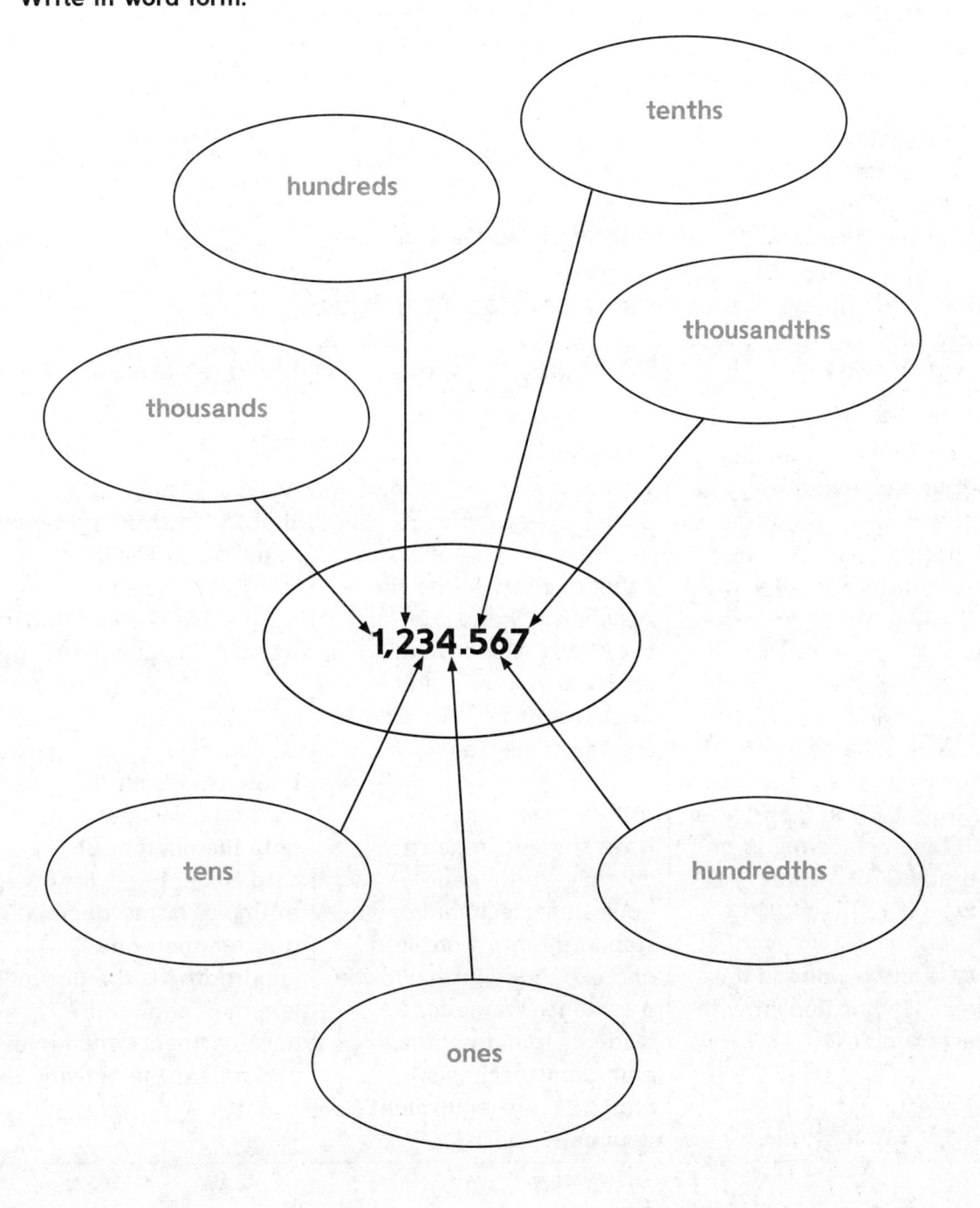

- hundreds
- tenths
- thousandths
- thousands
- **1,234.567**
- tens
- hundredths
- ones

Grade 5 • Chapter 1 *Place Value* 7

Lesson 7 Compare Decimals
English Learner Instructional Strategy

Language Structure Support: Communication Guide

Before the lesson, display a number line from 5.25 to 5.31 in increments of 0.01. Say, *Point to the end of the number line that is less.* Have students respond. Then ask, *Is it the left side or the right side?* Have students answer chorally. Then say, *Left is less.* Have students repeat chorally. Repeat for the greater/right side of the number line. During the lesson, display the following sentence frames to aid students in their responses:

____ is greater than ____.
____ is less than ____.
____ is equal to ____.
The digits in the ____ place are the same/different.

English Language Development Leveled Activities

Emerging Level	Expanding Level	Bridging Level
Number Sense Draw < and > on the board. Point to each symbol as you identify it. Draw a number line from 5.0 to 6.0. Label each tenth. Mark 5.4 and 5.2 on the line. Write 5.4 > 5.2. Say, *Five and four tenths is greater than five and two tenths.* Write 5.2 < 5.4 and read aloud the new comparison. Draw a number line from 0.75 to 0.85. Label each tenth. Place marks on 0.78 and 0.82. Write 0.78 ◯ 0.82 then ask, *Which symbol do I use?* Have students point to the correct symbol. Repeat with other examples.	**Making Connections** Write the words *equivalent decimals* and the Spanish cognate, *decimales equivalentes* on a classroom cognate chart. Write the decimals 0.3 and 0.30 on the board. Model shading in each decimal in a grid. Discuss that the two models are the same. Say, *Three tenths and thirty hundredths are equivalent decimals.* Have students repeat chorally. Write each equivalent decimal in standard form, word form, and expanded form. Provide a sentence frame for students to name other equivalent decimals: ____ and ____ are equivalent decimals.	**Academic Language** Create decimal cards by writing decimals (up to the thousandths) on individual index cards. Confidentially distribute a decimal card to each student. Have students work in bilingual pairs. Say, *You will try to guess your partner's number.* One student will ask comparison questions about the other student's decimal number until the number on the card is identified. For example, **Is the decimal greater than one hundredth?** Is the decimal less than one tenth? Then have partners switch roles and repeat the activity.

Teacher Notes:

NAME _____ DATE _____

Lesson 7 Four-Square Vocabulary
Compare Decimals

Write the definition for each math word. Write what each word means in your own words. Draw or write examples that show each math word meaning. Then write your own sentences using the words.

Definition	My Own Words
A number that has a digit in the tenths place, hundredths place, and beyond.	See students' examples.

decimal

My Examples	My Sentence
Sample answer: 0.5	Sample sentence: There are many different decimals between 0 and 1.

Definition	My Own Words
Decimals that have the same value.	See students' examples.

equivalent decimals

My Examples	My Sentence
Sample answer: 0.7 and 0.70 and 0.700 are all equivalent.	Sample sentence: Two decimals are equivalent if they have the same digits in the same place values, except for any zeros at the end of the decimal.

Lesson 8 Order Whole Numbers and Decimals

English Learner Instructional Strategy

Graphic Support: Word Web

Before the lesson, review *greatest* and *least*. Display a Word Web with *greatest* written in the center. Work with students to brainstorm words with similar meanings, such as *biggest, highest, largest*, and so on. Repeat the activity with *least*. Point out the common appearance of the suffix *-est*. Using realia, briefly explain the suffixes use in creating a superlative comparing three or more items. For example, display a small object and say, *The ____ is small.* Display a second, smaller object. Say, *The ____ is smaller than the ____.* Last, display a third, even smaller object. Say, *The ____ is smallest.*

Write 3.54 on the board. Have students suggest several other decimal numbers. Record them on the board. Then compare the new numbers to 3.54 using *less than, least, greater than*, and *greatest*. Have students repeat your comparisons chorally.

English Language Development Leveled Activities

Emerging Level	Expanding Level	Bridging Level
Word Knowledge	**Show What You Know**	**Building Oral Language**
Using four write-on/wipe-off boards, write a different whole number on each. For example: 23,947; 23,955; 39,748; 39,914. Distribute boards to four volunteers and have them stand in order from least to greatest. Ask, *Which is least? Which is greatest?* Allow students to respond by pointing. Have volunteers remain in their places and replace the comma with a decimal point in each number. For example: 23.947; 23.955; 39.748; 39.914. Explain that although the numbers are now decimals, the numbers are still in order from least to greatest.	Draw a labeled place-value chart on the board that includes ten thousand through thousandths. Have a student volunteer write one place value of a digit in the number 54.962. Say, *Write nine in the tenths place.* After the task is complete, have the student explain what he or she did using the following sentence frame: **I wrote ____ in the ____ place.** Be sure students are using the correct past tense form of *write*. Continue until the frame is filled. Repeat with other numbers.	Using a digital stop watch and a chart, record the amount of time it takes volunteers to complete a series of simple tasks, such as sharpening a pencil, getting a book out of a bag, tying a shoe, and so on. As students perform the tasks, record their times using decimal numbers to the hundredths or thousandths of a second, depending on what the stopwatch shows. After all times are recorded, work with students to categorize and order the times from least to greatest. Discuss the results as a group.

Teacher Notes:

T9 Grade 5 • Chapter 1 *Place Value*

NAME _____ DATE _____

Lesson 8 Note Taking
Order Whole Numbers and Decimals

Read the question. Write words you need help with and research each word. Use your lesson to write your Cornell notes. Write or draw math examples to explain your thinking. Share your examples with a classmate.

Building on the Essential Question	**Notes:**
How do you order whole numbers and decimals?	When comparing numbers, start comparing the numbers using the __greatest__ place value.
	Place value is the value given to a digit by its __position__ in a number.
	When you look at the place values of a decimal, each digit to the left has a value that is __ten__ times as much as that same digit would have in the place to its right.
	The hundreds place is __greater__ than the tens place.
Words I need help with:	The ten__ths__ place is __greater__ than the hundred__ths__ place.
See students' words.	You can also compare numbers by locating the numbers on a number line. A number line is a __line__ that represents numbers as __points__.
	When you look at a number line, each decimal number to the right of another decimal is __less__ than that number.
	After locating two decimals on a number line, the decimal to the __left__ is greater.

My Math Examples:
See students' examples.

Grade 5 • Chapter 1 *Place Value* **9**

Lesson 9 Problem-Solving Investigation Strategy: Use the Four-Step Plan

English Learner Instructional Strategy

Vocabulary Support: Frontload Academic Vocabulary

Write the word *plan* and the Spanish cognate, *el plan*, on a classroom cognate chart. Provide a concrete example, such as using a map to plan a route from one place to another.

Before the lesson, write Understand, Plan, Solve, and Check on separate index cards. Divide students into four groups and give each group one card. Have students brainstorm a math or non-math example for their group's word. Record the examples on the board. Then ask all students to help generate a list of synonyms for each word.

During the lesson, provide the following sentence frames to aid students in their participation: **I know ____. I need to find ____. My plan is to ____. The answer is ____. I can check my answer by ____.**

English Language Development Leveled Activities

Emerging Level	Expanding Level	Bridging Level
Word Knowledge	**Recognize and Act It Out**	**Academic Language**
Write the word *results* and the Spanish cognate, *los resultados* on a classroom cognate chart. Say, *I will buy lunch for my friend and me. Lunch will cost $8.50 a person. I have $20. Is it enough?* Ask students if they understand the problem. Have them give a thumbs-up or a thumbs-down. Ask students to help you plan. Say, *The plan is to estimate the total cost of two lunches.* Write the addition problem on the board and solve it. Say, *It will cost about $18 to buy two lunches, which is less than $20. The result of my estimate is that $20 is enough.*	Read aloud a problem from the lesson. Write Understand, Plan, Solve, and Check in a four column chart. Have students help you identify the information that is known and what problem needs to be solved, and write the information beneath Understand. Ask students how they would solve the problem. Allow them time to suggest different strategies and decide upon the best one. Write this plan in the chart. Have students help you solve the problem. After solving, check to determine if the result is reasonable. Record your work beneath Check.	Display the four problem-solving steps. Have bilingual pairs work together on a word problem from the lesson. Have one student in each pair read aloud the problem and identify what is known and what needs to be solved. The other student will plan how to solve the problem and carry out the plan to solve. His or her partner will check to see if the result is reasonable. Have pairs choose another word problem from the lesson and switch roles. Support the use of native language for clarification of problem-solving strategies.

Teacher Notes:

NAME _____ DATE _____

Lesson 9 Problem-Solving Investigation
STRATEGY: Use the Four-Step Plan

Use the four-step plan to solve each problem.

1. The **table** shows the number of **ounces** of butter **Marti** used in different recipes.
   She (Marti) has **6 ounces** of butter left.
   How many **ounces** of butter did **she** have at the **beginning**?

Recipe	Ounces of Butter
Pie	4
Cookie	8
Pasta	6

Understand	Solve
I know: I need to find:	
Plan	**Check**

2. At the end of their **3-day** vacation, the **Palmers** traveled a **total** of **530** miles.
   On the **third** day, they drove **75** miles.
   On the **second** day, they drove **320** miles.
   How many miles did **they** (the Palmers) drive the **first** day?

Understand	Solve
I know: I need to find:	
Plan	**Check**

Day	Miles Driven
1	
2	
3	

10 Grade 5 • Chapter 1 Place Value

Chapter 2 Multiply Whole Numbers

What's the Math in This Chapter?

Mathematical Practice 3: Construct viable arguments and critique the reasoning of others

Write the math problem 575 × 21 on the board. Tell students that you are going to round the two factors to estimate the product of the two whole numbers. Write 500 × 20 = 15,000 on the board. Say, *Is my estimation reasonable?* Allow time for students to think. Then have them turn and talk with a peer. Regroup and discuss their observations. The discussion goal should be for students to "construct viable arguments" proving that you rounded 575 incorrectly.

Ask, *What should I have done to estimate the product of the two whole numbers?* **Round each factor to its greatest place value. 21 rounds to 20 and 575 rounds to 600.**

Discuss with students how they critiqued your reasoning (rounding) and provided viable arguments as to why your estimation was not a reasonable answer to the original multiplication problem. Say, *All of you were applying Mathematical Practice 3.*

Display a chart with Mathematical Practice 3. Restate Mathematical Practice 3 and have students assist in rewriting it as an "I can" statement, for example: **I can construct an argument to show how to solve a problem**. Post the new "I can" statement.

Inquiry of the Essential Question:

What strategies can be used to multiply whole numbers?

Inquiry Activity Target: **Students come to a conclusion that the multiplication process is a strategy of repeated actions.**

As an introduction to the chapter, present the Essential Question to students. The inquiry graphic organizer will offer opportunities for students to observe, make inferences, and apply prior knowledge of problem solving strategies representing the Essential Question. As they investigate, encourage students to draw, write, and collaborate with peers to demonstrate their observations and thinking. Then have students present additional questions they may have to a peer to extend discussions.

Regroup students and restate Mathematical Practice 3 and the Essential Question. Pose questions to reflect on what has been learned to guide students in making connections between the Mathematical Practice and the Essential Question.

NAME _____ DATE _____

Chapter 2 Multiply Whole Numbers

Inquiry of the Essential Question:

What strategies can be used to multiply whole numbers?

Read the Essential Question. Describe your observations (I see..), inferences (I think...), and prior knowledge (I know...) of each math example. Write additional questions you have below. Then share your ideas and questions with a classmate.

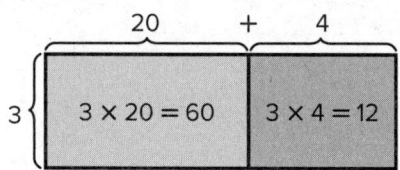

20 + 4 3 { [3 × 20 = 60] [3 × 4 = 12] }	I see …
	I think…
3 × 24 = (3 × 20) + (3 × 4) Find partial products. = 60 + 12 Multiply. = 72 Add.	I know…

21×10^3 = 21 × 1,000 ←	The power of 10 has three zeros.	I see …
		I think…
= 21,000 ←	The product has three zeros.	
		I know…

416 —(rounds to)→ 400 × 28 —(rounds to)→ × 30 ─────── 12,000		I see …
		I think…
		I know…

Questions I have…
_____
_____
_____

Grade 5 • Chapter 2 *Multiply Whole Numbers*   **11**

Lesson 1 Prime Factorization
English Learner Instructional Strategy

Vocabulary Support: Utilize Resources

Before the lesson, direct students to review the Glossary definition for *prime factorization* in both English and Spanish. Utilize other appropriate translation tools for non-Spanish speaking ELs.

During Math in My World, be sure students are correctly pronouncing the final /z/ sound in *factors* during the discussion.

Post and model using sentence frames such as the following to aid students in the discussion: **The prime factors of ___ are ___. ___ is a composite number. ___ is a prime number.** For bridging level students, provide a more complex sentence frame, such as: **I know ___ is a prime number because ___.**

English Language Development Leveled Activities

Emerging Level	Expanding Level	Bridging Level
Number Sense	**Memory Devices**	**Internalize Language**
Write the words *prime number* and *composite number* and the Spanish cognates, *número primo* and *número compuesto*, on a classroom cognate chart. Write a factor tree for 6. Say, *The factors of six are one, two, three, and six.* Write a factor tree for 5. Say, *Five has only two factors, one and itself. Five is a prime number.* Point to the 6 and say, *Six is a composite number. It has more than two factors.* Repeat with factor trees for other prime and composite numbers. Have students chorally say **prime** or **composite** to identify each number.	Write the word *factor* and the Spanish cognate, *factor*, on a classroom cognate chart. Display an image of a tree and the word *tree*. Point to and identify the trunk, branches, and leaves. Display a factor tree of the number 18. Say, *This is a factor tree.* Describe how the factor tree shows the factors of 18. Ask, *How are a real tree and a factor tree similar or different?* Discuss the similarities and differences with students. Have students help you create factor trees for other numbers and describe them.	Use composite numbers to create a set of number cards. Distribute the cards to bilingual pairs. Say, *Find the prime factorization.* Have pairs create a factor tree for their number on a write-on/wipe-off board. Distribute blank index cards to pairs. Say, *Write the prime factorization on the blank card.* Collect all cards and shuffle them. Redistribute the cards to individual students and have students find their match. Have pairs state the prime factorization using the following sentence frame: **The prime factorization of ___ is ___.**

Teacher Notes:

NAME _____ DATE _____

Lesson 1 Vocabulary Definition Map
Prime Factorization

Use the definition map to write a description and list characteristics about the vocabulary word or phrase. Write or draw math examples. Share your examples with a classmate.

My Math Vocabulary:

prime factorization

Description from Glossary:

A way of expressing a composite number as a product of its prime factors.

Characteristics from Lesson:

A factor is a number that is __multiplied__ by another number.

In the equation 3 × 2 = 6, the factors are __3__ and __2__.

A prime number is a whole number with exactly __two__ factors, 1 and itself.

3, 7, and 13 are __prime__ numbers.

A composite number is a whole number that has __more__ than __two__ factors.

12 is a __composite__ number. It has the factors: 1, __2__, 3, __4__, 6, and 12.

My Math Examples:
See students' examples.

12 Grade 5 • Chapter 2 *Multiply Whole Numbers*

Lesson 2 Inquiry/Hands On: Prime Factorization Patterns

English Learner Instructional Strategy

Language Structure Support: Tiered Questions

During the lesson, structure questions so students of different English proficiencies can answer in ways suitable to their level of understanding. For example, during the Build It lesson, you might point to the holes in the paper and ask, *How many?* Emerging students can answer with a gesture or a single word answer. For Expanding students, provide a sentence frame so they can answer in a complete sentence, such as, **There are ___ holes in the paper.** Encourage Bridging students to provide more complex answers by explaining the difference in the number of holes when the paper is folded compared to when it is unfolded.

During the Practice It part of the lesson, pair Emerging students with Expanding or Bridging students to complete the tables.

English Language Development Leveled Activities

Emerging Level	Expanding Level	Bridging Level
Modeled Talk	**Numbered Heads Together**	**Exploring Language Structure**
During Modeling the Math, write *times* on the board. As you say the word aloud, underline the final s and emphasize the /z/ sound. Write 3 × 3 × 3 × 3 on the board. As you model the expression with connecting cubes, say, *three times three times three times three*, and have students repeat after you. Monitor their pronunciation of the final /z/ sound and correct as needed.	Divide students into groups of four, and number the students in each group as one through four. Have the students in each group work together to answer Exercises 8–10 in the Apply It section of the lesson. Say, *Be sure everyone in your group understands the answers.* Come together again as a class. Call out a number from 1–4 randomly. Have students assigned to that number raise their hands. Call on a student to provide the answer to an exercise from the Apply It section.	Write *factorization* on the board. Below it, write *factorize + ation*. Below that write *factor + ize + ation*. Discuss the meanings of the suffixes (*-ize* means "to become something"; *-ation* means "the process of doing something") and how they change the meaning of the base word *factor*. Ask students for other words using the suffixes *-ize* and *-ation*. Next to the words on the board, create a factor tree for 18. Discuss how prime factorization is similar to and different from analyzing word parts.

Teacher Notes:

NAME _____ DATE _____

Lesson 2 Note Taking

Inquiry/Hands On: Prime Factorization Patterns

Read the question. Write words you need help with and research each word. Use your lesson to write your Cornell notes. Write or draw math examples to explain your thinking. Share your examples with a classmate.

Building on the Essential Question

How can you find patterns in prime factorization?

Words I need help with:
See students' words.

Notes:

A __prime__ number is a whole number with exactly __two__ factors, 1 and itself.

A __composite__ number is a whole number that has more than __two__ factors.

The numbers 2 and 3 are __prime__ numbers.

The number 6 is a __composite__ number.

Prime factorization is a way of expressing a __composite__ number as a product of its __prime__ factors.

The prime factorization of __6__ is 3 × 2.

The prime factorization of __12__ is 3 × 2 × 2.

The prime factorization of __24__ is 3 × 2 × 2 × 2.

The pattern I see is: __See students' examples.__

_____

_____ .

My Math Examples:
See students' examples.

Grade 5 • Chapter 2 *Multiply Whole Numbers* 13

Lesson 3 Powers and Exponents
English Learner Instructional Strategy

Sensory Support: Diagram

Write the words *exponent* and *base* and their Spanish cognates, *exponente* and *base*, on a classroom cognate chart. Utilize other appropriate translation tools for non-Spanish speaking ELs. Recreate the below diagram on chart paper. Use it as visual support during the lesson and post for future reference. (To color code, refer to the Teacher Edition.)

$$\text{BASE}^{exponent} \; \widehat{(4^3)} = 4 \times 4 \times 4$$

Use base 3 times

Gesture to the power and say, *Numbers expressed with exponents are called powers.* Point to the base and say, *The base is 4.* Gesture to the product and say, *The base is the number used as the factor.* Point to the exponent and say, *The exponent is 3.* Gesture to the product and say, *The exponent indicates how many times the base is used as a factor.*

English Language Development Leveled Activities

Emerging Level	Expanding Level	Bridging Level
Peer Modeling	**Memory Devices**	**Building Oral Language**
Pair emerging level students with an English proficient peer. Write expressions as products on index cards, for example $4 \times 4 \times 4 \times 4$ or $2 \times 2 \times 2 \times 2 \times 2 \times 2$. Distribute one card to each pair. Say, *On the back of the index card, rewrite the product as a power. Then find the value.* On the board, write the following sentence frames: **The base is ____. The exponent is ____. The value is ____.** Have the peer model using the sentence frames to describe the power written on their card. Then have the emerging level student echo the sentence frames.	Display images of a lamp and a mountain. Point to the base of each and say, *The base is the lowest, or bottom, part.* Write 4^3 and BASE^{exponent} on the board. Point to 4 and say, *The base is the lowest, or bottom, number of a power.* Write $4 \times 4 \times 4$. Circle the x in exponent and the times symbol. Stress the *ex* of exponent as you say, *The exponent indicates how many times the base is used. Four is used three times: four times four times four.* Stress the /z/ sound at the end of *times*. Display other powers. Have students name the bases and exponents.	Distribute two number cubes to each pair of bilingual students. Have pairs roll the cubes to create a power using one number as the base and the other as the exponent. Have pairs write the power on a piece of paper and then rewrite the power with the numbers' places reversed. For example, 4^3 and 3^4. Say, *Predict which power will have a greater value.* Have pairs write each power as a product and evaluate the values. Discuss and compare the values of each power. Ask, *Did you guess correctly? Which power is a greater value?*

Teacher Notes:

T14 Grade 5 • Chapter 2 *Multiply Whole Numbers*

NAME _____ DATE _____

Lesson 3 Vocabulary Chart

Powers and Exponents

Use the three-column chart to organize the vocabulary in this lesson. Write the word in Spanish. Then write the correct terms to complete each definition.

English	Spanish	Definition
base	base	In a <u>power</u>, the number used as a <u>factor</u>. In 10^3, the base is 10.
cubed	al cubo	A number raised to the <u>third</u> power; $10 \times 10 \times 10$, or 10^3.
exponent	exponente	In a power, the number of times the <u>base</u> is used as a <u>factor</u>. In 5^3, the exponent is 3.
power	potencia	A number obtained by raising a <u>base</u> to an <u>exponent</u>.
squared	al cuadrado	A number raised to the <u>second</u> power; 3×3, or 3^2.

Lesson 4 Multiplication Patterns
English Learner Instructional Strategy

Vocabulary Support: Modeled Talk

When presenting Example 1, gesture to each zero as you count aloud to reinforce the connection between the power of ten and the number of zeros. As you repeat the gestures, have students chorally say the power of ten and then count the zeros. Do this for each line of the table.

For Example 2, have a volunteer come to the board and, based on your model from Example 1, solve the problem aloud. Have the other students in the group follow along chorally as the student counts the zeros. Have a second volunteer come to the board to model Example 3. Provide sentence frames such as the following to aid in his or her model: **The basic multiplication fact is ____. There are ____ total zeros. The product is ____.**

English Language Development Leveled Activities

Emerging Level	Expanding Level	Bridging Level
Number Sense Write: 101 = 10; 102 = 100; 103 = 1,000. Say, *These are powers of ten and their products. Powers of ten have ten as the base. The exponent shows the number of zeros in the product.* Point to the exponent of each example and count the number of zeros in each value. Have students count aloud with you. Provide further examples of powers of ten. For each example, point to the exponent and ask, *How many zeros?* Have students respond chorally. Then have a volunteer come to the board and write the product.	**Listen and Write** Write the expression 2 × 3. Have students find the product. Ask a volunteer to write 6 to complete the expression. Write 2 × 30. Have students find the product and complete the expression. **60** Continue with 20 × 30. Ask, *Do you notice a pattern?* Provide a sentence frame to help students answer: **There are ____ zeros in the factors and ____ zeros in the product.** Continue writing similar series of expressions. Have students count the number of zeros aloud each time to find the product and complete each expression.	**Academic Language** Have students work in bilingual pairs to create expression cards. Ask them to write multiplication expressions using multiples and powers of ten on index cards. One student will find the product by multiplying mentally. Then he or she will explain the steps used to find the product to his or her partner. Provide the following sentence frames: **First, ____. Then, ____. Last, ____.** The partner will follow the steps to find the product and write it on the back of the card. Collect the cards and redistribute one to each pair to check the answers.

Teacher Notes:

NAME _____ DATE _____

Lesson 4 Concept Web
Multiplication Patterns

Use the concept web to write powers of 10 without exponents and the products of powers of 10.

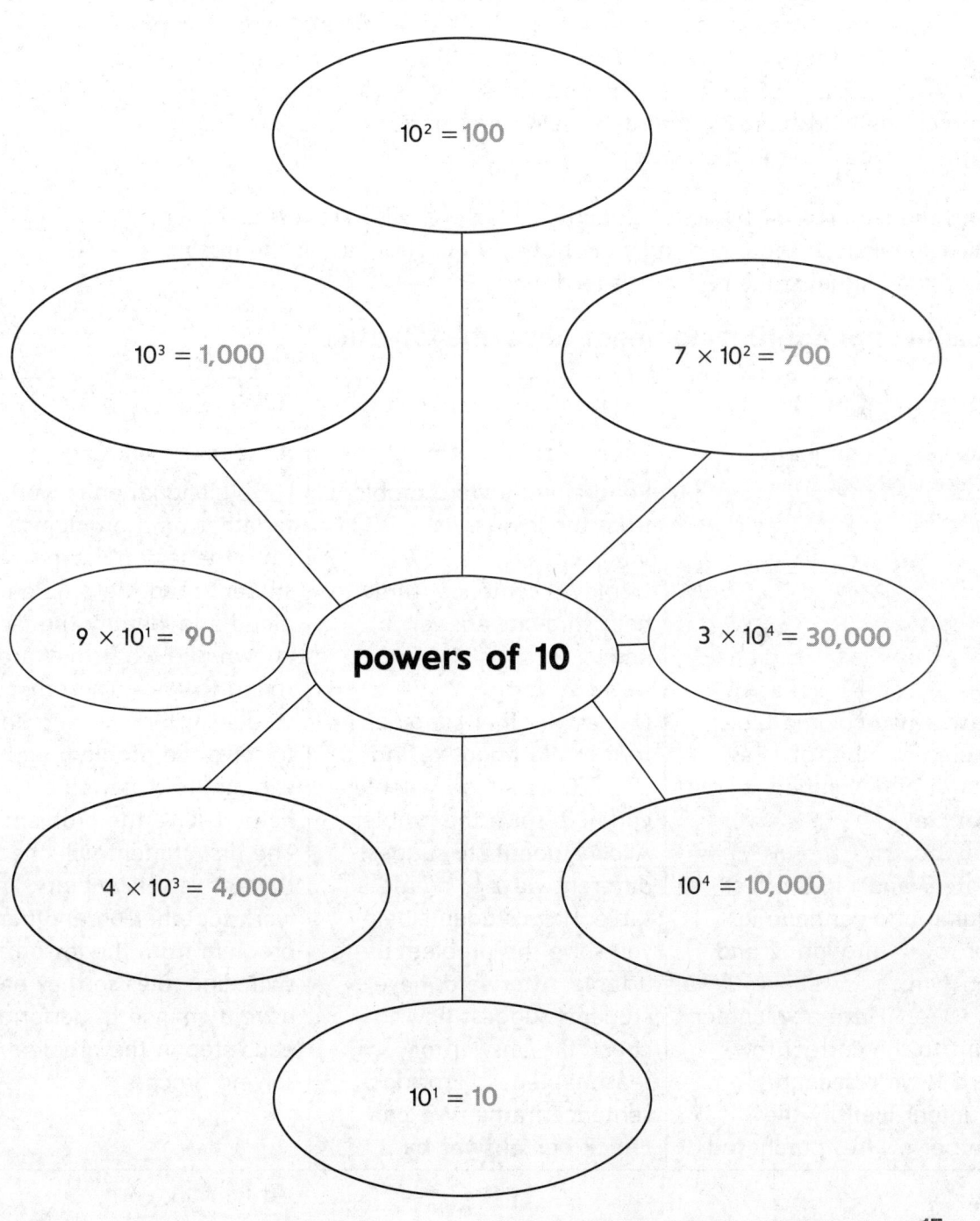

- $10^2 = 100$
- $10^3 = 1{,}000$
- $7 \times 10^2 = 700$
- $9 \times 10^1 = 90$
- **powers of 10**
- $3 \times 10^4 = 30{,}000$
- $4 \times 10^3 = 4{,}000$
- $10^4 = 10{,}000$
- $10^1 = 10$

Lesson 5 Problem-Solving Investigation Strategy: Make a Table

English Learner Instructional Strategy

Graphic Support: Tables

Write the word *table* and its Spanish cognate, *tabla*, on a classroom cognate chart. Provide concrete examples of tables included in non-math classroom resources, such as magazines, newspapers, or other nonfiction texts. Display a 2-column chart labeled *Math* and *Non-math*. Have students suggest math and non-math meanings of table, as well as different ways each is used. For example, one holds objects, the other holds information, and so on. Record students' suggestions in the chart.

During the lesson, pair bilingual students with those who are less proficient English speakers. If the emerging student has a question about the lesson, allow the bilingual student to ask it for them.

English Language Development Leveled Activities

Emerging Level	Expanding Level	Bridging Level
Internalize Language	**Recognize and Act It Out**	**Academic Language**
Write then say, *At a tire store, if you buy three tires, the fourth tire is free. A truck company bought tires at the store. They received three free tires. How many tires did they purchase?* Draw a two-column table labeled Number of Tires Bought and Number of Free Tires. Say, *This table will help me solve the problem.* Write 3 and 1 in the first column and continue for four rows (through 12 and 4). Ask, *Which row shows three free tires?* Have a volunteer point to the correct row. **third** Have a second volunteer identify the number of tires purchased. **9**	Read aloud a word problem from the lesson. Ask, *What information do we know?* Display a sentence frame to help students answer: **I know ___.** Ask, *What do we need to find out?* Display another sentence frame: **We need to find ___.** Discuss how a table can help solve the problem. Allow students to suggest different ways to create a table. Have students help you solve the problem using a table. Afterward, have students suggest how to check the answer for reasonableness. Provide a sentence frame: **We can check our answer by ___.**	Have bilingual pairs work together on a problem from the lesson. Have one student read the problem aloud and identify the facts known and what they are trying to solve. The other student will make a table to solve the problem and explain how the table helped solve the problem. The first student will check the answer. Have pairs work together on another problem from the lesson, switching roles so they each have a chance to perform each step in the problem-solving process.

Teacher Notes:

T16 Grade 5 • Chapter 2 *Multiply Whole Numbers*

NAME _____ DATE _____

Lesson 5 Problem-Solving Investigation
STRATEGY: Make a Table

Make a table to solve each problem.

1. **Betsy** is **saving** to buy a bird cage.
   She (Betsy) saves $1 the <u>first</u> week, $3 the <u>second</u> week,
   $9 the <u>third</u> week, and so on.
   How much **money** will she save in **5** weeks?

Understand	Solve
I know:	
I need to find:	

Plan	Check

Week	Money Saved
1	$1
2	$3
3	$5
4	
5	

2. **Kendall** is planning to buy a laptop for **$1,200**.
   Each month **she** <u>doubles</u> the amount she saved the <u>previous</u> month.
   If she saves **$20** the <u>first</u> month, in **how many months** will Kendall have
   enough money to buy the laptop?

Understand	Solve
I know:	
I need to find:	

Plan	Check
I will make a __table__.	

16 Grade 5 • Chapter 2 Multiply Whole Numbers

Lesson 6 Inquiry/Hands On: Use Partial Products and the Distributive Property

English Learner Instructional Strategy

Vocabulary Support: Review Vocabulary

Before the lesson, write the following monetary terms on separate index cards: *penny, nickel, dime, quarter, dollar.* Also create index cards showing each value: *1 cent, 5 cents, 10 cents, 25 cents, 1 dollar.* Distribute one index card per student, and give the remaining students play money representing each value (one coin or bill per student). Direct students to group themselves by values. For example, the students with pennies, the *penny* card, and the *1 cent* card form one group. Have students work together in their groups during the lesson.

Provide the following sentence frames to aid students during the Model the Math part of the lesson: **There are _____ dollars. There are _____ dimes. There are _____ pennies. The total value is _____ cents.**

English Language Development Leveled Activities

Emerging Level	Expanding Level	Bridging Level
Choral Responses	**Sentence Frames**	**Show What You Know**
Write and say, *partial products*. Have students chorally repeat. Underline *part* and say, *A partial products is **part** of the product.* As you model solving the Draw It and Try It exercises, pause to ask for answers at each step and give students a chance to respond chorally. Then write the answer and say it aloud. Have students chorally repeat. When you have partial products, point to each number as you say, *Fifty is one **part** of the product. Thirty-five is **part** of the product. They are the **partial products**.*	On the board, write 6 × 27. Display an area model on the board as well. Invite two volunteers to complete the area model and solve the problem using partial products. Have one student instruct the other in solving. Provide the following sentence frames for the student to use: **Write _____ in the _____ box. Multiply _____ times _____. The partial products are _____ and _____. _____ times _____ is _____.** Invite a new pair of volunteers to the board to solve a different problem using partial products.	Have students work in pairs. Distribute several index cards to each pair. On the board, write 3 × 68. Say, *On your index cards, write each step describing how you would solve this problem using partial products.* Then have pairs exchange cards. Say, *Follow the steps shown on the cards you were given to solve the problem.* Afterward, survey students to see which pairs did or did not solve the problem correctly. For incorrect answers, discuss which steps were incorrect or missing from their instructions.

Teacher Notes:

NAME _____ DATE _____

Lesson 6 Guided Writing

Inquiry/Hands On: Use Partial Products and the Distributive Property

How do you use partial products and the Distributive Property to multiply?

Use the exercises below to help you build on answering the Essential Question. Write the correct word or phrase on the lines provided.

1. Rewrite the question in your own words.
   See students' work

2. What key words do you see in the question?
   partial products, property, distributive

3. Decompose 15 into the sum of the tens and ones.
   15 = __10__ + __5__

4. Rewrite the multiplication expression.
   8 × 15 = __8__ × (__10__ + __5__)

5. The __distributive__ __property__ says that to multiply a sum by a number, you can multiply each addend by the same number and add the products.

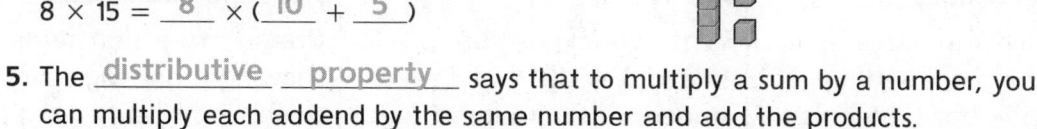

6. Use the Distributive Property to rewrite the multiplication expression.
   8 × 15 = __8__ × (__10__ + __5__) = (__8__ × __10__) + (__8__ × __5__)

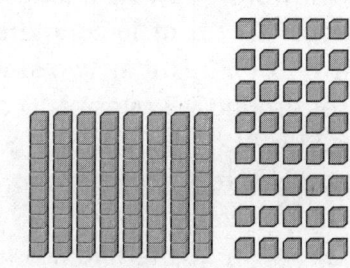

7. Find the product of 8 × 15.
   120

8. How do you use partial products and the Distributive Property to multiply?
   Break the two-digit number into tens and ones. Multiply the tens by the remaining factor. Multiply the ones by the remaining factor. The sum of the partial products will be the final product.

Lesson 7 The Distributive Property
English Learner Instructional Strategy

Sensory Support: Manipulatives

Before the lesson, write *distribute, distributed,* and *distributive* on the board. Write 5 × 4 on the board. Hand out five counters to four students. Say, *I distributed five counters to four students. How many counters did I distribute altogether?* Stress the /ed/ sound at the end of *distributed.* Have students help you find the product. Write 6 × 3 on the board. Have an expanding or bridging level volunteer hand out six counters to three students and say, **I distributed six counters to three students. How many counters did I distribute altogether?** Be sure the student correctly says the /ed/ ending of *distributed.* Say, *Distribute is a synonym of give or hand out.*

Write 5 × (20 + 3) on the board. Demonstrate the distributive property, stressing how the first part of the equation (5 ×) is "given" to both parts of the equation inside the parentheses.

English Language Development Leveled Activities

Emerging Level	Expanding Level	Bridging Level
Memory Devices	**Act It Out**	**Academic Language**
Ask students to name items in a single place setting (cup, plate, bowl, utensils). Write the words for the items on the board in a row, with + between each word. Say, *I need four place settings. I will multiply by four.* Put parentheses around the row of words. Write 4 × to the left of the row. Point to the first word and say, *I have one ____. I will multiply by four, and I will have four ____.* Below the word, write 4 × 1. Repeat for each item. Put parentheses around each expression and addition signs between them.	Write the expressions 3 × 16 and 3 × (10 + 6). Divide students into groups of four. Distribute counters to each group. In each group, have two students create 3 piles of 16 counters. Have the other two students create 3 piles of 10 counters and 3 piles of 6 counters. Say, *Count the total number of counters in your piles.* Write the findings next to each expression: 3 × 16 = 48 and 3 × (10 + 6) = (3 × 10) + (3 × 6) = 30 + 18 = 48. Provide a sentence frame for students to use in describing the expressions: **____ and ____ are equal.**	Create expression cards showing a one-digit number multiplied by a two-digit number on each card. Distribute one card to each bilingual pair. One student will use the Distributive Property to find the product mentally, describing each step. Provide sentence frames for students to use: **First, ____. Then, ____. Last, ____.** The other student will perform the steps as directed. For example, 7 × 23 = 7 × (20 + 3) = (7 × 20) + (7 × 3) = 140 + 21 = 161. Have students switch roles and repeat the activity.

Teacher Notes:

NAME _____ DATE _____

Lesson 7 Vocabulary Definition Map
The Distributive Property

Use the definition map to write a description and list characteristics about the vocabulary word or phrase. Write or draw math examples. Share your examples with a classmate.

My Math Vocabulary:

Distributive Property

Description from Glossary:
To multiply a sum by a number, you can multiply each addend by the same number and add the products.

Characteristics from Lesson:

A __property__ is a rule in mathematics that can be applied to all numbers.

A __product__ is the answer to a multiplication problem.

The order of operations states that you perform the operations in __parentheses__ before all other operations.

My Math Examples:
See students' examples

18 Grade 5 • Chapter 2 *Multiply Whole Numbers*

Lesson 8 Estimate Products

English Learner Instructional Strategy

Language Structure Support: Report Back

During the lesson, post and model using sentence frames such as the following to aid students in reporting back:

(Emerging level) ____ **rounds to** ____. **The product is about** ____.

(Expanding level) **When I round** ____ **up, I get** ____.
When I round ____ **down, I get** ____.

(Bridging level) **The estimate will be higher because** ____.
The estimate will be lower because ____.

Have students work in small groups for Independent Practice. Assign one problem to each group. Afterward, have a volunteer use the sentence frames to report back on how their group found the answer.

English Language Development Leveled Activities

Emerging Level	Expanding Level	Bridging Level
Number Sense	**Academic Vocabulary**	**Building Oral Language**
Model rounding using different place values. Write 267. Point to the tens place. Say, *I want to round to the tens place. Do I round up or down?* Have students respond by pointing up or down. Write 270. Say, *I want to round to the greatest place, the hundreds place. Do I round up or down?* Have students answer with a gesture. Write 300. Repeat the activity with 143. Repeat with other two-digit and three-digit numbers, rounding to different place values. Invite volunteers to come to the board and write the answers.	Display a clear container filled with paper clips. Ask, *How many paper clips are there?* Have students write their estimates on paper. Then ask several students to share their guesses. Provide the following sentence frame: **I estimate there ____ are paper clips.** Tell students the number of paper clips and compare it to the estimates. Write the multiplication expression 196 × 9 on the board. Say, *We will estimate the product for this expression.* Work with students to round the numbers and find an estimate. Compare the estimate to the exact product.	Say, *Things that are compatible go together well. Milk and cereal are compatible. What other things are compatible?* Provide a sentence frame for students to use: ____ **and** ____ **are compatible.** Record their suggestions on the board. Write the expressions 48 × 13 and 50 × 10. Point to 50 × 10 and say, *This expression is easier to solve because it uses compatible numbers.* Write more examples of expression pairs. Use compatible numbers in some expressions, but not in others. Have students identify which expressions have compatible numbers.

Teacher Notes:

NAME _____ DATE _____

Lesson 8 Note Taking

Estimate Products

Read the question. Write words you need help with and research each word. Use your lesson to write your Cornell notes. Write or draw math examples to explain your thinking. Share your examples with a classmate.

Building on the Essential Question	**Notes:**
How do you estimate products?	To round a number, find the __approximate__ value of the number.
	The symbol (=) means "is __equal__ to."
	The symbol (≈) means "is __about equal__ to."
	Round to the nearest ten.
	106 ≈ __110__
	47 ≈ __50__
	A __product__ is the answer to a multiplication problem.
	An estimate is a number close to an __exact__ value. An estimate indicates __about__ how much.
Words I need help with: See students' words.	Estimate the product by rounding each factor to the nearest ten.
	106 × 47 ≈ __110__ × __50__ = __5,500__
	__Compatible numbers__ are numbers in a problem that are easy to compute mentally.
	Even though 106 rounds to __110__, it is easier to mentally multiply by 100.
	Estimate the product using compatible numbers.
	86 × 23 ≈ __100__ × __20__ = __2,000__

My Math Examples:
See students' examples

Lesson 9 Multiply by One-Digit Numbers

English Learner Instructional Strategy

Collaborative Support: Share What You Know

Before the lesson, write *regroup* on the board with the prefix *re-* underlined. Point to the prefix and say, *The word part* **re-** *means "again or back." The word* **regroup** *means "group again."* Have students share any other words they know with the prefix *re-*, such as *review, return, rework,* and so on. Discuss how the prefix changes the meaning of the base word.

During the lesson, distribute base-ten blocks to students so they can use them when regrouping ones into tens and ones. Provide a sentence frame for students to report back on regrouping: **I can regroup ___ ones as ___ tens and ___ ones.**

English Language Development Leveled Activities

Emerging Level	Expanding Level	Bridging Level
Number Sense	**Listen and Write**	**Building Oral Language**
Write the expression 132 × 3 vertically. Say, *We can use the distributive property to find the product.* Rewrite the expression horizontally as 3 × (2 + 30 + 100). Have students find the product for each place value separately. Allow students to respond with a single-word answer or write the answer beside the expression. Record the corresponding digit in the correct place value for the vertical expression. Continue similarly with the other two place values. Add the products in the horizontal expression, and circle the value that completes each expression. **396**	Display the following sentence frame: ___ **times** ___ **is** ___. Write the expression 27 × 3. Model finding the product. Say, *First we multiply the ones. What is three times seven ones?* Have a volunteer answer using the sentence frame. Be sure he or she is correctly saying the /z/ sound at the end of times. Write 2 above the tens place and 1 in the product's ones place. Say, *Next we multiply tens. What is three times two tens?* Have another volunteer answer. Say, *Last we add the two tens we already had to six.* Write 8 in the product's tens place.	Create multiplication expression cards with factors of a three-digit and a one-digit number. Distribute one expression card to each bilingual pair. Have one student find the product by multiplying. Ask that he or she explain the steps used to find the product to his or her partner. Display sentence frames to help students describe the steps: **The first/ second/third/fourth step is ___.** The partner follows the steps and writes the product on the back of the card and checks using estimation. Have students switch roles and repeat the activity.

Teacher Notes:

NAME _____ DATE _____

Lesson 9 Multiple Meaning Word
Multiply by One-Digit Numbers

Complete the four-square chart to review the multiple meaning word or phrase.

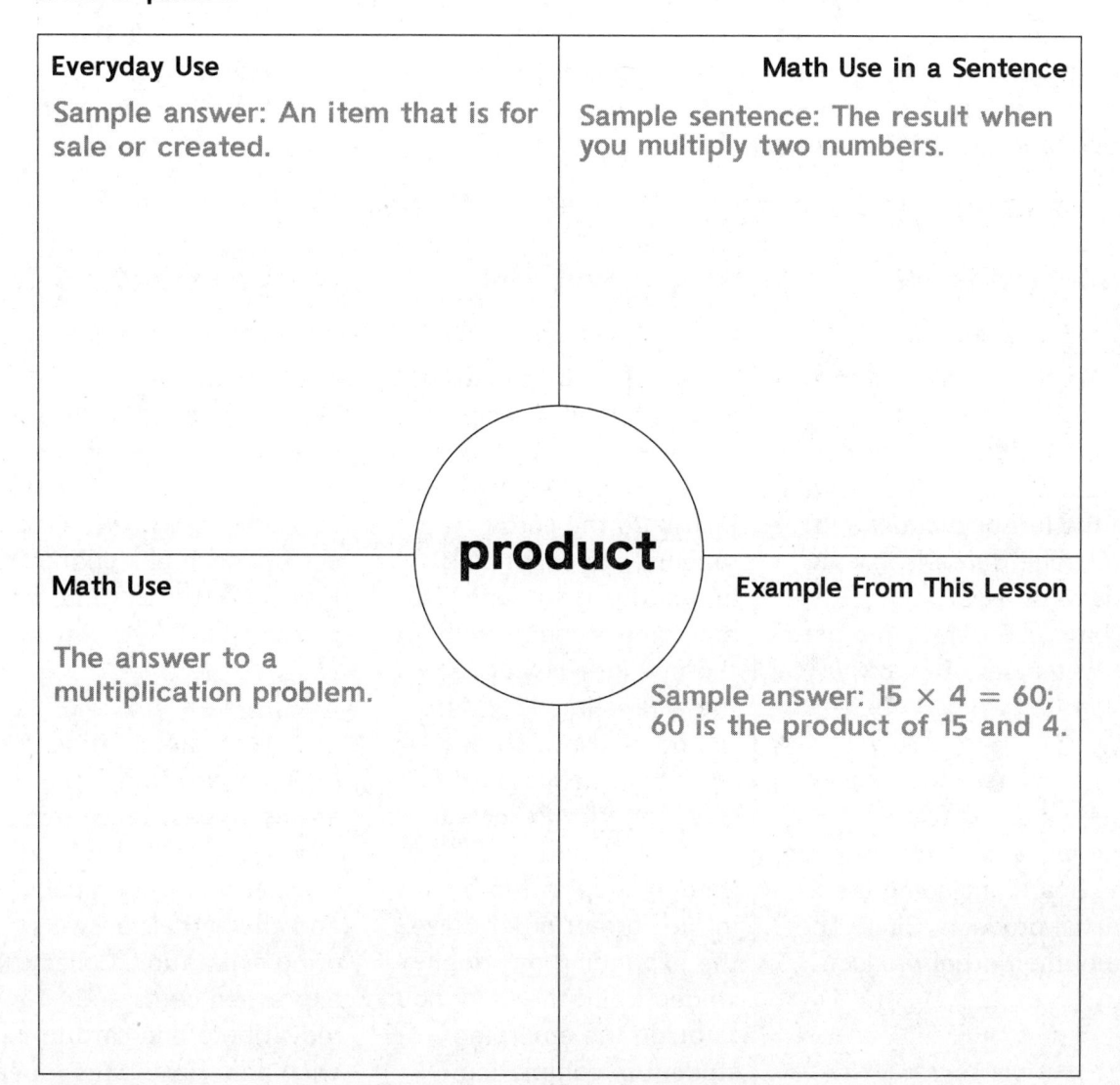

Everyday Use	**Math Use in a Sentence**
Sample answer: An item that is for sale or created.	Sample sentence: The result when you multiply two numbers.
Math Use	**Example From This Lesson**
The answer to a multiplication problem.	Sample answer: 15 × 4 = 60; 60 is the product of 15 and 4.

product

Write the correct term on each line to complete the sentence.

When using partial products to multiply, __add__ the partial products to find the __product__ of the __multiplication__ expression.

20 Grade 5 • Chapter 2 *Multiply Whole Numbers*

Lesson 10 Multiply by Two-Digit Numbers

English Learner Instructional Strategy

Language Structure Support: Choral Responses

For each step while you are modeling examples during the lesson, first describe what you will do, and then describe what you did. For example, say, *I am going to multiply the ones.* Have students repeat chorally. After you complete the step, say, *I multiplied the ones.* Have students repeat chorally again. Be sure students are correctly altering their pronunciation of the verb each time. In particular, listen for the /d/ sound applied to the past tense form. Be sure students are also saying the /z/ sound at the end of *ones*, and not pronouncing it as an /s/, as in *once*. Similarly, listen for correct pronunciations of the /z/ sound in *tens*, the /ed/ sound in *added*, and the /t/ sound in *checked*.

English Language Development Leveled Activities

Emerging Level	Expanding Level	Bridging Level
Word Knowledge Write the word *product* and the Spanish cognate, *producto*, on a classroom cognate chart. Write *partial* to the left of *product* and underline *part*. Display an image of a partially submerged object in a glass of water. Say, *Part of this object is in the water. It is partially in the water. It is not fully in the water.* Solve a two-digit by two-digit multiplication expression on the board, including the partial products. Circle and label the partial products. Say, *These are the partial products. They are part of the product, but not the total product.*	**Share What You Know** On separate index cards, write each step necessary to solve a two-digit multiplication expression. Distribute the cards randomly to expanding and bridging students. Pair emerging students with those who have cards. Write the expression 32 × 14 on the board. Have students work together to put the steps in order to solve the expression, and then have them line up at the board in the correct order. Have the expanding or bridging student read the card aloud to guide the emerging student in completing the step.	**Building Oral Language** Create multiplication expression cards with factors of a three-digit and a two-digit number written on each card. Distribute one card to each bilingual pair. Have one student find the product. Say, *Explain to your partner the steps you performed to find the product.* Monitor students' explanations to be sure they are using past-tense verbs. Have partners follow the steps to find the product and check their answers using estimation. Collect all expression cards and redistribute one card to each pair. Have pairs repeat the activity switching roles.

Multicultural Teacher Tip

EL students from Vietnam may have been taught a unique algorithm to check their answers after solving a multiplication problem. For example, here is how 473 × 12 = 5676 would be checked: First draw a large X. Add the digits of the top number in the problem (4 + 7 + 3) and write the result (14) at the top of the X. Add the digits of the bottom number (1 + 2) and write the result (3) at the bottom of the X. Multiply the top and bottom numbers of the X (14 × 3 = 42), add the digits of the product (4 + 2) and write the result (6) in the left space of the X. Finally, add the digits of the original answer (5 + 6 + 7 + 6 = 24), add the digits of the result (2 + 4), and write the number (6) in the right space of the X. If the numbers to the left and right of the X match, the answer is correct (6 = 6, so 5676 is the correct answer). Ask the student to model this algorithm for the class.

NAME _____ DATE _____

Lesson 10 Vocabulary Cognates

Multiply by Two-Digit Numbers

Use the Glossary to define the math word in English and in Spanish in the word boxes. Write a sentence using your math word.

product	**producto**
Definition The answer to a multiplication problem.	**Definición** Respuesta a un problema de multiplicación.

My math word sentence:

Sample answer: The product of 3 times 4 is 12.

estimate	**estimación**
Definition A number close to an exact value. An estimate indicates about how much.	**Definición** Número cercano a un valor exacto. Una estimación indica una cantidad aproximada.

My math word sentence:

Sample answer: The estimate of 23 times 4 is 80.

Chapter 3 Divide by a One-Digit Divisor

What's the Math in This Chapter?

Mathematical Practice 8: Look for and express regularity in repeated reasoning

Write the following pattern on the board: divide, multiply, subtract, bring down, divide, multiply, subtract, ____.

Ask, *What comes next in the pattern?* **bring down** Ask, *Have you ever seen this pattern or repeated process before?* Elicit from students this pattern shows the steps required to divide. Display a model showing the division process with two-digit dividends. Review the steps.

Discuss and model the steps used to divide are repeated calculations. Say, *When you solve a division problem, you are using **repeated reasoning**. If you remember to use these steps over and over to repeat the calculations, you will be able to divide greater numbers.*

Display a chart with Mathematical Practice 8. Restate Mathematical Practice 8 and have students assist in rewriting it as an "I can" statement, for example: **I can solve problems by using repeated calculations.** Have students draw or write examples of using repeated calculations. Post the examples and new "I can" statement.

Inquiry of the Essential Question:

What strategies can be used to divide whole numbers?

Inquiry Activity Target: **Students come to a conclusion that the division process is repeated calculations.**

As an introduction to the chapter, present the Essential Question to students. The inquiry graphic organizer will offer opportunities for students to observe, make inferences, and apply prior knowledge of division representing the Essential Question. As they investigate, encourage students to draw, write, and collaborate with peers to demonstrate their observations and thinking. Then have students present additional questions they may have to a peer to extend discussions.

Regroup students and restate Mathematical Practice 8 and the Essential Question. Pose questions to reflect on what has been learned to guide students in making connections between the Mathematical Practice and the Essential Question.

NAME _____ DATE _____

Chapter 3 Divide by a One-Digit Divisor

Inquiry of the Essential Question:

What strategies can be used to divide whole numbers?

Read the Essential Question. Describe your observations (I see...), inferences (I think...), and prior knowledge (I know...) of each math example. Write additional questions you have below. Then share your ideas and questions with a classmate.

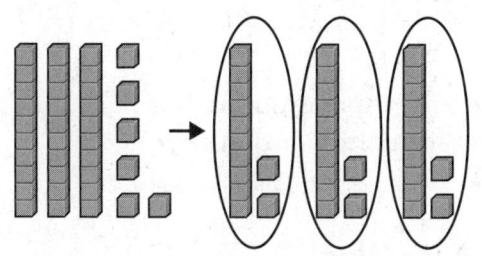

36 blocks can be arranged into 3 groups of 12 blocks each. So, 36 ÷ 3 = 12.

I see ...

I think...

I know...

	80	4	1
5	400	20	5

425 ÷ 5 = (400 + 20 + 5) ÷ 5
= 80 + 4 + 1
= 85

I see ...

I think...

I know...

```
    128
4)512
   -4
    11
   - 8
    32
   -32
     0
```

Step 1: Divide the hundreds.

Step 2: Divide the tens.

Step 3: Divide the ones.

I see ...

I think...

I know...

Questions I have...

_____

_____

_____

Lesson 1 Relate Division to Multiplication

English Learner Instructional Strategy

Graphic Support: Word Webs

During the New Vocabulary part of the lesson, display a word web with *unknown* written in the center. Underline the prefix un- and say, *This part of the word means "not." When un- is added to the beginning of a word, it changes the word's meaning. Unknown means "not known."* Have students brainstorm other words that use the prefix un- and record them in the word web. Have volunteers use the words in sentences to demonstrate meaning.

Display the following sentence frames to help students respond during the lesson: **What number times ____ is ____? A related multiplication fact is ____. The unknown number is ____.**

English Language Development Leveled Activities

Emerging Level	Expanding Level	Bridging Level
Word Knowledge Display images of families. Say, *A family is a group of related people. Families might share the same hair or eye color. Families might work together or play together.* Ask students to list common terms for family members (mother, uncle, grandfather, and so on) or name people in their own families. Write a fact family. Say, *This group of math facts is called a **fact family**. A fact family is a group of related facts that use the same numbers.* Display examples and nonexamples of fact families. Have students identify whether or not they are fact families.	**Show What You Know** Use counters to model each fact in a fact family. Write each fact on the board, gesture to it as you read it aloud, and have students repeat chorally. Say, *These facts all use the same three numbers. They are a **fact family**.* Distribute counters to students. Say, *Fact families use the same three numbers. You can use the same number of counters to model each fact.* Ask students to model the facts of a new fact family. Write the fact family on the board, and have students repeat chorally as you read aloud each fact.	**Building Oral Language** Write the word *variable* and the Spanish cognate, *variable*, on a classroom cognate chart. Pair students and assign each pair a number, such as 12, 20, 30, or 36. Say, *Find factors for your number and use them to write a fact family.* Ask pairs to rewrite their fact family on a blank index card, replacing one number with a variable. Have pairs exchange index cards and then solve to find the unknown. Provide a sentence frame for students to identify the variable: **The variable ____ is equal to ____.**

Teacher Notes:

NAME _____ DATE _____

Lesson 1 Vocabulary Cognates

Relate Division to Multiplication

Use the Glossary to define the math word in English and in Spanish in the word boxes. Write a sentence using your math word.

unknown	incógnita
Definition A missing value in a number sentence or equation.	**Definición** Valor desconocido en un enunciado numérico o una ecuación.

My math word sentence:
Sample answer: If you bought 4 tickets for $32, and want to know how much each individual ticket costs, the price of each ticket is the unknown. $4 \times ? = 32$

variable	variable
Definition A letter or symbol used to represent an unknown quantity.	**Definición** Letra o símbolo que se usa para representar una cantidad desconocida.

My math word sentence:
Sample answer: If you want to find how many $5 pizzas you can buy for $20, you have the variable p represent the number of pizzas. $5p = 20$

fact family	familia de operaciones
Definition A group of related facts using the same numbers.	**Definición** Grupo de operaciones relacionadas que tienen los mismos números.

My math word sentence:
Sample answer: $5 \times 4 = 20$, $4 \times 5 = 20$, $20 \div 4 = 5$, and $20 \div 5 = 4$ are all the same fact family.

Lesson 2 Inquiry/Hands On: Division Models

English Learner Instructional Strategy

Vocabulary Support: Activate Prior Knowledge

Write *equal groups* and the Spanish cognate, *grupos iguales,* on a classroom cognate chart. Provide several concrete models of the phrase's meaning, such as several equal groups of pencils, connecting cubes, students, and so on.

Display a KWL chart and, in the first column, record what students recall from previous grades about division. In the second column, record what students hope to learn during the lesson, including the use of models to represent and solve division problems.

After the lesson, display the following sentence frame and have students use it to describe what they learned during the lesson: **I learned that** ____. Use the third column of the KWL chart to record student responses.

English Language Development Leveled Activities

Emerging Level	Expanding Level	Bridging Level
Report Back Display the following sentence frames: ____ **equal groups.** ____ **left over.** Use base-ten blocks to represent 42, and then model 42 ÷ 6. Ask, *How many equal groups?* Allow students to report back with a single word answer, **seven** but then model using the sentence frame. If needed, count the groups aloud and then refer to the sentence frame as you say, *Seven equal groups.* Then ask, *How many are left over?* **none** Model using the sentence frame on the board and have students chorally repeat. Repeat the activity to model 52 ÷ 5.	**Non-transferrable Sounds** Divide students into pairs and distribute base-ten blocks and a number cube to each pair. Have one student in each pair use the base-ten blocks to create a dividend, and have the other student roll the cube to generate a divisor. Display the following sentence frames: ____ **equal groups.** ____ **left over.** Have students use the sentence frames to describe the division problem they modeled. Be sure students are correctly saying the /kw/ sound in *equal* and model pronunciation as needed.	**Round the Table** Divide students into groups of four and distribute base-ten blocks and a number cube to each group. Have one student in each group use the base-ten blocks to create a dividend. The student to his or her left rolls the number cube to generate a divisor. The next student to the left divides the blocks into the appropriate number of equal groups. Finally, the fourth student describes the quotient in terms of equal groups and blocks left over. Have groups repeat the activity starting with a different student. Circulate to monitor English grammar and usage.

Teacher Notes:

NAME _____ DATE _____

Lesson 2 Guided Writing

Inquiry/Hands On: Division Models

How do you model division?

Use the exercises below to help you build on answering the Essential Question. Write the correct word or phrase on the lines provided.

1. Rewrite the question in your own words.
   See students' work.

2. What key words do you see in the question?
   model, division

3. Identify the number modeled with the base-ten blocks. __48__

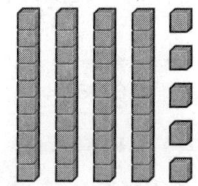

4. If you divide the tens into 3 equal groups, there will be __1__ ten in each group.

5. After dividing the tens into 3 equal groups, there will be __1__ ten remaining. If you regroup the remaining __1__ ten into ones, you will have __15__ ones altogether.

6. If you divide the 15 ones into 3 equal groups, there will be __5__ ones in each group.

7. Each group of tens and ones now has __1__ ten and __5__ ones.

8. So, 45 ÷ 3 = __15__.

9. How do you model division?
   Sample answer: Model the number being divided using base-ten blocks. Divide the tens into the number of equal groups for the division problem. Regroup any remaining tens as ones and then divide the ones. The number of blocks in each group is the answer to the division problem.

Lesson 3 Two-Digit Dividends
English Learner Instructional Strategy

Language Structure Support: Tiered Questions

During the lesson, be sure to ask questions according to students' level of English language proficiency. Use the following examples as a guide:

Emerging level: Ask simple questions that elicit one-word answers or allow the student to respond with a gesture, such as: *Which number is the remainder? Point to the dividend.* or *Do we add or subtract this number?*

Expanding level: Ask questions that elicit answers in the form of simple phrases or short sentences, such as: *What equation can we use to solve the problem?* or *What do we need to do next?*

Bridging level: Ask questions that require more complex answers, such as: *Explain how you know. What do we do if ____?* or *What steps do we need to take to solve the problem?*

English Language Development Leveled Activities

Emerging Level	Expanding Level	Bridging Level
Word Knowledge	**Show What You Know**	**Academic Language**
Write the words *dividend* in blue, *divisor* in green, and *quotient* in red, and their Spanish cognate's *dividend*, *divisor*, and *cociente*, on a classroom cognate chart. Say, *I can write ninety divided by ten in two different ways.* Model writing the problem using a division symbol and a division bracket. In both examples, use blue for the dividends, green for the divisors, and red for the quotient. As you model solving each equation, point to the words on the chart to reinforce how each word is represented in the equations. Have students repeat chorally.	Distribute write-on/wipe-off boards. Use a division bracket to write 43 ÷ 3 and narrate the steps as you model solving the equation. Say, *The quotient is 14 with a remainder of 1.* Have students use their write-on/wipe-off boards to solve another equation that results in a remainder. Have students model the division to find the quotient and remainder. Provide a sentence frame for students to share their answer: **The quotient is with a remainder of ____.** Be sure students are correctly saying the /kw/ sound in *quotient*.	Create index cards listing a one-digit divisor and a two-digit dividend on one side of the card. Distribute cards to pairs of bilingual students. Have one student read aloud the numbers using the sentence frame: **The divisor is ____, and the dividend is ____.** The other student will write and solve the problem. Ask the second student to say the answer with this sentence frame: **The quotient is ____, and the remainder is ____.** Ask students to check for reasonableness and record the quotient on the back of the card.

Teacher Notes:

T25 Grade 5 • Chapter 3 *Divide by a One-Digit Divisor*

NAME _____ DATE _____

Lesson 3 Vocabulary Chart
Two-Digit Dividends

Use the three-column chart to organize the vocabulary in this lesson. Write the word in Spanish. Then write the correct terms to complete each definition.

English	Spanish	Definition
dividend	dividendo	A number that is being <u>divided</u>.
divisor	divisor	The number that <u>divides</u> the <u>dividend</u>. The divisor tells you how many <u>groups</u>.
quotient	cociente	The result of a <u>division</u> problem.
remainder	residuo	The number that is <u>left</u> after one whole number is <u>divided</u> by another.
divisible or divide	divisible o dividir	Describes a number that can be divided into <u>equal</u> parts and has <u>no</u> remainder. or An operation on two numbers in which the first number is <u>split</u> into the same number of <u>equal</u> groups as the second number.

Grade 5 • Chapter 3 *Divide by a One-Digit Divisor* 25

Lesson 4 Division Patterns
English Learner Instructional Strategy

Collaborative Support: Act It Out

Have ELs do the following activity with an aide or older bilingual peer before the lesson. Ask students to suggest basic division facts, such as 42 ÷ 6 = 7. Write each suggestion on the board. Below it, write a multiple of ten division problem that uses the same basic fact, but without the quotient, such as 4200 ÷ 60. Demonstrate how to solve the equation by first crossing out the same number of zeros in both the dividend and divisor. Display the following sentence frame and model using it: **I crossed out ___ zeros.** Clearly pronounce the /t/ sound at the end of *crossed*. Be sure to listen that students are indicating past tense when they use the sentence frame. Then have the students help solve the simplified problem (420 ÷ 6 = 70). Repeat with students' other suggestions of basic division facts. Have volunteers model to solve.

English Language Development Leveled Activities

Emerging Level	Expanding Level	Bridging Level
Word Knowledge	**Partners Work**	**Academic Language**
Write the word *multiple* and the Spanish cognate, *múltiplo*, on a classroom cognate chart. Write 3 × 4 = 12. Point to 12 and say, *Twelve is a multiple of three, and twelve is a multiple of four.* Stress multiple each time, and have students chorally repeat the word. Write 10 × 1 = 10; 10 × 10 = 100; 10 × 100 = 1,000. Underline the 10 in each equation. Point to each product as you say, *These are multiples of ten.* Write 280; 2,800; 28,000. Say, *These are also multiples of ten.* Help students identify the other common factor of the numbers. **28**	Write the expression 28 ÷ 4. Have a volunteer write the quotient. **7** Write 280 ÷ 4 and 2,800 ÷ 4. Have volunteers write each quotient. **70; 700** Ask, *Do you notice a pattern?* Gesture to the quotients and say, *The number of zeros in the quotient is equal to the number of zeros in the dividend.* Underline the 28 and 4 in each expression. Say, *For each expression, the basic fact is 28 ÷ 4.* Have partners create similar patterns of three expressions each. Have a student from each pair use a sentence frame to identify the basic fact: **The basic fact is ___.**	Create expression cards by writing multiples of 10 of basic division facts on index cards. Distribute one card to each bilingual pair. Have one student find the quotient by dividing mentally. Ask that he or she explain the steps used to find the quotient to his or her partner. Provide the following sentence frames: **First ___. Then ___. Last ___.** The partner follows the steps and writes the quotient on a separate sheet of paper. Collect the cards, redistribute them, and have pairs switch roles.

Teacher Notes:

NAME _____ DATE _____

Lesson 4 Multiple Meaning Word
Division Patterns

Complete the four-square chart to review the multiple meaning word.

Everyday Use	Math Use in a Sentence
Sample answer: When something has several parts. A family can have multiple children in the family.	Sample sentence: When a number can be divided by another number without a remainder it is a multiple of that number.
Math Use	**Example From This Lesson**
A multiple of a number is the product of that number and any whole number.	Sample answer: 20 is a multiple of 5; 20 ÷ 5 = 4.

(center: **multiple**)

Write the correct numbers on each line to complete the sentences.

Multiples of __10__ are 10, 100, 1,000. If 24 ÷ 8 = 3, then 240 ÷ 8 = __30__.

Lesson 5 Estimate Quotients

English Learner Instructional Strategy

Sensory Support: Mnemonic Chant

Write then have students practice the following chant: **Four down to zero, round down. Five up to nine, round up.** Review rounding numbers using a ball or other round object. Write a three-digit number on the board, such as 437. Say, *I am going to round to the nearest ten.* Point to the number 7. Ask, *Do I round up or down?* Toss the ball up into the air and catch it as you say, *Five up to nine, round up.* Model rounding the number to 440. Then say, *I am going to round to the nearest hundred.* Point to the number 3. Ask, *Do I round up or down?* Drop the ball down from one hand into the other as you say, *Four down to zero, round down.* Model rounding the number to 400. Write a new three-digit number and hand a ball to each student. Repeat the above activity. Have the students first indicate how they would round by tossing the ball up or dropping the ball down. Then have them say the corresponding part of the chant.

English Language Development Leveled Activities

Emerging Level	Expanding Level	Bridging Level
Number Sense	**Academic Vocabulary**	**Building Oral Language**
Model rounding using the number 476. Write 476 and point to the tens place. Say, *We will round to the tens place.* Stress *round* and have students say **round** chorally. Be sure they are correctly saying the /aw/ sound. Model rounding to the tens place. **(480)** Point to the hundreds place. Say, *Now we will round to the greatest place, the hundreds place.* Have students repeat **round** chorally again. Model rounding to the hundreds place. **(500)** Repeat with other three-digit numbers, rounding to different place values for each number.	Write the word *estimate* and the Spanish cognate, *estimar*, on a classroom cognate chart. Write 273 ÷ 9. Say, *We will find an estimate for this expression.* Write 270 ÷ 9 below the original expression. Say, *I rounded 273 down to 270.* Have students find the quotient of 270 ÷ 9. Say, *The estimate is 30. The exact quotient of 273 ÷ 9 is 30 R3.* Write each quotient next to each expression. Ask, *Was 30 a good estimate?* Display a sentence frame for students: ___ **was/was not a good estimate because** ___. Repeat with other division expressions.	Review compatible numbers. Write examples of division expression pairs, with one using compatible numbers and the other not. Have students identify which expression is easier to solve. Display a sentence frame: ___ **uses compatible numbers.** Say, *You can use compatible numbers when estimating.* Provide an example, such as 237 ÷ 4. Model rounding 237 to 240. Say, *We round 237 to 240 because 24 is compatible with 4.* Write more expressions on the board and have students use rounding to create compatible numbers and estimate the answer.

Teacher Notes:

NAME _____ DATE _____

Lesson 5 Vocabulary Definition Map
Estimate Quotients

Use the definition map to write a description and list characteristics about the vocabulary word or phrase. Write or draw math examples. Share your examples with a classmate.

My Math Vocabulary:

compatible numbers

Characteristics from Lesson:

When finding compatible numbers, you may <u>round</u> the divisor, the dividend, or both the divisor **and** the dividend.

Description from Glossary:

Numbers in a problem that are easy to work with mentally.

Rounding means to find the <u>approximate</u> value of a number.

When finding compatible numbers, you may change the <u>dividend</u> to a number that is in the fact family with the divisor.

My Math Examples:
See students' examples

Grade 5 • Chapter 3 *Divide by a One-Digit Divisor*   **27**

Lesson 6 Inquiry/Hands On: Division Models with Greater Numbers

English Learner Instructional Strategy

Vocabulary Support: Anchor Chart

Divide students into four groups. Say, *Make an anchor chart showing what you know about division.* Explain that each chart should include a title at the top of the poster and definitions for math vocabulary related to division, such as *dividend, divisor, quotient, remainder, regrouping,* and so on. Suggest that students include an example of a long division problem with labels indicating the different elements. When the charts are completed, have groups display and describe their charts. Each group's description should include vocabulary words introduced up to this point in Chapter 3. Afterward, discuss how the anchor charts can help students better understand the steps needed to solve division problems with greater numbers.

English Language Development Leveled Activities

Emerging Level	Expanding Level	Bridging Level
Choral Responses	**Pairs Check**	**Basic Vocabulary**
Write *ones, tens, hundreds* on the board. Model solving 257 ÷ 2 using base-ten blocks. As you narrate the steps to solve and use each word listed on the board, point to it and have students chorally repeat. Clearly enunciate the final /z/ sound when indicating a plural, and be sure students are saying the sound correctly. Write a new problem on the board and invite a volunteer to model solving it with base-ten blocks. Narrate each step for the student, and have the other students chorally repeat after you.	Have students work in pairs to solve Apply It Exercises 6 and 7. Have one student complete Exercise 6 as the other student provides support and suggestions. Then have students switch roles to complete Exercise 7. Afterward, have each pair meet with another pair to compare answers. Once students have agreed on the correct answers, have them share with the class. Provide sentence frames to help students share their answers: **There were ___ newspapers in each box. A monkey's heart beats ___ times in one minute.**	Display a list of sequence words, such as *first, second, then, next,* and *last.* Have students work in pairs, and distribute index cards to each pair. Say, *Use the sequence words to describe how to use base-ten blocks to model division with greater numbers.* Have pairs write a step-by-step description on the index cards of how to model solving 432 ÷ 4. Then have pairs exchange cards and follow the steps using base-ten blocks to solve the problem.

Multicultural Teacher Tip

When EL students are calculating the mean, you may notice those from Europe and some parts of Latin America using a symbol similar to a long division bracket, but turned upside down. The dividend is written outside the bracket to the left, and the divisor is written inside the symbol. For example, 352 ÷ 8 is written as 352 ⌊8 . Calculations are made below the dividend.

T28 Grade 5 • Chapter 3 *Divide by a One-Digit Divisor*

NAME _____ DATE _____

Lesson 6 Note Taking

Inquiry/Hands On: Division Models with Greater Numbers

Read the question. Write words you need help with and research each word. Use your lesson to write your Cornell notes. Write or draw math examples to explain your thinking. Share your examples with a classmate.

Building on the Essential Question

How do you model division with greater numbers?

Words I need help with:
See students' words

Notes:

When you model the number 484 using base-ten blocks, you use __4__ hundreds flats, __5__ tens rods, and __2__ single cubes.

452 ÷ 4 = ?

To model the division, start by dividing the hundreds into __4__ equal groups, there will be __1__ hundred in each group.

After dividing the hundreds into 4 equal groups, there will be __0__ hundreds remaining.

If you divide the tens into 4 equal groups, there will be __1__ ten in each group.

After dividing the tens into 4 equal groups, there will be __1__ ten remaining. If you regroup the remaining __1__ ten into ones, you will have __12__ ones altogether.

If you divide the 12 ones into 4 equal groups, there will be __3__ ones in each group.

Each group has __1__ hundred, __1__ ten, and __3__ ones.

452 ÷ 4 = __113__

My Math Examples:
See students' examples

28 Grade 5 • Chapter 3 *Divide by a One-Digit Divisor*

Lesson 7 Inquiry/Hands On: Distributive Property and Partial Quotients

English Learner Instructional Strategy

Collaborative Support: Show What You Know

Write the phrases *Distributive Property* and *partial quotient* and their Spanish cognates, *propiedad distributiva* and *cocientes parciales*, on a classroom cognate chart. Write 5 × 57 on the board and draw a blank area model next to it. Say, *We can we use the Distributive Property and partial products to find the answer.* Invite a volunteer to come to the board, and have the other students guide him or her in solving the problem using the area model. Write *488 ÷ 4* on the board and draw a blank bar diagram next to it to use for solving. Say, *We can we use the Distributive Property and partial quotients to find the answer.* Model solving the problem using the bar diagram, and then ask students to describe similarities in the steps used to solve both problems. Display a sentence frame for students to use: **In both problems we ____.**

English Language Development Leveled Activities

Emerging Level	Expanding Level	Bridging Level
Exploring Language Structures	**Listen and Write**	**Partners Work**
On one side of the board, write *I am ____*. On the other side, write *You are ____*. Invite a student to the board to solve 248 ÷ 2 using a bar diagram. As the student works, say, *You are dividing. What are you doing?* Emphasize *You* and point to the corresponding sentence frame. Then point to the other sentence frame as you prompt the student to answer: **I am dividing.** Use the sentence frames to describe other steps as the student solves: **I am adding. I am using partial quotients.** and so on. Write a new problem and invite another student to solve.	Divide students into pairs. Say, *I'm going to tell you a word problem. Take notes. Then work with your partner to solve the problem.* Then share the following word problem: *Each month I saved the same amount of money. After 6 months, I had $450. How much did I save each month?* Give students time to complete the task. Display sentence frames to help students share their answers: **Each month you saved ____. I know this is the answer because ____.**	Write *375 ÷ 5 = ____* on the board. Pair students and say, *Work with your partner to create a real-world math problem for this number sentence.* Allow students time to complete the task, and then have partners exchange word problems with others. Direct partners to read the math problem they received and correct any mistakes in spelling or grammar. Then have them solve the problem and write a complete sentence describing the answer. Ask volunteers to read their answers.

Teacher Notes:

NAME _____ DATE _____

Lesson 7 Concept Web

Inquiry/Hands On: Distributive Property and Partial Quotients

Use the concept web to identify the partial quotients and partial sums involved in the division problem.

- 30 ÷ _3_ = _10_
- 600 ÷ _3_ = _200_
- 9 ÷ _3_ = _3_
- **639 ÷ 3 = ?**
- 639 = 600 + _30_ + _9_
- 200 + _10_ + _3_ = _213_

Grade 5 • Chapter 3 *Divide by a One-Digit Divisor* 29

Lesson 8 Divide Three- and Four-Digit Dividends

English Learner Instructional Strategy

Collaborative Support: Think-Pair-Share

For Independent Practice, pair emerging students with expanding or bridging students. Have students complete the exercises on their own, then pair up and share/compare solutions. For Problem Solving, randomly assign one of the Exercises 14–16 to each pair and have them work together to solve Have the more proficient English speaker coach their partner to report the solution back to you. Display the following sentence frames and have students utilize those most appropriate for their exercise.

We divided ____ by ____ to find the answer.
Each game system costs ____.
The cars travel ____ yards per minute.
One kangaroo weighs ____ pounds.
We used ____ to check our answer.

English Language Development Leveled Activities

Emerging Level	Expanding Level	Bridging Level
Look, Listen, and Identify Write *dividend, divisor, quotient*, and *remainder* on the board. Model solving 521 ÷ 3 using a division bracket. **173 R2** As you say each word listed on the board, identify its counterpart in the division problem. Have students chorally repeat each word after you say it. Model solving another division problem with a remainder using a bracket. Have volunteers approach the board to identify each component of the problem. Students can either point to the component as they say the word aloud, or they can point to it after you say the word.	**Recognize and Act It Out** Display the following sentence frames: **The dividend is ____. The divisor is ____.** Write 343 ÷ 3. Ask students to identify the dividend and divisor. Display 343 cubes using base-ten blocks. Say, *I have 343 blocks. I will divide by distributing 343 blocks to 3 students.* Record each step as it is acted out. First distribute the hundreds, then the tens, and then the ones. Afterward, say, *Each student has 114 cubes and one cube remains.* Ask a student to write the quotient. **114 R1** Have students use blocks to solve another, similar division problem.	**Academic Language** Display the following sentence frames: **The divisor is ____. The dividend is ____. The quotient is ____. The remainder is ____.** Using problems found in the lesson, create expression cards with one-digit divisors and three- or four-digit dividends. Distribute cards to student pairs. Have one student read aloud the components of the problem. Have the other student write the division problem on another sheet of paper, solve it, and then identify the quotient and any remainder. Ask students to check their answers for reasonableness.

Teacher Notes:

NAME _____ DATE _____

Lesson 8 Vocabulary Definition Map

Divide Three- and Four-Digit Dividends

Use the definition map to write a description and list characteristics about the vocabulary word or phrase. Write or draw math examples. Share your examples with a classmate.

My Math Vocabulary:

place value (in division)

Characteristics from Lesson:

When dividing a three- or four-digit number, divide from <u>greatest</u> place value to <u>least</u> place value.

Divide each place value by the <u>divisor</u> to find the partial quotients. Multiply the partial quotient and divisor to find the product.

Subtract the <u>product</u> from the <u>dividend</u> and compare to the divisor before moving to the next place value.

Description from Glossary:

The value given to a digit by its position in a number.

My Math Examples:
See students' examples

30 Grade 5 • Chapter 3 *Divide by a One-Digit Divisor*

Lesson 9 Place the First Digit
English Learner Instructional Strategy

Language Structure Support: Report Back

As you work through the Guided Practice Exercises 1–2 with students, allow them to participate and report back in a manner compatible with their level of English proficiency. Allow emerging students to respond with single-word answers or gestures.

Expanding students should be able to respond with simple phrases and short sentences. Provide sentence frames, such as the following, to aid as they report back: **There are not enough ___. I need to regroup ___. The quotient is ___ with a remainder of ___.**

For bridging students, provide sentence frames that prompt more complex answers, such as: **I had to regroup because ___. The answer is reasonable because ___.**

English Language Development Leveled Activities

Emerging Level	Expanding Level	Bridging Level
Number Sense Write 48 ÷ 6 = 8. Point to each number in the equation and identify it as *divisor*, *dividend*, or *quotient*. Have students chorally repeat each word after you. Next write 35 ÷ 7 = *x*. Point to each aspect of the equation again and identify it as before. Ask, *Which is not a number: the divisor, dividend, or quotient?* Have students answer **quotient** chorally. Say, *The quotient is unknown, so we put a letter there. The letter is called a variable.* Repeat with a new equation with a variable as the dividend or divisor.	**Recognize and Act It Out** Write 215 ÷ 3. Display 215 cubes using base ten blocks. Say, *I have 215 blocks. I will divide by distributing 215 blocks to 3 students.* Record each step as it is acted out. Say, *I start with the hundreds. I have two hundreds, but there are three students. I must regroup the hundreds.* Model regrouping 2 hundreds as 20 tens, and then distribute the total of 21 tens. Distribute the ones, and then say, *Each student has seventy-one cubes and two cubes remain.* Have students act out a similar division problem using blocks to find the quotient and remainder.	**Academic Language** Display the following sentence frames: **The divisor is ___. The dividend is ___. The quotient is ___. The remainder is ___.** Using expressions from the lesson, create division cards listing a one-digit divisor and a four-digit dividend. Distribute cards to student pairs. Have one student read aloud the components of the problem. Have the other student write the division problem on another sheet of paper, solve it, and then identify the quotient and any remainder. Ask students to check their answers for reasonableness.

Teacher Notes:

NAME _____ DATE _____

Lesson 9 Multiple Meaning Word
Place the First Digit

Complete the four-square chart to review the multiple meaning word.

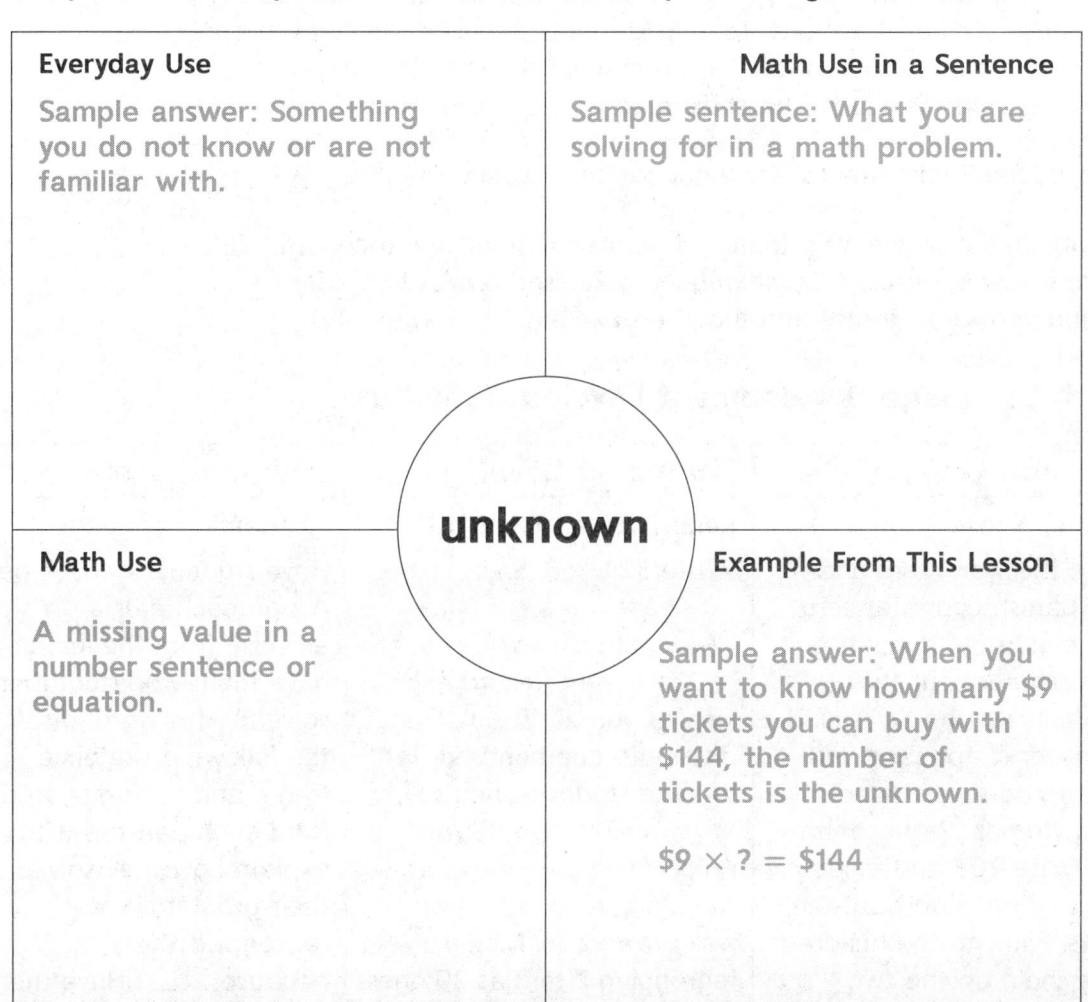

Everyday Use	Math Use in a Sentence
Sample answer: Something you do not know or are not familiar with.	Sample sentence: What you are solving for in a math problem. |￼
Math Use	**Example From This Lesson**
A missing value in a number sentence or equation.	Sample answer: When you want to know how many $9 tickets you can buy with $144, the number of tickets is the unknown. $9 × ? = $144

(center circle: **unknown**)

Write the correct term on the line to complete the sentence.

Alexander has 138 cards to organize into sheets that hold 6 cards each. He wants to **know** how many sheets he will need to place every card into a sheet. The unknown is the number of __sheets__.

Grade 5 • Chapter 3 *Divide by a One-Digit Divisor* **31**

Lesson 10 Quotients with Zeros
English Learner Instructional Strategy

Graphic Support: Organized List

Before Guided Practice, work with students to create a list on chart paper of the steps needed to solve a division problem having quotients with zeros. Display Exercise 1. As you work with students to solve the problem, record each step on the chart. To help students describe the steps in order, display sentence frames containing ordinal words: **The first step is** _____. **The second step is** _____. **The third step is** _____. and so on.

Have students utilize the list for Independent Practice.

For Problem Solving, have a bilingual peer or aide review the words: *fish, aquarium, tank, music, CDs, neighbors, rake,* and *leaves* to clarify meaning. Provide a translation tool if a peer/aide is unavailable.

English Language Development Leveled Activities

Emerging Level	Expanding Level	Bridging Level
Number Sense Write the word *zero* and the Spanish cognate, *cero*, on a classroom cognate chart. Throughout the lesson, stress the /z/ sound whenever it appears, and listen to be sure students are saying /z/ rather than /s/. Write 203 and model the number using base-ten blocks. Point to the hundreds place, hold up the two hundred base-ten blocks, and ask, *How many hundreds?* Have students answer two hundreds chorally. Point to the tens place. Show your empty hands. Ask, *How many tens?* Have students answer **zero tens** chorally. Finish with the ones place.	**Recognize and Act It Out** Write 324 ÷ 3. Say, *I have 324 base-ten blocks. I will divide by distributing them to 3 students.* Record each step as you act it out. First distribute one hundred flat to each student. Then say, *I will now distribute the tens. I have two tens, but there are three students. I must regroup.* Model regrouping 2 tens as 20 ones. Ask, *What do I write in the tens place?* Have students answer **zero** chorally. Distribute the ones and say, *Each student has one hundred and eight cubes.* Repeat with another division problem from the lesson.	**Academic Language** Have students work in pairs. Assign each pair a division exercise from the lesson and have them work together to solve it. Afterward, display the following sentence frame and have one student from each pair use it to explain how they solved their problem: **We regrouped the _____ because _____.** Have the other student use the following sentence frame to further explain the process: **We wrote zero in the _____ place because _____.**

Teacher Notes:

NAME _____ DATE _____

Lesson 10 Note Taking

Quotients with Zeros

Read the question. Write words you need help with and research each word. Use your lesson to write your Cornell notes. Write or draw math examples to explain your thinking. Share your examples with a classmate.

Building on the Essential Question	Notes:
How do you find quotients with zeros?	428 ÷ 4 = ?

Notes:

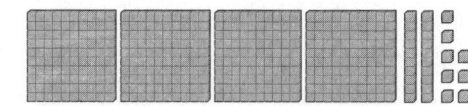

428 ÷ 4 = ?

To model the division, start by dividing the hundreds into __4__ equal groups, there will be __1__ hundred in each group.

After dividing the hundreds into 4 equal groups, there will be __0__ hundreds remaining.

You have __2__ tens. If you divide the tens into 4 equal groups, there will be __0__ tens in each group. If there are not enough tens to divide, place a __0__ in the quotient.

There will be __2__ tens remaining. If you regroup the __2__ tens into ones, you will have __28__ ones altogether.

If you divide the ones into 4 equal groups, there will be __7__ ones in each group.

Each group has __1__ hundred, __0__ tens and __7__ ones.

So, 428 ÷ 4 = __107__

Words I need help with:
See students' words.

My Math Examples:
See students' examples.

32   Grade 5 • Chapter 3 *Divide by a One-Digit Divisor*

Lesson 11 Inquiry/Hands On: Use Models to Interpret the Remainder

English Learner Instructional Strategy

Vocabulary Support: Sentence Frames

Display the following sentence frames to help students participate during the Build It and Try It sections of the lesson:

(Emerging) ____ cubes; ____ left over; ____ plates

(Expanding) ____ cubes on each plate. ____ cube is left over. ____ students are left over.

(Bridging) **Each food bank will receive** ____. ____ **students are left over, therefore** ____.

Be sure to create and display similar sentence frames for the remaining lessons in the chapter.

English Language Development Leveled Activities

Emerging Level	Expanding Level	Bridging Level
Number Recognition	**Look, Listen, and Identify**	**Turn & Talk**
As students work through the Practice It exercises, help them with difficult or unfamiliar vocabulary, such as: *picnic, reunion, volleyballs,* or *centerpieces,* by directing students to translation tools. Guide students in recognizing that the greater number in each problem is the dividend, and the lesser number is the divisor. Allow students to simply identify the amount of the remainder rather than interpreting its meaning in relation to the word problem.	On the board, write out Apply It Exercise 6. Read the problem aloud. Ask, *What do we need to find out?* Invite a volunteer to underline the relevant sentence in the problem. Ask, *How many calculators are there altogether?* Have another volunteer come forward to circle the answer. Continue in this manner as you work with students to solve the problem. Repeat the activity with Apply It Exercises 7 and 8.	Read aloud Apply It Exercise 9. Direct students to turn to a partner and discuss the answer. After partners have had time to talk, ask students to share their answers. Display the following sentence frames to help students in the discussion: **One way to interpret the remainder is** ____. **Another way to interpret the remainder is** ____.

Teacher Notes:

T33 Grade 5 • Chapter 3 *Divide by a One-Digit Divisor*

NAME _____ DATE _____

Lesson 11 Guided Writing

Inquiry/Hands On: Use Models to Interpret the Remainder

How do you model and interpret the remainder?

Use the exercises below to help you build on answering the Essential Question. Write the correct word or phrase on the lines provided.

1. Rewrite the question in your own words.
   See students' work.

2. What key words do you see in the question?
   remainder, model, interpret

3. There are 75 people on a field trip; a row on the trolley seats 6 people. If each row is filled completely with people from the field trip before moving to the next row, how many rows will have people from the field trip sitting on them? Identify the division expression that will solve the problem. 75 ÷ 6

4. If you divide the tens into 6 equal groups, there will be __1__ ten in each group.

5. After dividing the tens into 6 equal groups, there will be __1__ ten remaining. If you regroup the remaining __1__ ten into ones, you would have __15__ ones altogether.

6. If you divide the 15 ones into 6 equal groups, there will be __2__ ones in each group AND there will be __3__ ones remaining.

7. How many rows will be filled completely with people from the field trip? How many will be filled partially?
   12 completely full, 1 partially filled

8. So, __12__ + __1__ = __13__ describes the number of rows that will have people from the field trip sitting on them.

9. How do you model and interpret the remainder?
   Model the division. Identify what each part of the quotient means.
   Interpret the remaining parts of the division problem in terms of the problem situation.

Grade 5 • Chapter 3 *Divide by a One-Digit Divisor* 33

Lesson 12 Interpret the Remainder
English Learner Instructional Strategy

Sensory Support: Manipulatives

Distribute base-ten blocks to students. During the lesson, have them use the manipulatives to visually represent the solution to each division story problem, including the remainder.

After solving each problem, have students share their interpretation of the remainder according to their proficiency level. For example, emerging students might point to base-ten blocks representing the remainder and say **remainder** or **This is the remainder**. Provide sentence frames for expanding and bridging students as they extend their interpretation of the remainder: **The remainder is ___. The remainder shows that ___.** or **I interpret the remainder to mean ___.**

English Language Development Leveled Activities

Emerging Level	Expanding Level	Bridging Level
Word Knowledge Write the word *interpret* and the Spanish cognate, *interpretar*, on a classroom cognate chart. Say, *To interpret something is to give it meaning.* Write a simple sentence in Spanish, such as *El sol es caliente*. Say, *If I don't speak Spanish, I don't know what this means. I will give the meaning in English. I will interpret the sentence.* Write the sentence in English: The sun is hot. Using a word problem from the lesson, solve the division problem, and then interpret the remainder. Say, *I gave meaning to the remainder. I interpreted the remainder.*	**Recognize and Act It Out** Say, *I have twenty-three apples. A bag can hold up to five apples. What is the least number of bags I need to carry all twenty-three apples?* Use counters to model the solution. Make four groups of five counters and one group of three counters. Count the piles. Write 23 ÷ 5. Have a student write the quotient. **4 R3** Say, *We can also find the answer by interpreting the remainder. The remainder means a fifth bag is needed.* Repeat with a new story problem. Provide a sentence frame for students to interpret the remainder: **The remainder means ___.**	**Internalize Language** Say, *Some fifth graders are going on a field trip.* Have pairs roll two 0–5 number cubes to create a two-digit dividend representing the total number of students. Say, *The teacher split the students into groups.* Have pairs roll one cube and record the number as the divisor representing the number of students in a full group. Have students solve the equation. Provide sentence frames to help students interpret the remainder: **There are ___ total groups. There are ___ full groups. The partial group has ___ students. There is no partial group.**

Teacher Notes:

T34 Grade 5 • Chapter 3 *Divide by a One-Digit Divisor*

NAME _____ DATE _____

Lesson 12 Note Taking

Interpret the Remainder

Read the question. Write words you need help with and research each word. Use your lesson to write your Cornell notes. Write or draw math examples to explain your thinking. Share your examples with a classmate.

Building on the Essential Question

How can you interpret the remainder of a division problem?

Words I need help with:
See students' words.

Notes:

The fifth grade classes collected 347 cans for the local food bank. The cans were packed into four boxes in a way that each box contains the same number of cans.

347 ÷ 4 = ?

Start by dividing the hundreds into __4__ equal groups, there will be __0__ hundreds in each group.

Regroup the __3__ hundreds into tens and you have __34__ tens altogether.

If you divide the tens into 4 equal groups, there will be __8__ tens in each group, with __2__ tens remaining.

Regroup the remaining __2__ tens into ones and you have __27__ ones altogether.

If you divide the ones into 4 equal groups, there will be __6__ ones in each group with __3__ ones remaining.

347 ÷ 4 = __86__ R __3__

Each box contains __86__ cans and there are __3__ cans **remaining** that are not in a box.

My Math Examples:
See students' examples.

Grade 5 • Chapter 3 *Divide by a One-Digit Divisor*

Lesson 13 Problem-Solving Investigation Strategy: Determine Extra or Missing Information

English Learner Instructional Strategy

Vocabulary Support: Utilize Resources

As students work through the Problem Solving exercises, be sure to remind students that they can refer to the Glossary or the Multilingual eGlossary for help, or direct students to other translation tools if they are having difficulty with non-math language in the problems. Point out signal words and phrases that frequently appear in story problems, such as *What is the total, How many more than, How much will be left, for each* and help students understand that these words and phrases will show them which kinds of equations are needed for solving. Have students record these signal words and phrases in a math journal so they will have them to refer to when solving other story problems.

English Language Development Leveled Activities

Emerging Level	Expanding Level	Bridging Level
Academic Vocabulary Display $137 of play money and images of a sweatshirt and a t-shirt. Say, *A shop has a sale on clothing. Sweatshirts are $15 off and t-shirts are $5 off. How many sweatshirts can I buy with $137?* Write the sale information next to the images. Circle the t-shirt. Say, *I do not need the information about t-shirts. That is* **extra** *information.* Have students say **extra** chorally. Write ? − $15 = sale price by the sweatshirt. Say, *I don't know the cost of a sweatshirt. I cannot solve this problem. I am* **missing** *information.* Have students say **missing** chorally.	**Recognize and Act It Out** Read aloud a problem from the lesson. Have students help identify what is known and what needs to be found and write that information on the board. Ask, *Does the problem have any* **extra** *information? Does it have any* **missing** *information?* Display the following sentence frames to help students answer: **The missing information is ____. The extra information is ____.** Ask students if the problem can be solved with the information given. If the problem has no missing information, have students help you solve the problem.	**Academic Language** Have pairs work together on problems from the lesson. One student will read aloud the word problem and identify what is known and what needs to be found. The other student will identify any extra or missing information. If there is no missing information, have students solve the problem and check the answer for reasonableness. Provide the following sentence frame for students to report back on their work: **We could/could not solve the problem because ____.**

Teacher Notes:

T35 Grade 5 • Chapter 3 *Divide by a One-Digit Divisor*

NAME _____ DATE _____

Lesson 13 Problem-Solving Investigation

STRATEGY: Determine Extra or Missing Information

Determine if there is extra or missing information. Then solve the problem, if possible.

1. **Jayden** is downloading songs onto **his** MP3 player.
   One song is <u>**5 minutes**</u> long, another is <u>**2 minutes**</u> long, and a third is <u>**between**</u> the lengths of the other two songs.
   What is the <u>**total**</u> length of **all** <u>**three**</u> songs?

Understand	Solve
I know:	
I need to find:	
Plan	**Check**
Facts that are important to solve the problem are:	
The extra or missing information is:	

2. Room 220 and Room 222 are having a canned food drive.
   Room 222 collected <u>**346**</u> **cans** and **Room 220** collected <u>**278**</u> **cans**.
   How many <u>**more**</u> **cans** has **Room 222** collected <u>**than**</u> **Room 220**?

Understand	Solve
I know:	
I need to find:	
Plan	**Check**
Facts that are important to solve the problem are:	
The extra or missing information is:	

Grade 5 • Chapter 3 *Divide by a One-Digit Divisor* **35**

Chapter 4 Divide by a Two-Digit Divisor

What's the Math in This Chapter?

Mathematical Practice 8: Look for and express regularity in repeated reasoning

Write the math problem 11)̄121 on the board. Ask, *How could we use estimation and rounding to help solve this problem?* Elicit from students that they could round the divisor to 10 and the dividend to 120. Display this new division problem and have students solve. Say, *What is 120 divided by 10?* **12**

Model solving 11)̄121 emphasizing the steps: dividing, multiplying, and subtracting until the remainder is 0 and the quotient is 11. Explain to students that they **repeated** steps while dividing.

Ask, *Is the quotient of 11 close to our estimate?* **yes** Discuss how estimating a quotient allowed them to get an idea of what the answer should be and check for reasonableness. Say, *When we get into the habit of estimating the quotient before solving, we are using estimation as a repeated strategy.*

Display a chart with Mathematical Practice 8. Restate Mathematical Practice 8 and have students assist in rewriting it as an "I can" statement, for example: **I can use repetitive strategies to help solve difficult problems.** Post the new "I can" statement.

Inquiry of the Essential Question:

What strategies can I use to divide by a two-digit number?

Inquiry Activity Target: **Students come to a conclusion that dividing larger numbers uses the same division process.**

As an introduction to the chapter, present the Essential Question to students. The inquiry graphic organizer will offer opportunities for students to observe, make inferences, and apply prior knowledge of division representing the Essential Question. As they investigate, encourage students to draw, write, and collaborate with peers to demonstrate their observations and thinking. Then have students present additional questions they may have to a peer to extend discussions.

Regroup students and restate Mathematical Practice 8 and the Essential Question. Pose questions to reflect on what has been learned to guide students in making connections between the Mathematical Practice and the Essential Question.

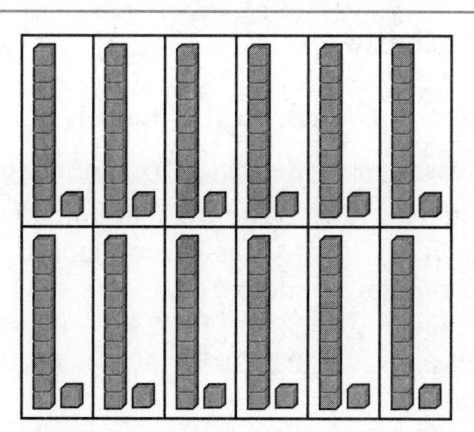

NAME _____ DATE _____

Chapter 4 Divide by a Two-Digit Divisor

Inquiry of the Essential Question:

What strategies can I use to divide by a two-digit number?

Read the Essential Question. Describe your observations (I see...), inferences (I think...), and prior knowledge (I know...) of each math example. Write additional questions you have below. Then share your ideas and questions with a classmate.

$$14\overline{)364}$$
$$\underline{-28}$$
$$84$$
$$\underline{-84}$$
$$0$$
with quotient 26

Step 1 Divide the tens.
Step 2 Multiply, subtract, and compare.
Step 3 Bring down the ones.
Step 4 Divide the ones.

I see ...

I think...

I know...

Each group contains 1 ten and 1 one.
So, 132 ÷ 12 = 11.

I see ...

I think...

I know...

Use rounding to adjust the quotient with a two-digit divisor.

Step 1 Use compatible numbers to estimate.
Step 2 Try the estimate.
Step 3 Adjust the estimate as needed.
Step 4 Divide using the division algorithm.

I see ...

I think...

I know...

Questions I have...

_____

_____

36 Grade 5 • **Chapter 4** *Divide by a Two-Digit Divisor*

Lesson 1 Estimate Quotients
English Learner Instructional Strategy

Collaborative Support: Partners Work/Pairs Check

Assign Exercises 2-7 in Independent Practice. Have students work in pairs. For the first problem, have one student coach the other in finding the estimate. For the second problem, have students switch roles. When pairs have finished the second problem, have them get together with another pair and check answers.

Provide the following sentence frames:
What is your estimate for exercise _____?
Our estimate is _____.
How did you round your numbers?
We rounded our numbers by _____.

When both pairs have agreed that their estimates are reasonable, ask them to shake hands and continue working in their original pairs for the next two problems, exchanging roles as before. Then have new sets of pairs check answers.

English Language Development Leveled Activities

Emerging Level	Expanding Level	Bridging Level
Word Knowledge Display a word web with *estimate* written in the center and *about, almost, around,* and *close to* written in the surrounding ovals. Say, *When you estimate, you make a careful guess.* Write 83 ÷ 8 and model estimating the answer. Say, *First I round 83 to 80. Then I solve. The quotient of 83 ÷ 8 is around 10.* Reword the answer several times using synonyms from the word web. Point to the word and say, *The answer is about/almost/around/close to 10.* Have students repeat chorally. Be sure students correctly pronounce the "ou" as /aw/ in *about* and *around*.	**Phonemic Awareness** Model the pronunciation change between the noun and verb forms of *estimate* (noun: -mət, verb: -māt). Write 140 ÷ 22 and say, *I will estimate the answer.* Model rounding 22 to 20 to find the estimated answer. Say, *My estimate of 140 ÷ 20 is 7.* Write another division equation, such as 218 ÷ 11. Display the following sentence frames so a volunteer can explain how to estimate the answer and share his or her estimate: **I can estimate the answer by rounding _____. My estimate is _____.** Repeat the activity with similar division equations.	**Building Oral Language** Write the following division expressions on individual index cards: 360 ÷ 42; 552 ÷ 54; 237 ÷ 83; 425 ÷ 34; 595 ÷ 28. Have students work in pairs. Distribute one card to each pair and have them estimate an answer. Display the following sentence frame: **Our estimate is higher/lower than the actual quotient because _____.** Have one student in each pair use the sentence frame to compare their estimate to the actual quotient. Redistribute the cards and repeat the activity so the other student in each pair has a chance to use the sentence frame.

Teacher Notes:

NAME _____ DATE _____

Lesson 1 Note Taking

Estimate Quotients

Read the question. Write words you need help with and research each word. Use your lesson to write your Cornell notes. Write or draw math examples to explain your thinking. Share your examples with a classmate.

Building on the Essential Question How do you estimate quotients?	**Notes:** When you __round__ a number, you find the approximate value of a number. 37 rounded to the nearest ten is __40__. 849 rounded to the nearest hundred is __800__. When you find an __estimate__, you find a number close to an exact value. Using **rounded numbers,** an estimate for 849 ÷ 37 is __800__ ÷ __40__ = __20__.
Words I need help with: See students' words	__Compatible__ numbers are numbers in a problem that are easy to work with mentally. 8 ÷ 4 is __2__ 4 ÷ 4 is __1__ 84 ÷ 4 is __21__ Using **compatible numbers,** an estimate for 849 ÷ 37 is __840__ ÷ __40__ = __21__.
My Math Examples: See students' examples	

Grade 5 • Chapter 4 *Divide by a Two-Digit Divisor* **37**

Lesson 2 Inquiry/Hands On: Divide Using Base-Ten Blocks

English Learner Instructional Strategy

Language Structure Support: Tiered Questions

During the Model the Math and Build It sections of the lesson, utilize tiered questions as formative assessment. Emerging students may be able to respond only with gestures or single-word answers. Ask questions such as: *Is this multiplication or division? Do we need to regroup? How many tens rods? How many ones? Are these equal groups?* For Expanding students, ask questions that can be answered with short phrases or simple sentences: *What do we do first? Where do these tens rods go?* For Bridging students, ask questions that require more elaborate answers, such as: *How do you know this is a division problem? Why did we regroup the remaining tens rods into 30 ones?*

English Language Development Leveled Activities

Emerging Level	Expanding Level	Bridging Level
Activate Prior Knowledge	**Round the Table**	**Turn & Talk**
Write the word *regroup* and the Spanish cognate, *reagrupar,* on a classroom cognate chart. On the board, write 231 ÷ 3. Use 2 hundreds blocks, 3 tens rods, and 1 one to represent 231. Write and say: *I need to regroup.* Ask, *What do I need to do?* Have students answer chorally, **Regroup**. Model regrouping 231 as 23 tens rods and 1 one. Say, *I regrouped.* Emphasize the final /t/ sound indicating past tense. Ask, *What did I do?* Have students answer chorally, **Regrouped**. Model the remainder of the problem.	Place students into 4 groups. Assign a Practice It exercise to each group. Have one student write the problem on a large piece of paper. Then have students work jointly to solve the problem by passing the paper around the table. Each student will perform one step in drawing a model to find the quotient. Direct each member of the group to write with a different color to ensure all students participate in solving the problem. Provide sentence frames to help students participate: **I will regroup ____ into ____. There are ____ groups of ____.**	Say, *How is dividing by a two-digit divisor similar to dividing by a one-digit divisor? How is it different? Turn to the student nearest you and discuss your answer.* Give students a chance to discuss their ideas. Then come together again as a group and ask volunteers to share their answers. Have students write in their math journals about the benefits of using models to solve both types of division problems.

Teacher Notes:

NAME _____ DATE _____

Lesson 2 Guided Writing

Inquiry/Hands On: Divide Using Base-Ten Blocks

How do you model division using base-ten blocks?

Use the exercises below to help you build on answering the Essential Question. Write the correct word or phrase on the lines provided.

1. Rewrite the question in your own words.
   See students' work
   _____

2. What key words do you see in the question?
   division, model, base-ten blocks

3. What numbers are modeled with each base-ten block example below?

   a. 10   b. 10

4. How would you model 168 using the least amount of base-ten blocks?
   __1__ hundred, __6__ tens, and __8__ ones

5. You cannot divide the hundreds into 12 equal groups. So, regroup the hundreds into tens, and you have __16__ tens altogether.

6. Divide the tens into 12 equal groups. There will be __1__ ten in each group. Regroup the remaining __4__ tens into ones, and you will have __48__ ones altogether.

7. Divide the ones into 12 equal groups. There will be __4__ ones in each group.

8. 168 ÷ 12 = __14__

9. How do you model division?
   Model the number being divided using base-ten blocks. Divide the hundreds into the number of equal groups for the division problem. Regroup as needed and then divide the tens and ones. The number of blocks in each group is the answer to the division problem.

Lesson 3 Divide by a Two-Digit Divisor

English Learner Instructional Strategy

Vocabulary Support: Sentence Frames

During Model the Math, provide the following sentence frames to aid students in their participation:

I rounded ____ to the nearest ____.
I divided ____ by ____.
My estimate is ____.
The quotients all have ____ or ____ digits.

During the Problem Solving part of the lesson, certain words and phrases may be difficult for some students to understand, particularly emerging or expanding level students. Utilize appropriate translation tools for ELs to clarify terms, such as: *flag, store, area, width, length, uploads, pictures, album, scroll, represent, sleep, per,* and *standard procedure.*

English Language Development Leveled Activities

Emerging Level	Expanding Level	Bridging Level
Number Sense	**Show What You Know**	**Language Structure**
Write 367 ÷ 15. Point to the dividend and say, *The dividend.* Identify each place value by pointing to it and saying, *Ones, tens, hundreds.* Ask, *Which place value is greatest?* Have students answer chorally, **hundreds**. Round to 400. Say as you point to each, *The divisor. Ones, tens.* Ask, *Which place value is greatest?* Have students answer chorally, **tens**. Round to 20. Estimate the quotient. **20** Say, *This is an estimate.* Find the exact quotient. **24 R7** Use a number line to show that 20 ≈ 24 R7.	Use a division bracket to write 367 ÷ 15. Display the following sentence frames for students to identify the dividend and divisor: **The dividend ____ is ____. The divisor is ____.** Have students estimate a quotient. Record it near the expression. Tape two vertical pieces of string between the digits in the dividend and below to help students align digits as they solve. Have a volunteer solve the equation. Provide sentence frames for students to answer: **The quotient is ____ with a remainder of ____.** Repeat with another division problem.	On the board, write *added, subtracted, multiplied,* and *divided.* Using division expressions from the lesson, assign one apiece to students working in pairs. Distribute an index card to each pair. Say, *As you solve your equation, write sentences that describe the process using the words on the board.* If necessary, provide sentence frames such as, **First we multiplied ____. Then we subtracted ____.** and so on. Afterward, have pairs read aloud from their index cards. Ensure students pronounce the /əd/ and /d/ sounds to indicate past tense.

Teacher Notes:

Lesson 3 Multiple Meaning Word
Divide by a Two-Digit Divisor

Complete the four-square chart to review the multiple meaning word.

Everyday Use	Math Use in a Sentence
Sample answer: The amount left over after an event. For example, the remainder of the pizza after dinner was 3 slices.	Sample sentence: The amount that is left after you divide two numbers.
Math Use	**Example From This Lesson**
The number that is left after one whole number is divided by another.	Sample answer: 21 ÷ 4 = 5 R1 The remainder of 21 ÷ 4 is 1.

(center: **remainder**)

Write the correct term on the line to complete the sentence.

The remainder of a division problem will always be ___less___ than the divisor.

Lesson 4 Adjust Quotients
English Learner Instructional Strategy

Language Structure Support: Cognates

Write the word *adjust* and its Spanish cognate, *ajustar*, on a classroom cognate chart. Provide a concrete example of the meaning by adjusting a piece of clothing or an object on a desk or shelf. Say, *When I adjust my sweater, I move it to make it more comfortable. When I adjust something, I improve it or make it better.* During Independent Practice, have students work in pairs. Assign one problem for each pair to work on together. Have them first estimate an answer, and then place the estimate as the first digit in the ones or tens place as they solve for the exact quotient. Afterward ask, *Did you need to adjust your first estimate?*

Provide sentence frames for students to use when answering:
Our estimate was ____.
We adjusted our first estimate to ____.

English Language Development Leveled Activities

Emerging Level	Expanding Level	Bridging Level
Number Sense Write the expressions 4,187 ÷ 23 and 4,000 ÷ 20. Ask, *Which is easier to solve?* Allow time to answer, then point to 4,000 ÷ 20. Say, *This expression is easier to solve. It uses compatible numbers. It uses numbers that are easy to solve in your head.* Emphasize compatible. Write additional expression pairs; one using compatible numbers and one that does not. Have students identify which expression is easier to solve by pointing and saying, **compatible** Then model using rounding to create compatible numbers for estimating.	**Academic Vocabulary** Write 122 ÷ 23 and say, *We are going to find an estimate for this expression.* Write 120 ÷ 20 below the first expression. Say, *I rounded to make compatible numbers.* Have students find the quotient. **6** Say, *The estimate is 6.* Say, *Now we can try to divide 122 by the estimate 6.* Have students try the estimate. Students should discover that the estimate 6 is too high. Have them *adjust* the estimate to 5. Then have students solve. **5 R7** Repeat with another, similar division equation.	**Academic Language** Distribute three or four number cubes to student pairs. Ask pairs to roll two cubes and create a two-digit divisor. Have them roll three or four cubes again to create a three- or four-digit dividend. Have one student estimate the quotient using compatible numbers. Have the other student solve, adjusting the quotient as needed. During the process of finding the quotient, have students record the steps and check the answer for reasonableness. Have pairs switch roles and repeat the activity. Have students share the steps they recorded.

Teacher Notes:

NAME _____ DATE _____

Lesson 4 Vocabulary Cognates
Adjust Quotients

Use the Glossary to define the math word in English and in Spanish in the word boxes. Write a sentence using your math word.

estimate	estimación
Definition A number close to an exact value. An estimate indicates about how much.	**Definición** Número cercano a un valor exacto. Una estimación indica una cantidad aproximada.

My math word sentence:
Sample answer: The estimate of 27 ÷ 5 is 30 ÷ 5 = 6.

quotient	cociente
Definition The result of a division problem.	**Definición** Resultado de un problema de división.

My math word sentence:
Sample answer: The quotient of 30 ÷ 5 is 6.

Lesson 5 Divide Greater Numbers

English Learner Instructional Strategy

Graphic Support: Charts

To help students during the Talk Math part of the lesson, create a 3-column chart labeled *Before, During,* and *After*. Model solving a division equation from the lesson. During each step of the solving process, record examples of how estimation is being used. Afterward, write the following sentence frames below the chart:

Before solving we estimate to ____. (get an idea what the answer should be)
During solving we estimate to ____. (to decide what digit to place in the quotient)
After solving we use our original estimate to ____. (check for reasonableness)

Have students use the sentence frames as you lead a discussion of how estimation is used when solving a division problem.

English Language Development Leveled Activities

Emerging Level	Expanding Level	Bridging Level
Synthesis	**Recognize and Act It Out**	**Academic Language**
Write *reasonable* and its Spanish cognate, *razonable*, on a classroom cognate chart. Write 27,351 ÷ 92. Work with students to find an estimate. Round 27,351 to 27,000 and 92 to 90 and solve. Write 300. *Say, I found a quotient of 297 R27. My friend found a quotient of 29 R27.* Write both quotients near the expression. Ask, *Which answer is reasonable?* Allow students to answer with a gesture. Circle 297 R27 and 300. Say, *This quotient is close to our estimate. 297 R27 is a reasonable answer.* Repeat the activity with other division expressions.	Choose a division problem from the lesson and recreate it on the floor. Write the digits on separate pieces of paper and line up the digits to create the dividend and divisor. Use masking tape to create the division bracket. Have students identify the dividend and divisor and estimate a quotient. Record the estimate. Use a marker, paper, and masking tape to model solving the equation. Write digits on separate pieces of paper and include masking tape arrows when bringing down place values. Have pairs repeat the activity with 76,912 ÷ 92.	Use division expressions found in the lesson to create expression cards. Distribute cards to student pairs. Display the following sentence frames for one student to read aloud the numbers: **The divisor is ____. The dividend is ____.** Have the other student estimate the quotient. Have pairs work together to find the exact quotient. Display the following sentence frames to have students describe their answers: **Our estimate was ____. The quotient is ____. The remainder is ____. Our answer is reasonable because ____.**

Multicultural Teacher Tip

Depending on where a student was originally taught, you may find he or she writes decimal numbers and greater numbers using slightly different notations. In the US, numbers are separated into groups of three place values by commas (3,252,689), but in Latin American countries, the groups may be separated by points (3.252.689) or spaces (3 252 689), and in Mexico it may be a combination of a comma and apostrophe (3'252,689) or a comma and semicolon (3;252,689). Similarly, decimals in the US use points (3.45) while Latin American countries use a comma (3,45).

NAME _____ DATE _____

Lesson 5 Concept Web

Divide Greater Numbers

Use the concept web to identify the parts of a division sentence.

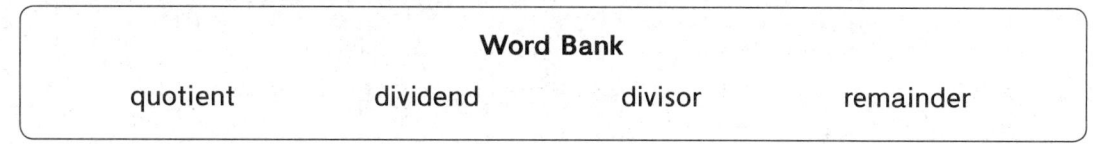

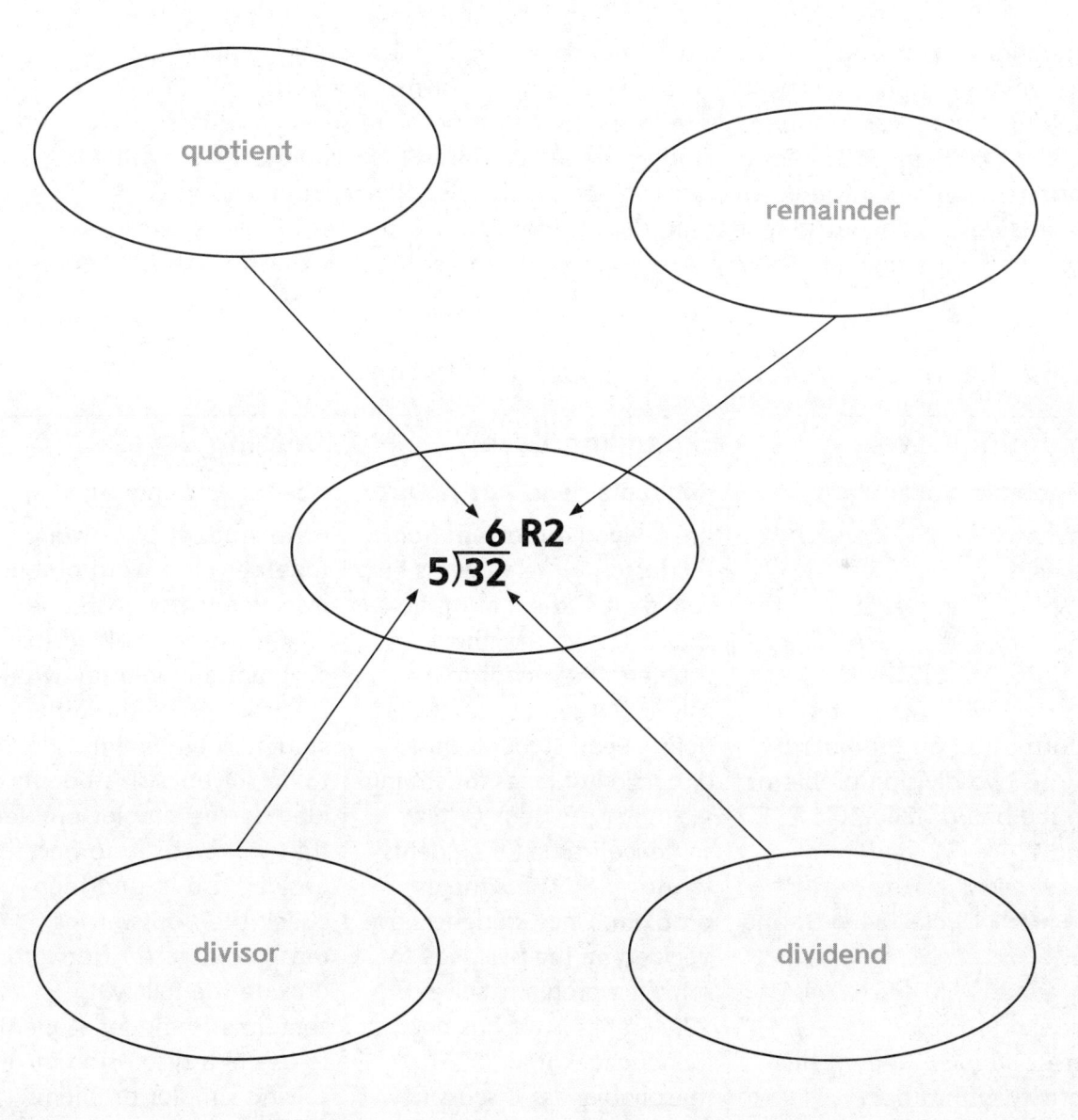

Word Bank
quotient   dividend   divisor   remainder

- quotient
- remainder
- 6 R2
- 5)32
- divisor
- dividend

Grade 5 • Chapter 4 *Divide by a Two-Digit Divisor*   **41**

Lesson 6 Problem-Solving Investigation Strategy: Solve a Simpler Problem

English Learner Instructional Strategy

Vocabulary Support: Build Background Knowledge

Before the lesson, review the comparative ending *-er*. Remind students that this suffix adds the meaning "more" to base words. Write *simpler* on the board and say, *If one thing is simpler than another, it is more simple.* Have students brainstorm other examples of comparatives with the suffix *-er*, such as *bigger, smaller, hotter, colder* and so on.

Throughout the lesson, display and fill-in a KWL chart. Start by completing the first column during a review of what students have learned previously about problem solving. For example, write in the four-step problem solving process: *Understand, Plan, Solve, Check*. During the lesson, fill in the middle column with information about solving a simpler problem. Display sentence frames to help students describe what they learned during the lesson and record the answers in the chart's third column: **Solving a simpler problem is ____.** or **During the lesson I learned ____.**

English Language Development Leveled Activities

Emerging Level	Expanding Level	Bridging Level
Academic Vocabulary Say, *How many basketballs can I buy with $124? The price of one basketball is $46, but the store is having a sale. Each basketball now costs $15 less.* Write this information on the board. Write two division problems on the board: 124 ÷ (46 − 15) and 124 ÷ 31. Say, *Both expressions can solve our problem.* Circle 124 ÷ 31 and say, *This problem requires fewer steps to solve. This problem is simpler to solve.* Stress *simpler*. Repeat the activity with other expressions. Have students say **simpler** to identify which expression is simpler to solve.	**Recognize and Act It Out** Read aloud a problem from the lesson. Have students help you identify the information that is known and the problem that needs to be solved. Write both on the board. Lead students in a discussion of ideas for solving a simpler problem. Display a sentence frame for students to use: ____ **is a simpler problem.** Once students have decided on the best way to solve the problem, solve it. When an answer has been found, check it for reasonableness. Discuss how using a simpler problem made it easier to solve the original problem.	**Academic Language** Have student pairs work together on a word problem from the lesson. Ask one student to read aloud the problem and identify what is known. Have the other student identify what needs to be solved. Ask students to identify the simpler problem that will be used to find the answer, use it, and then check their answer for reasonableness. Afterward, provide the following sentence frame for students to use in a discussion of solving simpler problems: ____ **was a simpler problem to solve because** ____.

Teacher Notes:

T42 Grade 5 • Chapter 4 *Divide by a Two-Digit Divisor*

NAME _____ DATE _____

Lesson 6 Problem-Solving Investigation

STRATEGY: Solve a Simpler Problem

Using a simpler problem, solve each problem below.

1. Mr. Santiago has a flight from New York to Paris that covers a distance of **3,640** **miles** in **7** **hours**. If the **plane** travels at the same speed per hour, how many **miles** will **it** have traveled after **4 hours**?

   plane

Understand	Solve
I know:	
I need to find:	
Plan	**Check**
Speed plane traveled in 1 hour:	
Speed plane traveled in 4 hours:	

2. **Josh** watches **720** television **shows** in **one year**. If **he** watches the **same** number of shows **each month**, how many shows does he watch in **5 months**?

   television show

Understand	Solve
I know:	
I need to find:	
Plan	**Check**
Shows watched in 1 month:	
Shows watched in 5 months:	

Chapter 5 Add and Subtract Decimals

What's the Math in This Chapter?

Mathematical Practice 2: Reason abstractly and quantitatively

Write 638 − 321 and $6.38 − $3.21 vertically on the board. Distribute write-on/wipe-off boards to students and have them copy the two problems. Say, *Use what you know about subtraction and money to solve these two problems.* Encourage students to use multiple representations to solve the problem including manipulative money and place value charts. Provide time for students to solve the two problems and check their solutions with a peer. Regroup and discuss their solutions.

Draw two place-value charts and write the whole numbers correctly aligned in the chart. Use play money or images of dollar bills and coins on the board to represent $6.38 − $3.21. Model proper subtraction of both problems showing the answers to be 317 and $3.17 and discuss decimal placement. Say, *Using place value and tools like manipulative money helps us to* **quantify** *digits and subtract decimals correctly.*

Display a chart with Mathematical Practice 2. Restate Mathematical Practice 2 and have students assist in rewriting it as an "I can" statement, for example: **I can reason quantitatively to make connections.** Post the new "I can" statement.

Inquiry of the Essential Question:

How can I use place value and properties to add and subtract decimals?

Inquiry Activity Target: **Students come to a conclusion that adding and subtracting decimals is similar to addition and subtraction they have done before.**

As an introduction to the chapter, present the Essential Question to students. The inquiry graphic organizer will offer opportunities for students to observe, make inferences, and apply prior knowledge of addition and subtraction representing the Essential Question. As they investigate, encourage students to draw, write, and collaborate with peers to demonstrate their observations and thinking. Then have students present additional questions they may have to a peer to extend discussions.

Regroup students and restate Mathematical Practice 2 and the Essential Question. Pose questions to reflect on what has been learned to guide students in making connections between the Mathematical Practice and the Essential Question.

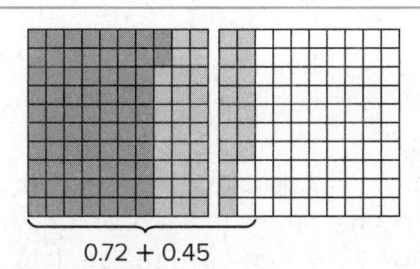

NAME _____ DATE _____

Chapter 5 Add and Subtract Decimals

Inquiry of the Essential Question:

How can I use place value and properties to add and subtract decimals?

Read the Essential Question. Describe your observations (I see...), inferences (I think...), and prior knowledge (I know...) of each math example. Write additional questions you have below. Then share your ideas and questions with a classmate.

$\begin{array}{r}8.72\\-3.05\end{array}$ ⟶ rounds to ⟶ $\begin{array}{r}8.7\\-3.1\\\hline 5.6\end{array}$

I see ...

I think...

I know...

0.72 + 0.45

0.72 + 0.45 = 1.17

I see ...

I think...

I know...

1.5 + 6.4 + 3.5 = 1.5 + 3.5 + 6.4	Commutative Property
= (1.5 + 3.5) + 6.4	Associative Property
= 5 + 6.4	Add.
= 11.4	Add.

I see ...

I think...

I know...

Questions I have...

_____
_____
_____

Grade 5 • Chapter 5 *Add and Subtract Decimals* 43

Lesson 1 Round Decimals
English Learner Instructional Strategy

Language Structure Support: Tiered Questions

Throughout Guided/Independent Practice, utilize tiered questions as formative assessment. Emerging students may be able to respond only with gestures or single-word answers. Ask questions such as: *Show me the tenths place. Do you look at the digit to the right or to the left? Do you drop this digit, or do you round up this digit?*

For expanding students, ask questions that can be answered with short phrases or simple sentences: *What do you do first? What place are you rounding to?*

For bridging students, ask questions that require more elaborate answers, such as: *How do you know that you need to round up? Why did you drop this digit?*

English Language Development Leveled Activities

Emerging Level	Expanding Level	Bridging Level
Word Recognition Provide concrete examples for the non-math meanings of *round* using realia from the classroom. Then review the meaning of *round* in a math context. Draw a large number line from 2.3 to 2.4 on the board and mark increments of one hundredths. Place your finger on 2.32. Ask, *Is 2.32 closer to 2.3 or 2.4?* Allow students to answer by pointing. Say, *It is closer to 2.3, so we* **round down**. Place your finger on 2.38 and ask, *Is 2.38 closer to 2.3 or 2.4?* Have students answer. Say, *It is closer to 2.4 so we* **round up**. Repeat with other decimals between 2.3 and 2.4.	**Act It Out** Write *decimal* and its Spanish cognate, *decimal*, on a classroom cognate chart. Write numbers 0–9 on sheets of paper, and draw a decimal point on another sheet. Distribute the sheets to students. Say a decimal number, such as 8.67. Have students direct each other to stand with their papers at the front of the class to show the number. Provide sentence frames, such as: *The ___ is left/right of the ___.* or *The decimal goes between ___ and ___.* Model rounding the number. Repeat the activity with a new decimal.	**Building Oral Language** Write 4.73 and say, *We will round to the tenths place.* Ask a student to identify the tenths place digit. Underline 7 and have the student point to and identify the place value one digit to the right. Ask, *Will we round seven up or keep it the same?* **same** Say, *Three is less than five. We will keep seven the same.* Erase the 3 to drop the digit. Write 6.92 and have a student model rounding it, using the example you just modeled. Repeat with other decimal examples, having students model how to use place value terms when rounding decimals.

Teacher Notes:

NAME _____ DATE _____

Lesson 1 Vocabulary Cognates
Round Decimals

Use the Glossary to define the math word in English and in Spanish in the word boxes. Write a sentence using your math word.

decimal	decimal
Definition A number that has a digit in the tenths place, hundredths place, and beyond.	**Definición** Número que tiene al menos un digito en la posición de los décimos o centesimos, o en cualquier posición posterior.

My math word sentence:
Sample answer: 7.5 and 0.23 and 1.006 are all decimals.

rounding	redondear
Definition To find the approximate value of a number.	**Definición** Hallar el valor aproximado de un número.

My math word sentence:
Sample answer: 23 rounded to the nearest ten is 20, 2.3 rounded to the nearest one is 2.

Lesson 2 Estimate Sums and Differences

English Learner Instructional Strategy

Collaborative Support: Small Groups

During the Vocabulary Review, divide students into three groups. Assign each group one of the review words, *sum, difference,* or *round.* Have students work together in their group to provide a math example demonstrating the meaning of the word. Afterward, have an expanding or bridging student present the example. If necessary, provide a sentence frame such as: **Our example shows ___ because ___.**

During Model the Math, allow students to suggest food items for the menu that reflect their native culture. If the foods are unfamiliar to others in the group, have students explain what the food is made of or how it tastes. If possible, look up the foods on the Internet using a classroom computer to provide a visual example.

English Language Development Leveled Activities

Emerging Level	Expanding Level	Bridging Level
Developing Oral Language Write 12.32 and 15.96. Point to the decimal and say, *decimal.* Have students repeat chorally. Point to each whole number place as you say, *ones, tens.* **ones, tens** Listen for correct pronunciation of the /z/ sound at the end of each word. Say, *Whole numbers are left of the decimal.* Point to the decimal number places as you say, *tenths, hundredths.* Have students repeat chorally. Listen for correct pronunciation of the /ths/ sound. Say, *Decimal numbers are right of the decimal.* Then model rounding both decimals to whole numbers. Have students add.	**Recognize and Act It Out** Draw a number line from 2.0 to 3.0 on the board, marking increments of one tenth. Write several decimal numbers to the hundredths between 2.0 and 3.0 individually on sticky notes. Give each student a sticky note. Have them round their numbers to the nearest whole number and place their notes under the corresponding number on the board. Provide the following sentence frame for students to explain their reasoning: **I rounded to ___ because ___.** Repeat the activity with a new number line and set of sticky notes.	**Internalize Language** Have students work in pairs. Provide several problems showing addition and subtraction of decimals. Say, *Estimating can help you find a reasonable answer.* Model estimating the answers, but as you work through the problems, have some answers that are incorrect. Help students vocalize the errors in estimation by providing a sentence frame: **The estimate is incorrect because ___.** Have pairs create new decimal addition and subtraction problems and then trade problems with other pairs to estimate the solutions.

Teacher Notes:

Student page

NAME _____ DATE _____

Lesson 2 Vocabulary Definition Map
Estimate Sums and Differences

Use the definition map to write a description and list characteristics about the vocabulary word or phrase. Write or draw math examples. Share your examples with a classmate.

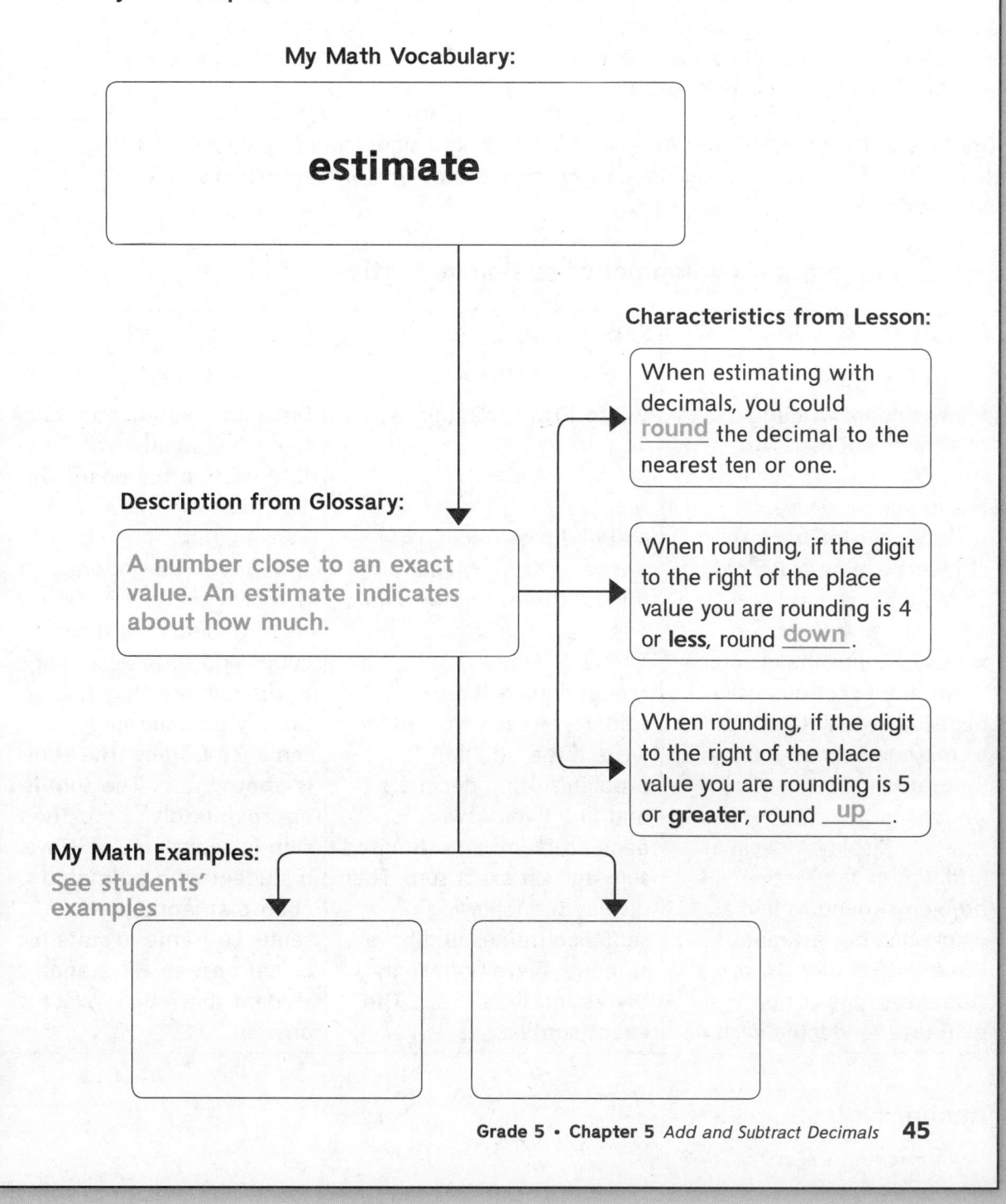

My Math Vocabulary:

estimate

Description from Glossary:

A number close to an exact value. An estimate indicates about how much.

Characteristics from Lesson:

When estimating with decimals, you could <u>round</u> the decimal to the nearest ten or one.

When rounding, if the digit to the right of the place value you are rounding is 4 or **less**, round <u>down</u>.

When rounding, if the digit to the right of the place value you are rounding is 5 or **greater**, round <u>up</u>.

My Math Examples:
See students' examples

Grade 5 • Chapter 5 *Add and Subtract Decimals*   **45**

Lesson 3 Problem-Solving Investigation Strategy: Estimate or Exact Answer

English Learner Instructional Strategy

Vocabulary Support: Signal Words/Phrases

Have a student read aloud a word problem from the lesson. As he or she reads, list words or phrases from the problem that signify whether an estimate or an exact answer is needed to solve the problem, such as: *about, approximately,* or *how much/many*. Discuss with students how recognizing this kind of language can help indicate whether an estimate or an exact answer is needed when solving a word problem.

Read aloud Exercises 1–5 and have students highlight signal words/phrases then write estimate or exact answer next to the problem. Students then solve independently or in pairs.

English Language Development Leveled Activities

Emerging Level	Expanding Level	Bridging Level
Word Recognition Show students a baggie filled with counters. Ask, *About how many counters are in the bag? We can guess, or estimate.* **Have students estimate.** Record their numbers. Say, *Now I want to know the exact number.* **Count the counters.** Write the exact number next to the estimates. Point to the numbers as you say, *The estimate is close to, or about the same as the exact number.* **Display a decimal addition or subtraction problem. Round to find a whole number estimate. Find the exact solution. Have students say exact or estimate to identify each.**	**Listen and Write** Write 10.8 + 9.3 and say, *To find an* **estimate** *for the sum, we round each number. An estimate tells us* **about** *how much the answer will be.* **Model finding the estimated sum.** Then say, *To find the* **exact** *sum, we do not round.* Model finding the exact sum. Have each student write three addition problems using decimals and find two answers for each problem: an estimated sum and an exact sum. Then display the following sentence frames and have students share both sums: **My estimate is _____. The exact sum is _____.**	**Number Game** Distribute one number cube to each student. Write 12.7 − 5.3 on the board. Say, *Roll the cube. If you get an even number, find the exact sum. If you get an odd number, find an estimate.* Have students roll their cubes and solve according to the number they roll. Display the following sentence frames: **The sum is about _____. The sum is approximately _____. The sum is exactly _____.** Have a student who estimated choose an appropriate sentence frame to state his or her answer. Have another student share their exact answer.

Teacher Notes:

T46 Grade 5 • Chapter 5 *Add and Subtract Decimals*

NAME _____ DATE _____

Lesson 3 Problem-Solving Investigation

STRATEGY: Estimate or Exact Answer

Determine whether you need an estimate or exact answer to solve each problem.

1. A restaurant can make **95** dinners <u>each</u> night.
   The restaurant has been **sold out** for **7** <u>nights</u> in a row.
   How many dinners were sold during <u>this</u> week?

Understand	Solve
I know: I need to find:	
Plan There are 7 nights in 1 week. **Sold out** means, "all 95 dinners were sold that night."	**Check**

2. A **family** is renting a cabin for **$59.95** <u>a</u> <u>day</u> for 3 days.
   <u>About</u> how much will **they** (the family) pay for the cabin?

Understand	Solve
I know: I need to find:	
Plan How much will 1 day cost? **About** how much will 1 day cost? **About** how much will 3 days cost?	**Check**

46 Grade 5 • Chapter 5 *Add and Subtract Decimals*

Lesson 4 Inquiry/Hands On: Add Decimals Using Base-Ten Blocks

English Learner Instructional Strategy

Collaborative Support: Round the Table

Place students into multilingual groups of 4 or 5. Assign a Practice It Exercise to each group. Have one student draw a place-value table on a large piece of paper. Then have students work jointly to solve the problem using base-ten blocks placed onto the paper. Each student will perform one step in adding the decimals. Direct each member of the group to write with a different color to ensure all students participate in solving the problem. If necessary, provide a step-by-step list for groups to follow, such as 1) Model the first decimal number with base-ten blocks and set the blocks in the appropriate column. 2) Model the second decimal number with base-ten blocks and set it in the appropriate column. 3) Regroup as needed. 4) Write the solution. 5) Use estimation to check the answer for reasonableness. Assign a second Practice It Exercise and have groups repeat the activity.

English Language Development Leveled Activities

Emerging Level	Expanding Level	Bridging Level
Report Back Display the following sentence frames: ____ **ones.** ____ **tenths.** ____ **hundredths.** Use base-ten blocks and a place-value table to model 2.4 + 0.66. As you model, pause to ask, *How many [ones/tenths/hundredths]?* Allow students to report back with a single word answer, but then model using the sentence frames. If needed, count the blocks aloud and then refer to the sentence frame as you clearly enunciate the answer, for example, *Two ones.* Have students chorally repeat. Repeat the activity to model 1.54 + 0.78.	**Sentence Frames** Divide students into pairs and distribute base-ten blocks to each pair. Assign Apply It Exercise 9, and have the students in each pair work together to solve. Afterward, ask volunteers to explain each of the steps they completed to solve the problem. Display sentence frames to help students participate: **First, we** ____. **Then, we** ____. **Last, we** ____. Display another sentence frame for students to use in sharing their answer: **Valerie now has** ____ **in her bank at home.**	**Public Speaking Norms** Write *formal* and *informal* on several slips of paper and place them in a container. Briefly discuss with students the difference between formal and informal language. Write 2.84 + 0.38 on the board. Invite two students to the board to solve the problem using base-ten blocks and a place-value table. Have one student draw a slip of paper from the container. Based on the slip drawn, have the student use formal or informal language to guide the other student in solving the problem. Repeat the activity with a new problem and different pair of students.

Teacher Notes:

T47 Grade 5 • Chapter 5 *Add and Subtract Decimals*

NAME _____ DATE _____

Lesson 4 Guided Writing

Inquiry/Hands On: Add Decimals Using Base-Ten Blocks

How do you add decimals using base-ten blocks?

Use the exercises below to help you build on answering the Essential Question. Write the correct word or phrase on the lines provided.

1. Rewrite the question in your own words.
   See students' work.

2. What key words do you see in the question?
   add, decimal, base-ten blocks

3. When using base-ten blocks, which decimals are modeled below, 1.0, 0.1, or 0.01?

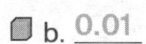

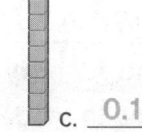

   a. 1.0   b. 0.01   c. 0.1

4. When adding two decimals, you can model each decimal using base-ten blocks, and then __combine__ the base-ten blocks.

5. When you combine two decimals, both modeled using base-ten blocks, you may need to regroup. __10__ hundredths blocks (flats) can be regrouped as 1 __tenths__ block (rod). 10 __tenths__ blocks (rods) can be regrouped as 1 __ones__ block (ones cube).

6. How do you add using base-ten blocks?
   Model each decimal being added using base-ten blocks. Combine all the base-ten blocks. Regroup as needed.

Lesson 5 Inquiry/Hands On: Add Decimals Using Models

English Learner Instructional Strategy

Vocabulary Support: Anchor Chart

Divide students into four groups. Say, *Make an anchor chart showing what you know about decimals and adding decimals.* Explain that each chart should include a title at the top of the poster and definitions for math vocabulary related to decimals, such as *tenths, hundredths, period,* and so on. Suggest that students include an example of a decimal addition problem solved with base-ten blocks or another type of model. Direct students to label different elements in their charts with appropriate math vocabulary. When the charts are completed, have groups display and describe their charts. Afterward, discuss how the anchor charts can help students better understand the steps needed to solve decimal addition problems using models.

English Language Development Leveled Activities

Emerging Level	Expanding Level	Bridging Level
Look, Listen, and Identify Use 10-by-10 grids to model solving 1.3 + 0.6. As you shade the grids to solve, pause after each step to emphasize what you have done. Point to the grid and describe the step, for example, *I shaded 30 squares.* Then ask, *How many did I shade?* Have students answer chorally, **30.** Count aloud as you indicate each column, and then say, *I shaded three tenths. How many tenths?* **3** Continue in this manner as you complete the problem. Repeat with a second decimal addition problem.	**Exploring Language Structure** Display the following sentence frames: **I will shade ____. I am shading ____. I shaded ____.** Model using 10-by-10 grids to solve 1.23 + 0.06. As you shade the grids for each step, use the sentence frames to describe what you do. For example, say, *I will shade one whole grid.* As you shade, say, *I am shading the whole grid.* Afterward, say, *I shaded the whole grid.* After you have solved the problem, write new problems and have students solve them using grids. Have students use the sentence frames to narrate their actions.	**Money Sense** Have students work in pairs. Distribute manipulative coins and 10-by-10 grids to each pair. Say, *Create two piles of coins. Each pile should total less than one dollar.* Direct students to write the values of the piles as decimal numbers. For example, *$0.67 and $0.85.* Say, *Using the grids to model the addition, find the total value of the two piles.* Afterward, gather the grids and redistribute them randomly to new pairs. Say, *Check the model you received for correctness.* If there are any wrong answers, work with students to identify where a mistake was made.

Teacher Notes:

T48 Grade 5 • Chapter 5 *Add and Subtract Decimals*

NAME _____ DATE _____

Lesson 5 Note Taking

Inquiry/Hands On: Add Decimals Using Models

Read the question. Write words you need help with and research each word. Use your lesson to write your Cornell notes. Write or draw math examples to explain your thinking. Share your examples with a classmate.

Building on the Essential Question	Notes:
How do you model adding decimals?	You can model decimals by __shading__ squares in a 10-by-10 grid. The decimal __0.01__ is modeled by shading one square. The decimal __0.1__ is modeled by shading ten squares. The decimal __1.0__ is modeled by shading all the squares. When adding two decimals, you can model each decimal by __shading__ squares on 10-by-10 grids, and then count the __combined__ squares to find the __sum__.
Words I need help with: See students' words.	
My Math Examples: See students' examples.	

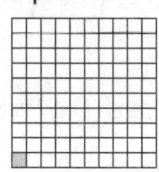

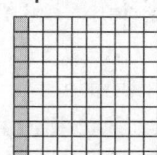

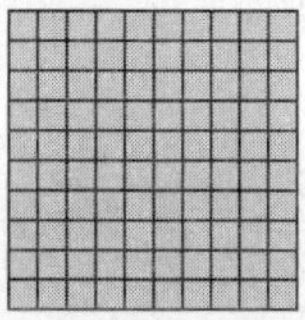

48 Grade 5 • Chapter 5 Add and Subtract Decimals

Lesson 6 Add Decimals
English Learner Instructional Strategy

Language Structure Support: Report Back

Divide students into three groups. Assign each group one of the Problem Solving exercises from the lesson. Give students time to solve their assigned problem, and then have a volunteer from each group report back with the solution. Display the following sentence frames to aid students in reporting back:

The total cost of the puzzle and batteries is ____.
The athlete swam four laps in ____ seconds.
Terrance bicycled a total of ____ miles.

As students report back, listen closely for the /s/ sound indicating plural for *laps* and *seconds* versus the /z/ sound in *batteries* and *miles*. If necessary, say these words aloud to reinforce correct pronunciation and have students repeat chorally.

English Language Development Leveled Activities

Emerging Level	Expanding Level	Bridging Level
Number Sense	**Listen and Write**	**Building Oral Language**
Students may be confused by the term *point*. They may be more familiar with *point* as meaning *the tip* or *to point with one's finger*. Write 3.17 on the board. As you write, read aloud and emphasize point. Say, *The decimal point tells us that this is a whole number plus a little more. Point to the point.* Have students point to the point. Next, write a decimal addition problem. Model lining up the decimal points before beginning to add. Allow students to make suggestions of decimals to be added and read the problems and solutions aloud.	Draw a large point on a sticky note. Write a series of up to five numbers on the board, such as 3764. Put the point after the 3 to create 3.764. Display the following sentence frame to have students identify the place value of each number: **The ____ is in the ____ place.** Move the decimal point and have them identify the changed place value of each number. Next, write a new series of numbers, such as 5641. Say, *This number is 56 point four one.* Have a volunteer place the decimal point in the correct place. Repeat the activity with new numbers.	Ask students to use number cubes to generate numbers with two decimal places. Have them write the number and then combine it with a partner's number to create an addition problem. Have partners come to the board, write the problem, read it aloud, and solve. Be sure students are lining up the decimals prior to adding. Have students repeat the activity by generating numbers with up to three decimal places.

Teacher Notes:

NAME _____ DATE _____

Lesson 6 Concept Web
Add Decimals

Use the concept web to identify the place values of the decimal.

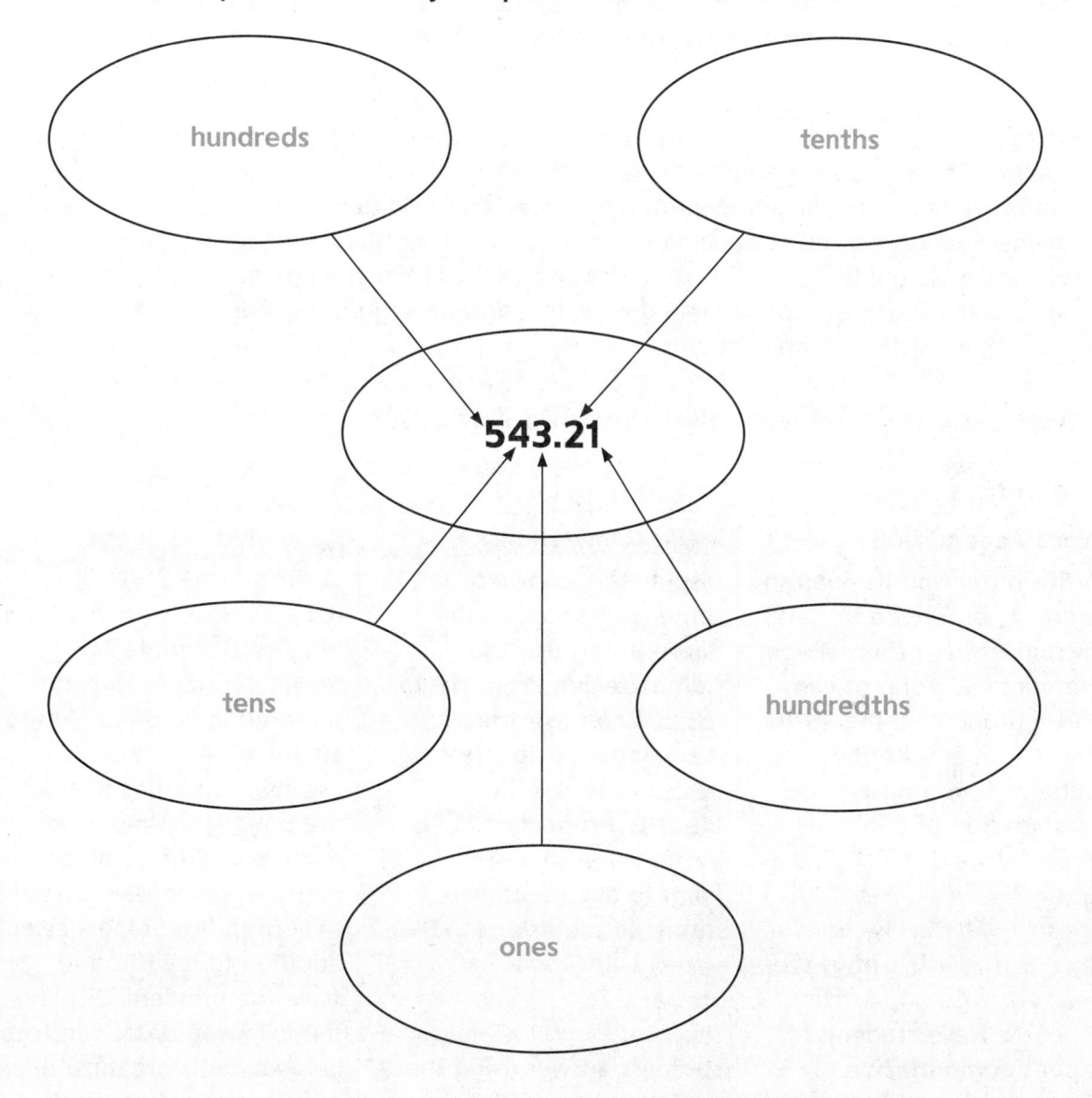

- hundreds
- tenths
- **543.21**
- tens
- hundredths
- ones

Lesson 7 Addition Properties
English Learner Instructional Strategy

Vocabulary Support: Frontload Academic Vocabulary

Write: *Associative Property of Addition, Commutative Property of Addition,* and *Identity Property of Addition*. Underline *Property* in each phrase. Discuss the non-math meaning of *property*, and then clarify that in a math context, *property* means "a trait" or "quality." Say, *Knowing the properties of addition can make adding numbers easier.*

Divide students into three groups. Distribute one of the following My Vocabulary Cards to each group: Associative Property of Addition; Commutative Property of Addition; Identity Property of Addition. Include the online Spanish version of each card if applicable. Using the example on the card as a guide, have each group create two additional examples of their assigned property on chart paper. Afterward, have groups share their examples with the other groups.

English Language Development Leveled Activities

Emerging Level	Expanding Level	Bridging Level
Word Recognition Write *order* and its Spanish cognate, *orden*, on to stand in front of the group. Divide them into a group of two and a group of three. Write $2 + 3 = 5$. Shuffle the students into a new combination of two and three. Write $3 + 2 = 5$. Say, *The order of addends doesn't matter. We still have 5 students. This is the called the Commutative Property.* Have students repeat **commutative property** chorally. Model the Associative Property in a similar way, but use three groups of students and rearrange them.	**Make Connections** Display the following sentence frames on the board: **I can use the Commutative Property to ____. I can use the Associative Property to ____. I can use the Identity Property to ____.** Write $(14 + 7) + 0 + 3$. Point to each sentence frame as you ask, *How can you use an Addition Property to help you find the sum mentally?* Have students answer using the sentence frames. Repeat with an addition problem that uses decimal numbers, such as $2.5 + (0.5 + 1.8) + 0$.	**Academic Language** Write $(15 + 9.2) + 1.8 + (0 + 5)$. Have students work in pairs. Distribute My Vocabulary Cards for the three addition properties to each pair. Ask students to use the properties to solve the equation. Then have each pair write a short paragraph describing how they used the properties of addition to rewrite and solve the problem. Display the following order words to help students organize their explanations: *first, next, then, last.*

Teacher Notes:

T50   Grade 5 • Chapter 5 *Add and Subtract Decimals*

NAME _____ DATE _____

Lesson 7 Vocabulary Chart

Addition Properties

Use the three-column chart to organize the vocabulary in this lesson. Write the word in Spanish. Then write the correct terms to complete each definition.

English	Spanish	Definition
property	propiedad	A __rule__ in mathematics that can be applied to all numbers.
Associative Property	propiedad asociativa	Property that states that the way in which numbers are __grouped__ does not change the sum.
Commutative Property	propiedad conmutativa	Property that states that the __order__ in which numbers are added does not change the __sum__.
Identity Property	propiedad de identidad	Property that states that the sum of any number and 0 __equals__ the number.

50 Grade 5 • Chapter 5 *Add and Subtract Decimals*

Lesson 8 Inquiry/Hands On: Subtract Decimals Using Base-Ten Blocks

English Learner Instructional Strategy

Graphic Support: Venn Diagram

Display a Venn diagram. Label once side *Adding with Base-Ten Blocks* and the other side *Subtracting with Base-Ten Blocks*. Provide an example of each, such as 1.25 + 0.6 and 1.5 − 0.4. Then ask students to compare and contrast how each was solved. Display the following sentence frames for students to use as they compare similarities and differences: **In both problems** ____. **When you add** ____, **but when you subtract** ____. **Subtracting is different because** ____. **Addition is different because** ____. Record student answers in the appropriate areas of the diagram.

During the Try It and Talk About It parts of the lesson, allow Emerging students to partner with Expanding or Bridging students who share their native language. Have the Emerging student participate in the discussion by suggesting answers to his or her partner using their native language. Then have the more proficient English speaker translate the answer to English.

English Language Development Leveled Activities

Emerging Level	Expanding Level	Bridging Level
Act It Out	**Listen and Identify**	**Numbered Heads Together**
Distribute base-ten blocks to students. Have each student use the blocks to model a decimal number of their choice to the tenths place between 1.0 and 2.0. Invite two students to the board with their blocks. Guide them in using the blocks to model a subtraction problem to find the difference between their decimal numbers. Display sentence frames to help them describe the steps: **Take away** ____ **one. Take away** ____ **tenths.** ____ **are left. The difference is** ____. Invite a new pair of students to the board to solve another problem.	Distribute base-ten blocks to students. Have each student use the blocks to model a decimal number of their choice to the hundredths place between 1.0 and 2.0. Invite two students to the board with their blocks. Direct them to use the blocks to model a subtraction problem to find the difference between their decimal numbers. As they work together to solve, ask them to identify elements in their models. For example, *Which blocks show tenths? Which blocks show hundredths?* Invite another pair to the board to solve another problem.	Have students get into groups of four. Ask the students in each group to number off as 1–4. Have the students in each group work together to solve Apply It Exercise 10. Afterward, display the following sentence frames: **First we used base-ten blocks to model** ____. **Then we modeled** ____. **We had to regroup** ____. **We removed** ____. **The remaining piece of wood is** ____. Choose numbers from 1–4 to designate which student in each group will use the sentence frames to describe their group's answer.

Teacher Notes:

NAME _____ DATE _____

Lesson 8 Note Taking

Inquiry/Hands On: Subtract Decimals Using Base-Ten Blocks

Read the question. Write words you need help with and research each word. Use your lesson to write your Cornell notes. Write or draw math examples to explain your thinking. Share your examples with a classmate.

Building on the Essential Question	Notes:
How do you model subtraction using base-ten blocks?	The decimal <u>1.63</u> is modeled below.

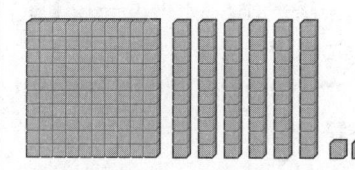

The decimal 0.45 is modeled using <u>4</u> tenths blocks (rods) and <u>5</u> hundredths blocks (ones cubes).

When you subtract two decimals, start by removing base-ten blocks from the <u>least</u> place value.

1.63 − 0.45

There are not 5 hundredths in the base-ten blocks for 1.63. So, you regroup 1 <u>tenths</u> block (rod) as <u>10</u> hundredths blocks (ones cubes).

After regrouping, there will be <u>13</u> hundredths blocks (ones cubes) altogether.

Subtract <u>5</u> hundredths, and there are <u>8</u> hundredths remaining.

Next, subtract the tenths place value. After the regrouping, there are <u>5</u> tenths blocks (rods).

Subtract <u>4</u> tenths, and <u>1</u> tenth remains.

After subtracting 0.45 from 1.63, <u>1</u> one, <u>1</u> tenth, and <u>8</u> hundredths remain.

Words I need help with:
See students' words.

My Math Examples:
See students' examples.

Grade 5 • Chapter 5 *Add and Subtract Decimals* 51

Lesson 9 Inquiry/Hands On: Subtract Decimals Using Models

English Learner Instructional Strategy

Vocabulary Support: Utilize Resources

On the board, create a list of words related to subtracting decimals using models that students can refer to and use during the lesson. Include the following words and add any others that seem appropriate: *decimal, period, ones, tenths, hundredths, regroup, shade, grid*.

As students work through the Apply It Exercises, be sure to remind them that they can refer to the Glossary or the Multilingual eGlossary for help, or direct students to other translation tools if they are having difficulty with nonmath language in the problems, such as *megabyte, memory, trails, weekend,* or *hiking*.

English Language Development Leveled Activities

Emerging Level	Expanding Level	Bridging Level
Number Recognition	**Pairs Share**	**Turn & Talk**
As you work through the Model the Math and Build It parts of the lesson, monitor students' comprehension by prompting them to identify different aspects of the math and the model. For example, point to 0.47 and ask, *Which number is in the tenths place?* Students may answer by pointing to the four or by saying **four**. After shading to model 0.47, point to the grid and ask, *Is this 0.47 or 1.32?* Again, allow students to answer verbally or with a gesture. Continue in this manner as you complete the lessons.	Randomly assign Practice It Exercises 1–7 to student pairs. After they have had time to work through their problems, ask one student in each pair to share the steps they followed to reach a solution. Provide sentence frames to help students participate: **We shaded ____ squares. We crossed out ____ squares. ____ were left, so the answer is ____ .**	Write 2.65 − 1.24. Ask, *I will use a model to solve. How many grids do I need?* Direct students to turn to a partner and discuss the answer. After partners have had time to talk, call on a student to answer. Encourage the student to use a complete sentence. Display three grids, and ask, *How many squares do I need to shade?* Have students turn to a partner again to discuss the answer, and then call on a different student to answer. Continue in this manner as you solve the problem using the grids. Repeat with a new problem.

Teacher Notes:

NAME _____ DATE _____

Lesson 9 Guided Writing

Inquiry/Hands On: Subtract Decimals Using Models

How do you model decimal subtraction?

Use the exercises below to help you build on answering the Essential Question. Write the correct word or phrase on the lines provided.

1. What key words do you see in the question?
   model, decimal, subtraction

2. The decimal __1.7__ is modeled in the shaded squares below.

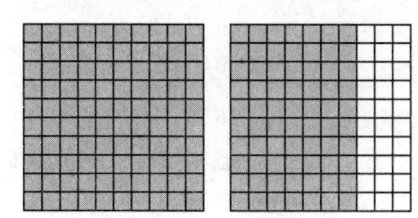

3. When subtracting a decimal from 1.7, you __cross out__ enough squares that represent the decimal on the model for 1.7.

4. To subtract 0.45, cross out __45__ squares.

5. The __difference__ is represented by the number of shaded squares that remain.

6. 1.7 − 0.45 = __1.25__

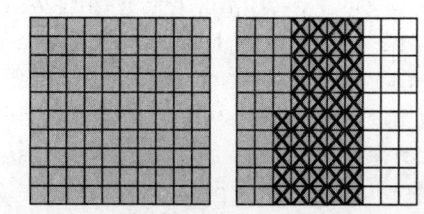

7. How do you model decimal subtraction?
   Model the decimal being subtracted from by shading 10-by-10 grids. Cross out shaded squares on the grid to represent the decimal being subtracted. The shaded squares that remain represent the difference.

Lesson 10 Subtract Decimals
English Learner Instructional Strategy

Graphic Support: KWL Chart

Write *inverse operations* and its Spanish cognate, *operaciones inversas*, on a classroom cognate chart. Display a KWL chart. In the first column, record what students recall from previous grades about inverse operations. In the second column, record what students hope/want to learn during the lesson; include wanting to know how the use of inverse operations will apply to subtracting decimals.

After the lesson, display the following sentence frame and have students use it to describe what they learned during the lesson: **I learned that inverse operations can help to ___.** Use the third column of the KWL chart to record student responses.

English Language Development Leveled Activities

Emerging Level	Expanding Level	Bridging Level
Word Recognition Write 3.8 − 2.1 vertically on the board. Name the place value in each addend and have students repeat chorally. Model solving to find the difference and write it on the board. **1.7** Say, *We will check our answer.* Emphasize *check* and have students repeat chorally. Be sure they are correctly saying the /ch/ and /k/ sounds in check. Write 1.7 + 2.1 vertically on the board and solve it to model checking the previous answer. Provide students with additional problems to solve and check.	**Developing Oral Language** Write several examples of subtraction problems using decimals. Work with students to solve them. Ask students to identify the inverse operation they would use to check the difference in each answer. Provide the following sentence frame: **I can check the answer by adding ___ and ___.** Have students perform the inverse operations and then explain how they know the answer is correct or incorrect using the following sentence frame: **I know the answer is correct/incorrect because ___.**	**Signal Words/Phrases** Have a student read aloud a word problem from the lesson. As he or she reads, list words and phrases from the problem that signify subtraction. Discuss with students how these words and phrases can help them recognize that subtraction is needed to solve the word problem. Ask a different student to read aloud another word problem from the lesson while a volunteer writes the words/phrases from the problem that signifies subtraction. Have students search through the My Homework Exercises for signal words/phrases and add to the list.

Multicultural Teacher Tip

ELs may use an alternative algorithm when solving subtraction problems. In particular, Latin American students may have been taught the *equal additions method* of subtraction instead of the traditional US method of "borrowing" from the column to the left when the top number is less than the bottom number. In the *equal additions method,* a problem such as 35 − 18 solved vertically would start with ten ones added to the top number (15 − 8) and then one ten is added to the bottom number (30 − 20), to get 7 and 10, or 17. Similarly, 432 − 158 would be solved as 12 − 8, 130 − 60, and 400 − 200 (4 + 70 + 200 = 274).

NAME _____ DATE _____

Lesson 10 Concept Web

Subtract Decimals

Use the concept web to identify the inverse operation of each operation shown in the concept web.

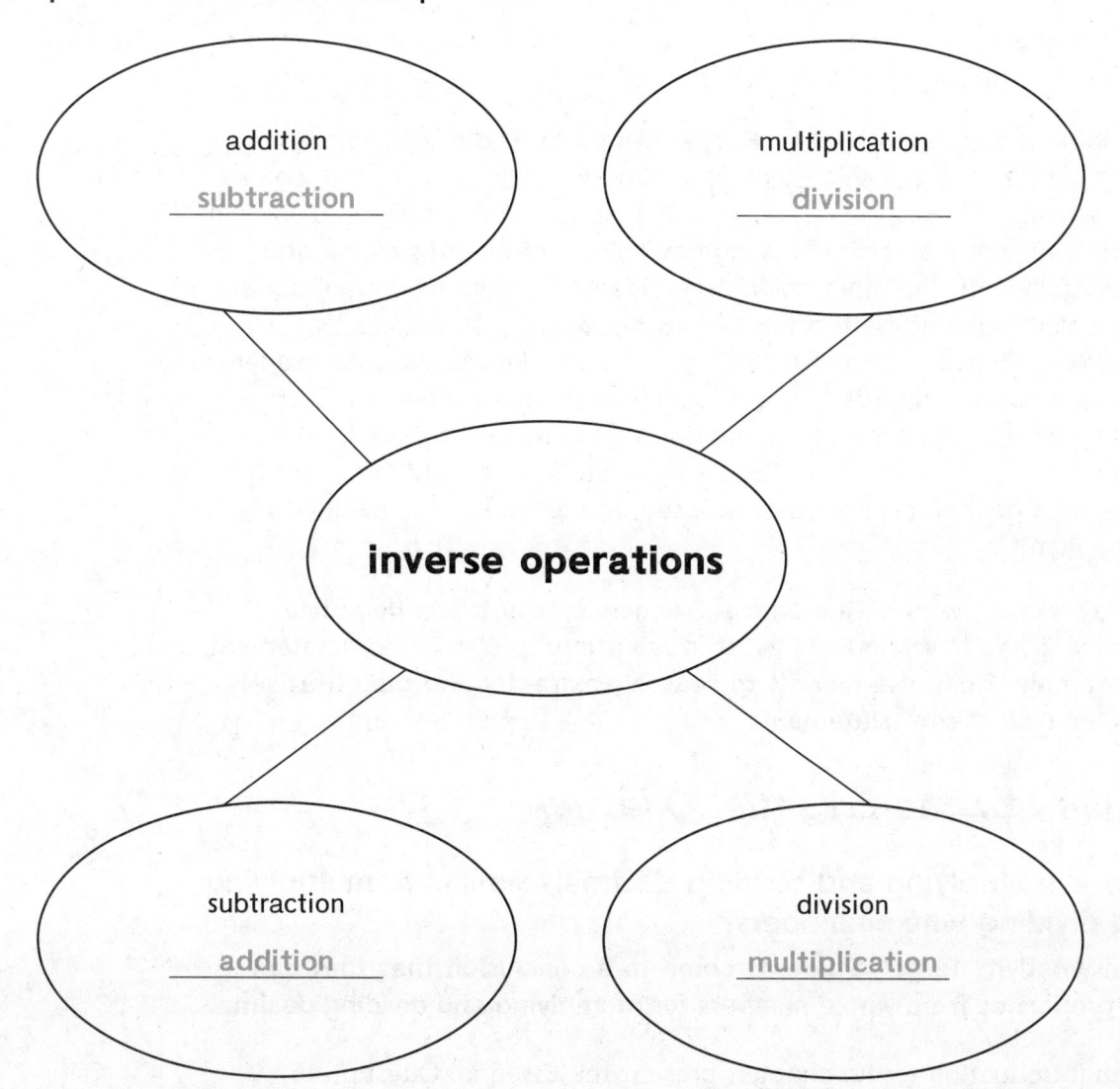

- addition
  - subtraction
- multiplication
  - division

inverse operations

- subtraction
  - addition
- division
  - multiplication

Chapter 6 Multiply and Divide Decimals

What's the Math in This Chapter?

Mathematical Practice 2: Reason abstractly and quantitatively

Write 0.5 × 0.7 on the board. Ask, *How can we solve this problem?* **We need to multiply.** Distribute write-on/wipe-off boards and have students try to find a solution on their own. Have students share their answers.

Distribute a copy of Work Mat 8: First Quadrant Grid to each student along with 2 different colored crayons. Say, *We are going to use tools to help us reason abstract problems more easily.* Instruct students to color in 5 tenths of the grid vertically using one color and 7 tenths of the grid horizontally with the other color. Ask students to count how many squares were colored in with both colors of crayons. **35** Say, *The product of 0.5 and 0.7 is 0.35. How is that problem like 5 × 7 = 35?* Provide time for students to think and then discuss. Students should make the connection that multiplying decimals is just like multiplying whole numbers. Ask, *How do we know where to put the decimal point?* Students should infer that the placement of the decimal point is based on the sum of the decimal places in the factors.

Display a chart with Mathematical Practice 2. Restate Mathematical Practice 2 and have students assist in rewriting it as an "I can" statement, for example: **I can use models to reason abstractly and quantitatively.** Post the new "I can" statement.

Inquiry of the Essential Question:

How is multiplying and dividing decimals similar to multiplying and dividing whole numbers?

Inquiry Activity Target: **Students come to a conclusion that they can use the properties from whole numbers for multiplying and dividing decimals.**

As an introduction to the chapter, present the Essential Question to students. The inquiry graphic organizer will offer opportunities for students to observe, make inferences, and apply prior knowledge of properties representing the Essential Question. As they investigate, encourage students to draw, write, and collaborate with peers to demonstrate their observations and thinking. Then have students present additional questions they may have to a peer to extend discussions.

Regroup students and restate Mathematical Practice 2 and the Essential Question. Pose questions to reflect on what has been learned to guide students in making connections between the Mathematical Practice and the Essential Question.

NAME _____ DATE _____

Chapter 6 Multiply and Divide Decimals

Inquiry of the Essential Question:

How is multiplying and dividing decimals similar to multiplying and dividing whole numbers?

Read the Essential Question. Describe your observations (I see..), inferences (I think...), and prior knowledge (I know...) of each math example. Write additional questions you have below. Then share your ideas and questions with a classmate.

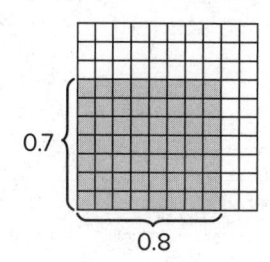

[grid diagram with 0.7 and 0.8 labeled]

I see …

I think…

I know…

```
    1
  0.13   ← 2 decimal places
×  2.5   ← 1 decimal place
  ─────
    65
 +260
  ─────
 0.325   ← 2 + 1 = 3 decimal places
```

I see …

I think…

I know…

34.7 ÷ 6
↓   ↓
 36 ÷ 6 ← Round 34.7 to 36 since 36 and 6 are compatible numbers.

Since 36 ÷ 6 = 6, 34.7 ÷ 6 is **about 6**

I see …

I think…

I know…

Questions I have…

_____

_____

_____

Lesson 1 Estimate Products of Whole Numbers and Decimals

English Learner Instructional Strategy

Language Structure Support: Report Back

During Model the Math, provide the following sentence frames to help groups report back: **We shaded** _____ **square(s). We shaded** _____ **column(s). More squares were shaded to show** _____. As students report back, be sure they are saying the /sh/ sound. If necessary, model the correct pronunciation of *shaded* and *show*.

As you work through the Guided Practice problems, prompt emerging students' participation by asking either/or questions that can be answered with a gesture or a single word. For example, asking *Do we round up or down?* allows the student to answer by pointing up or down or saying **up** or **down**.

English Language Development Leveled Activities

Emerging Level	Expanding Level	Bridging Level
Activate Prior Knowledge	**Number Sense**	**Internalize Language**
Refer to the classroom cognate chart as you review *estimate*, *decimal*, and *value*. Choose a problem from the Independent Practice section and use it to model estimating. As you describe each step in the process, emphasize the three Review Vocabulary terms for this lesson. Have students repeat each word chorally. Listen for correct pronunciation, and have students repeat if necessary until they are saying each word appropriately.	Show a group of 10 items plus a group of two additional ones. You can use pencils, paper clips, sticky notes, and so on. Point to the group of ten and say, *These ten together make one whole group.* Point to the other group and say, *These are 2 extra. They make two-tenths of a group.* Write 1.2. Ask, *If we had four times as many pencils, about how many groups would we have?* Write 1.2 × 4 and have students find the estimate. Provide a sentence frame to help students share their answer: **My estimate is** _____.	Divide students into two teams, A and B. Team A writes a decimal number, such as 18.7. Team B writes a whole number, such as 6. Team A rounds 18.7 to 19 and then multiplies 19 by 6 to estimate the product. Team B vocalizes the necessary steps to check the answer using repeated addition and performs the steps. Have teams switch roles. Play continues back and forth as time permits.

Multicultural Teacher Tip

Because many word problems involve prices and/or determining changes in monetary value, ELs will benefit from an increased understanding of American coins and bills. A chart or other kind of graphic organizer visually comparing coin and bill values and modeling how to write dollars and cents in decimal form would help these students. You may also want to have ELs describe the monetary systems of their native countries. Identifying similarities or differences with the American system can help familiarize students with dollars and cents.

NAME _____ DATE _____

Lesson 1 Review Vocabulary Chart
Estimate Products of Whole Numbers and Decimals

Use the three-column chart to organize the review vocabulary in this lesson. Write the word in Spanish. Then write the correct terms to complete each definition.

English	Spanish	Definition
decimal	decimal	A number that has a __digit__ in the tenths place, hundredths place, and beyond.
estimate	estimación	A number __close__ to an exact value. An estimate indicates __about__ how much.
place value	valor posicional	The value given to a digit by its __position__ in a number.
product	producto	The answer to a __multiplication__ problem.

Grade 5 • Chapter 6 *Multiply and Divide Decimals*   **55**

Lesson 2 Inquiry/Hands On: Use Models to Multiply

English Learner Instructional Strategy

Language Structure Support: Tiered Questions

During the lesson, be sure to ask questions according to the ELs' level of English comprehension. For example, ask Emerging students simple questions that elicit one-word answers: *How many squares? How many rows? How many did I shade? What does this grid show? Does this show tenths or hundredths? How many tenths/hundredths?*

For Expanding students, ask more complex questions that elicit simple phrases or short sentences: *How do we find the product? How do I represent 0.7 on the grid? How many tenths in one whole?*

For Bridging students, ask questions that require more complex answers: *Why do we need three grids? How can we check the reasonableness of our answer?*

English Language Development Leveled Activities

Emerging Level	Expanding Level	Bridging Level
Choral Responses	**Sentence Frames**	**Listen, Write, and Read**
Distribute 10-by-10 grids to students and model solving 0.5 × 3 using grids as students follow along. As you model each step, first describe what you will do, and then describe what you did. For example, say, *I will shade five rows.* Have students repeat chorally and then shade their grid(s). After you complete the step, say, *I shaded five rows.* Have students repeat chorally again. Be sure students are correctly altering their pronunciation of the verb each time by applying the /ed/ sound to the past tense form. Repeat with a new problem.	On the board, write: 0.___ × ___. Give a volunteer two number cubes to roll. Use the numbers rolled to fill in the blanks on the board. Have students guide you as you model solving the multiplication problem using 10-by-10 grids. Provide sentence frames to help students participate: **We need ___ grids. Shade ___ rows in each grid. The total amount shaded is ___. ___ times ___ is ___.** Have another volunteer roll the cubes to generate digits for a new problem.	Have students work in small groups. On the board, write 0.4 × 3. Say, *Write a step-by-step description of how you would solve this math problem using models.* Distribute index cards to groups for writing their descriptions. After groups have had time to complete the task, call on volunteers to read aloud their group's description. Then have groups exchange index cards and follow the steps to solve the problem. Have groups present their solutions, and discuss the steps that were followed.

Teacher Notes:

NAME _____ DATE _____

Lesson 2 Guided Writing

Inquiry/Hands On: Use Models to Multiply

How do you use models to multiply decimals?

Use the exercises below to help you build on answering the Essential Question. Write the correct word or phrase on the lines provided.

1. Rewrite the question in your own words.
   See students' work.

2. What key words do you see in the question?
   models, multiply, decimals

3. You can model decimals by __shading__ squares in a 10-by-10 grid. The decimal __0.7__ is modeled in the shaded squares to the right.

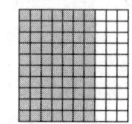

4. To model the multiplication of a whole number and a decimal, use as many grids as the value of the __whole__ number. Then model the __decimal__ on each grid.

5. To model the multiplication of 0.7 and 3, you will need __3__ grids. Model __0.7__ on each of the grids.

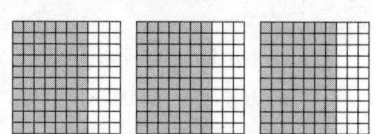

6. To find the product of a whole number and a decimal, combine the shaded squares onto one model. The total number of shaded squares is __210__. The model below represents the product, which is __2.1__.

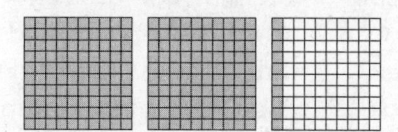

7. How do you model multiplication of whole numbers and decimals?
   Model the decimal being multiplied by shading squares on 10-by-10 grids. Repeat until you have modeled the decimal as many times as the whole number. Combine all the shaded squares onto a new model. The shaded squares on the new model represent the product.

56 Grade 5 • Chapter 6 *Multiply and Divide Decimals*

Lesson 3 Multiply Decimals by Whole Numbers

English Learner Instructional Strategy

Vocabulary Support: Sentence Frames

During Talk Math, provide tiered instruction and sentence frames that will allow students of varying proficiency levels to participate. For emerging students, ask questions that prompt gestures or one word answers as you model the estimation: *Do I round up or down? What is the estimate? Is six greater or less than the exact answer?* For emerging students, display the following sentence frames to help them participate: **The estimate is ___. We rounded ___ up/down. Our estimate is greater/less than the exact answer.** For bridging students, display a sentence frame that requires a fuller answer: **I know the estimate is ___ than the exact answer because ___.**

English Language Development Leveled Activities

Emerging Level	Expanding Level	Bridging Level
Phonemic Awareness Write *multiply* and its Spanish cognate, *multiplier*, on a classroom cognate chart. Write and say *who* and *whose*. Cross out the w in each word as you emphasize the pronunciation and have students chorally repeat. Say, *The w is silent.* Repeat with *whole*. Say, *Whole means "having all parts."* Write 6 × 3.54. Point to each number in the equation as you ask, *Is this a whole number or a decimal?* Have students answer chorally. Then model solving the equation. Repeat with a similar multiplication problem.	**Making Connections** Show 4 quarters. Have students identify the value of 4 quarters. Say, *$1.00 is equal to 4 quarters.* Write 0.25 × 4 and model solving. Vocalize the steps and write the answer as $1.00. Ask, *How many places appear to the right of the decimal in 0.25?* **(2 places)** Explain that two places will also appear to the right of the decimal in the answer. Use play coins to model $0.65. Say, *How much do I have if I have five times as much?* Model solving 0.65 × 5. Have students read the answer using the following sentence frame: **You have ___ dollars and ___ cents.**	**Show What You Know** Write the vertical multiplication problem 0.25 × 5. Model multiplying as whole numbers (25 × 5) and then ask, *Where do we place the decimal?* Display a sentence frame for students to use when answering: **Count ___ places from right to left in the product.** Count aloud as you model placing the decimal. Use grocery advertisements to find items that cost less than 1 dollar. Then have pairs multiply to find the cost of 3 items. Have students describe the process to a peer and show how they determined the total cost.

Teacher Notes:

NAME _____ DATE _____

Lesson 3 Four-Square Vocabulary

Multiply Decimals by Whole Numbers

Write the definition for each review math word. Write what each word means in your own words. Draw or write examples that show each math word meaning. Then write your own sentences using the words.

multiplication

Definition	My Own Words
An operation on two numbers to find their product. It can be thought of as repeated addition.	See students' examples.
My Examples	**My Sentence**
Sample answer: 2 × 3.6 means 3.6 + 3.6.	Sample sentence: The result of multiplication is a product.

factor

Definition	My Own Words
A number that is multiplied by another number.	See students' examples.
My Examples	**My Sentence**
Sample answer: In the expression 2 × 3.6, both 2 and 3.6 are factors.	Sample sentence: When you multiply a decimal and a whole number, both the decimal and the whole number are factors.

Grade 5 • Chapter 6 *Multiply and Divide Decimals*   **57**

Lesson 4 Inquiry/Hands On: Use Models to Multiply Decimals

English Learner Instructional Strategy

Collaborative Support: Partner Reading

Have students work in pairs to solve the Apply It exercises. Have partners take turns reading aloud information from each problem. Tell students to listen closely to what their partners say and politely suggest corrections for any mistakes in pronunciation or usage as necessary.

Display the following sentence frames to help partners work together to solve the problems and share their answers: **We need to multiply ____ and ____. We need to shade ____ squares in ____ rows. The total amount shaded is ____ squares. Each square represents ____. ____ times ____ is ____. The area of the window is ____.**

English Language Development Leveled Activities

Emerging Level	Expanding Level	Bridging Level
Developing Oral Language Write and say, *Hundredth.* Display a 10-by-10 grid. Point to one square and say, *One hundredth.* Have students chorally repeat. Add an *-s* to *Hundredth* on the board. Indicate a row of squares in the grid and say, *Ten hundredths.* Have students chorally repeat. Repeat with *Tenth(s).* As you work through the Model the Math and Draw It lessons, point to each word on the board as you use it and clearly enunciate the /th/ and /ths/ endings. Have students chorally repeat. Listen for the correct pronunciation of the ending sounds and remodel as needed.	**Show What You Know** Have students work in pairs. Randomly assign one of Practice It Exercises 6–9 to each pair. Have students work together to solve the exercise. Then invite a pair to come to the board and show how they solved the problem using grids. Provide sentence frames to help students describe the process: **First we shaded ____. Then we shaded ____. We counted ____. The answer is ____ hundredths.** Have pairs work on a different Practice It Exercise, and invite a new pair to the board to show how they solved it.	**Academic Vocabulary** Display a Venn diagram. Label one side *Multiplying Decimals* and the other side *Multiplying Whole Numbers.* Model multiplying 7 × 8 using a 10-by-10 grid. Then model multiplying 0.7 × 0.8 using another 10-by-10 grid. Lead a discussion of the differences and similarities between the two models. Display sentence frames to help students participate: **Each square in the grid represents ____. Both models ____. Multiplying decimals is different because ____.**

Teacher Notes:

T58 Grade 5 • **Chapter 6** *Multiply and Divide Decimals*

NAME _____ DATE _____

Lesson 4 Note Taking

Inquiry/Hands On: Use Models to Multiply Decimals

Read the question. Write words you need help with and research each word. Use your lesson to write your Cornell notes. Write or draw math examples to explain your thinking. Share your examples with a classmate.

Building on the Essential Question

How do you use models to multiply decimals?

Notes:

To model the multiplication of 0.4 and 0.7, shade a __rectangle__ that has a __length__ of 0.7 and a __width__ of 0.4.

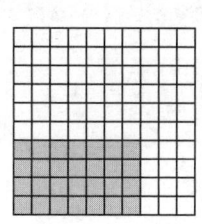

The __product__ is found by counting the total number of shaded squares on the rectangle.

The total number of shaded squares for 0.4 × 0.7 is __28__ squares. The product is __0.28__.

Words I need help with:

See students' words.

My Math Examples:

See students' examples.

58 Grade 5 • Chapter 6 *Multiply and Divide Decimals*

Lesson 5 Multiply Decimals
English Learner Instructional Strategy

Collaborative Support: Peers/Mentors

During Model the Math, be sure each group contains students of varying levels of English proficiency. Allow same-native language expanding and bridging students to translate any unfamiliar words for the emerging students as groups complete the task. When groups have completed the multiplication and counted the decimals placed in each factor, have emerging students say how many decimal places are in the product. If emerging students are comfortable speaking aloud, have the other students in the group say the number aloud and encourage the emerging student to repeat it.

English Language Development Leveled Activities

Emerging Level	Expanding Level	Bridging Level
Number Sense Write the decimal 2.75 on the board. Gesture to the decimal point and ask, *What is this?* Have students answer **decimal** chorally. Count aloud the number of decimal places to the right of the decimal point. Ask, *How many places to the right of the decimal?* Then write and say, *Two places.* Display the following phrase frame: ____ **places.** Write several other decimals and have students use the frame to identify the number of decimal places each has. Then choose two of the decimal numbers to use for modeling multiplication.	**Academic Vocabulary** Write and solve a decimal multiplication problem. Have students count chorally the decimal places in each factor and the product. Record the numbers to the right of the problem. Say, *Adding the decimal places in each factor gives us the decimal places in the product.* Repeat with another decimal multiplication problem, but do not place the decimal. Display the following sentence frames for students to explain where it goes: **One factor has ____ decimal places. The other factor has ____ places. So the product has ____ decimal places.**	**Number Game** Divide students into two teams, A and B. Distribute a write-on/wipe-off board to each team. Designate a student to be referee and give him or her a calculator. Team A writes a decimal number. Team B writes a decimal number. Both teams multiply the decimal numbers together. The first team to solve the problem and then read it aloud along with the solution receives a point after the referee checks for accuracy. Continue until both teams have 5 points.

Teacher Notes:

NAME _____ DATE _____

Lesson 5 Vocabulary Note Taking
Multiply Decimals

Read the question. Write words you need help with and research each word. Use your lesson to write your Cornell notes. Write or draw math examples to explain your thinking. Share your examples with a classmate.

Building on the Essential Question

How do you multiply decimals?

Words I need help with:

See students' words.

Notes:

Multiplication with decimal numbers is similar to multiplication with __whole__ numbers.

When you multiply decimals, __multiply__ as with whole numbers.

```
   0.25              25
 ×  3.1            × 31
                    25
                 + 750
                   775
```

Count the decimal places in each decimal factor.
There are __2__ decimal places in 0.25.
There is __1__ decimal place in 3.1.

The total number of decimal places in the factors is __equal__ to the number of decimal places in the product of the decimals.

There are a total of __3__ decimal places in 0.25 × 3.1.
There will be __3__ decimal places in the product of 0.25 × 3.1.
The decimal point will be placed __3__ decimal places from the __right__ in 775.

0.25 × 3.1 = __0.775__

My Math Examples:

See students' examples.

Grade 5 • Chapter 6 *Multiply and Divide Decimals* **59**

Lesson 6 Multiply Decimals by Powers of Ten

English Learner Instructional Strategy

Graphic Support: Tables

Draw a three-column table titled *Multiplying by Powers of Ten*. Label the first two columns *Power of Ten* and *Number of Places*. Leave the third column unlabeled.

In the rows of the first column, write in *10, 100, 1,000,* and *10,000*. Work with students to complete the second column with the corresponding number of places the decimal will move to the right when multiplied by the power of ten. Display a sentence frame to help students: **Move the decimal ___ places to the right.**

Then write a decimal number in the header for the third column, such as 12.53. Have students multiply the decimal number by each power of ten in the first column using the rule in the second column. Record each answer in its appropriate row. Have students scribe the table into their math journals for future reference.

English Language Development Leveled Activities

Emerging Level	Expanding Level	Bridging Level
Synthesis	**Recognize and Act It Out**	**Partner Work**
Write a decimal number prominently on the board. Use a paper circle taped to the board as the decimal point. Say, *When you multiply a decimal by 10, you move the decimal point one place to the right.* Move the paper circle one place to the right. Display the following sentence frame: ___ **zeros = ___ places.** Write a new decimal number and model multiplying by 100. Count the zeros and say, *Two zeros equals two places.* Move the decimal point two places. Have students use the sentence frame to guide you in solving other powers of ten problems.	Write each number from 0 to 9 on large pieces of paper. Create a decimal point out of a piece of black paper. Have three or four students hold up numbers and the paper circle to model a decimal number. Say, *Multiply the decimal by 10.* The student holding the decimal point will move one place to the right. Repeat with other decimal numbers and have students act out multiplying them by 100 and 1,000. Make additional papers with 0s to annex zeros.	Distribute number cubes to pairs and have them create and solve multiplication problems. Student A will roll the number cubes to create a decimal number. Student B will choose a power of 10 (10, 100, or 1,000). Then student A will multiply to find the product. Repeat as time permits, or until each student has multiplied a decimal by 10, 100, and 1,000. Afterward, have students explain how to use zeros to determine how many places to move the decimal. Provide a sentence frame: **I move the decimal ___ places when there are ___ zeros.**

Teacher Notes:

NAME _____ DATE _____

Lesson 6 Concept Web

Multiply Decimals by Powers of Ten

Use the concept web to write each exponent as a power of 10 and each power of ten as an exponent.

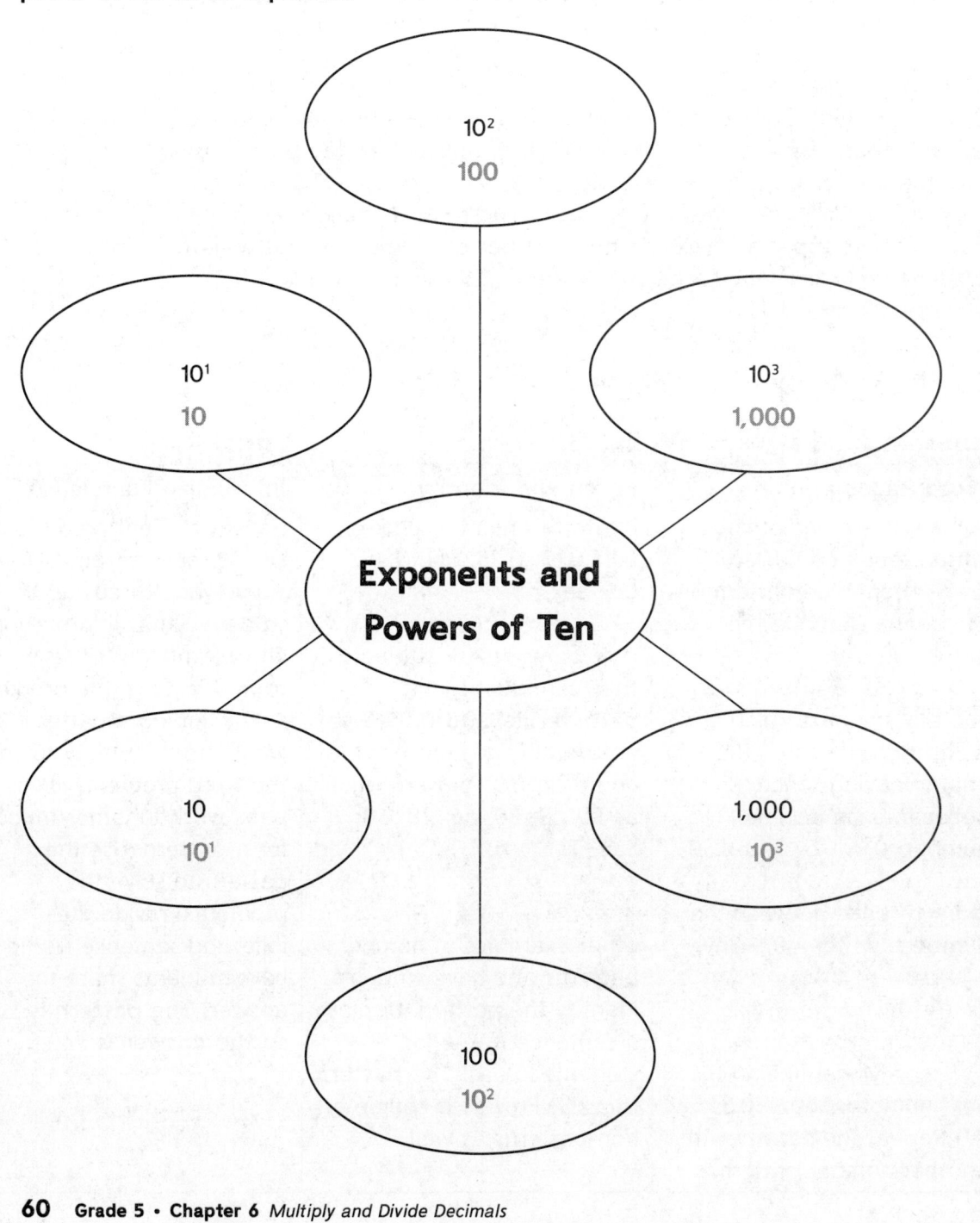

- 10^2
  100

- 10^1
  10

- 10^3
  1,000

Exponents and Powers of Ten

- 10
  10^1

- 1,000
  10^3

- 100
  10^2

60 Grade 5 · Chapter 6 *Multiply and Divide Decimals*

Lesson 7 Problem-Solving Investigation Strategy: Look for a Pattern
English Learner Instructional Strategy

Language Structure Support: Modeled Talk

Display the following sentence frames to help students describe the problem-solving steps: **I know ____. I need to find out ____. My plan is ____. I will ____ to solve the problem. I will ____ to check my answer.**

As you work through the Practice lessons, model using each sentence frame so students can hear standard English pronunciation. After saying each sentence, ask a question that will allow a volunteer to respond using the same sentence frame. For example, ask, *What do you know? What do you need to find out? What is your plan?* and so on. Listen for correct pronunciation, especially with sounds or letter combinations that do not transfer from other languages, such as the initial sounds /ch/ and the silent-k in /kn/.

English Language Development Leveled Activities

Emerging Level	Expanding Level	Bridging Level
Word Recognition	**Listen and Identify**	**Internalize Language**
Show a piece of clothing with a repeated pattern or draw a repeated pattern on the board. Describe the pattern and say, *This has a pattern.* Have students chorally say **pattern.** Then write this pattern on the board, leaving space between each decimal number: 0.0, 0.5, 0.10, 1.50. Say, *This has a pattern too.* In each space between the numbers, write + 0.5. Say, *The pattern is to add five tenths each time. What number is next in the pattern?* Model finding the next number: 1.50 + 0.5 = 2.0 Repeat the activity with another number pattern.	Write this number pattern: 0.25, 0.3, 0.35, 0.4, 0.45, 0.5. Say, *These numbers follow a rule. They are part of a pattern.* Ask students if they can identify the pattern rule. **add 0.05** Give an example of numbers that do not form a pattern, such as 30, 18, 30, 66, 28, 64. Say, *These numbers do not follow a specific rule. They are not a pattern.* Give other examples of number patterns and have students identify the pattern. Display a sentence frame for students to use: **The pattern is ____.** Provide a couple of non-patterns as well.	Have a student read aloud a word problem from My Homework. Discuss with students what information is known and what they need to find to solve the problem. Ask if finding a pattern would be a strategy to solve the word problem. Have pairs work together to look for a pattern. Use the pattern to solve the problem. Provide the following sentence frame to have students share the answer: **The pattern is ____ so the answer is ____.**

Teacher Notes:

NAME _____ DATE _____

Lesson 7 Problem-Solving Investigation

STRATEGY: Look for a Pattern

Look for a pattern to solve each problem.

1. Every year, **Victoria** receives **$30** for her birthday, plus **$2** for **each year** of **her** (Victoria's) **age**.
   Lacey received **$20** for her birthday and **$4** for **each year** of **her** (Lacey's) **age**.
   In **2013**, Victoria is **10**, and Lacey is **6**.
   In what year will **they** both receive the **same** amount of money?

Understand	Solve
I know: I need to find:	<table><tr><th>Year</th><th>Victoria</th><th>Lacey</th></tr><tr><td>2013</td><td>30 + 2 × 10</td><td>20 + 4 × 6</td></tr><tr><td>2014</td><td>30 + 2 × 11</td><td>20 + 4 × 7</td></tr><tr><td>2015</td><td></td><td></td></tr><tr><td>2016</td><td></td><td></td></tr><tr><td>2017</td><td></td><td></td></tr></table>
Plan I will use a table and look for a _pattern_.	**Check**

2. **Trent** lifts weights **7 days a week**.
   He spends **18 minutes** lifting weights on **Monday**, **29 minutes** on **Tuesday**, **40 minutes** on **Wednesday**, and **51 minutes** on **Thursday**.
   If this pattern continues, how **many minutes** will Trent lift weights on **Saturday**?

Understand	Solve
I know: I need to find:	
Plan I will make a _table_ and look for a pattern.	**Check**

Grade 5 • Chapter 6 Multiply and Divide Decimals **61**

Lesson 8 Multiplication Properties
English Learner Instructional Strategy

Graphic Support: Venn Diagrams

On a large piece of chart paper, write *Addition Properties*. Underneath, write *Associative, Commutative,* and *Identity* with examples of each addition property. On the other side of the paper, write *Multiplication Properties*. Underneath, write *Associative, Commutative,* and *Identity* with examples of each multiplication property using decimal numbers.

Display a Venn diagram. Label one side *Associative Property of Addition* and the other side *Associative Property of Multiplication*. Fill in the diagram comparing the two properties. Display sentence frames to help students in the discussion: **The properties are similar because ____. The properties are different because ____.** Be sure students are correctly saying the /z/ sound in properties to indicate plural. Display new Venn diagrams and repeat the exercise for both the Commutative and Identity Properties.

English Language Development Leveled Activities

Emerging Level	Expanding Level	Bridging Level
Word Recognition Distribute the My Vocabulary Cards for *Associative Property of Multiplication, Commutative Property of Multiplication,* and *Identity Property of Multiplication* to each student. Provide examples of each property using decimal numbers. As you model each property, have students identify the example by displaying the corresponding My Vocabulary Card. Repeat with several examples of each property.	**Recognize and Act It Out** List these multiplication properties on the board: *Associative, Commutative, Identity*. Write $(16 \times 0.25) \times 1 \times 3.5$. Point to each of the properties of multiplication and ask students how it might be used to find the product. Provide the following sentence frame to help students respond: **The ____ property can be used to ____.** Repeat with other equations, such as $2 \times (0.75 \times 13.5) \times 1$.	**Academic Language** Divide students into three groups. Assign each group one of the Properties of Multiplication: Associative, Commutative, or Identity. Have groups write a definition in their own words and provide an example for their assigned properties. Have a volunteer from each group share the definition and write the example on the board. Afterward, discuss the definitions and examples.

Teacher Notes:

Student page

NAME _____ DATE _____

Lesson 8 Vocabulary Definition Map
Multiplication Properties

Use the definition map to write a description and list characteristics about the vocabulary word or phrase. Write or draw math examples. Share your examples with a classmate.

My Math Vocabulary:

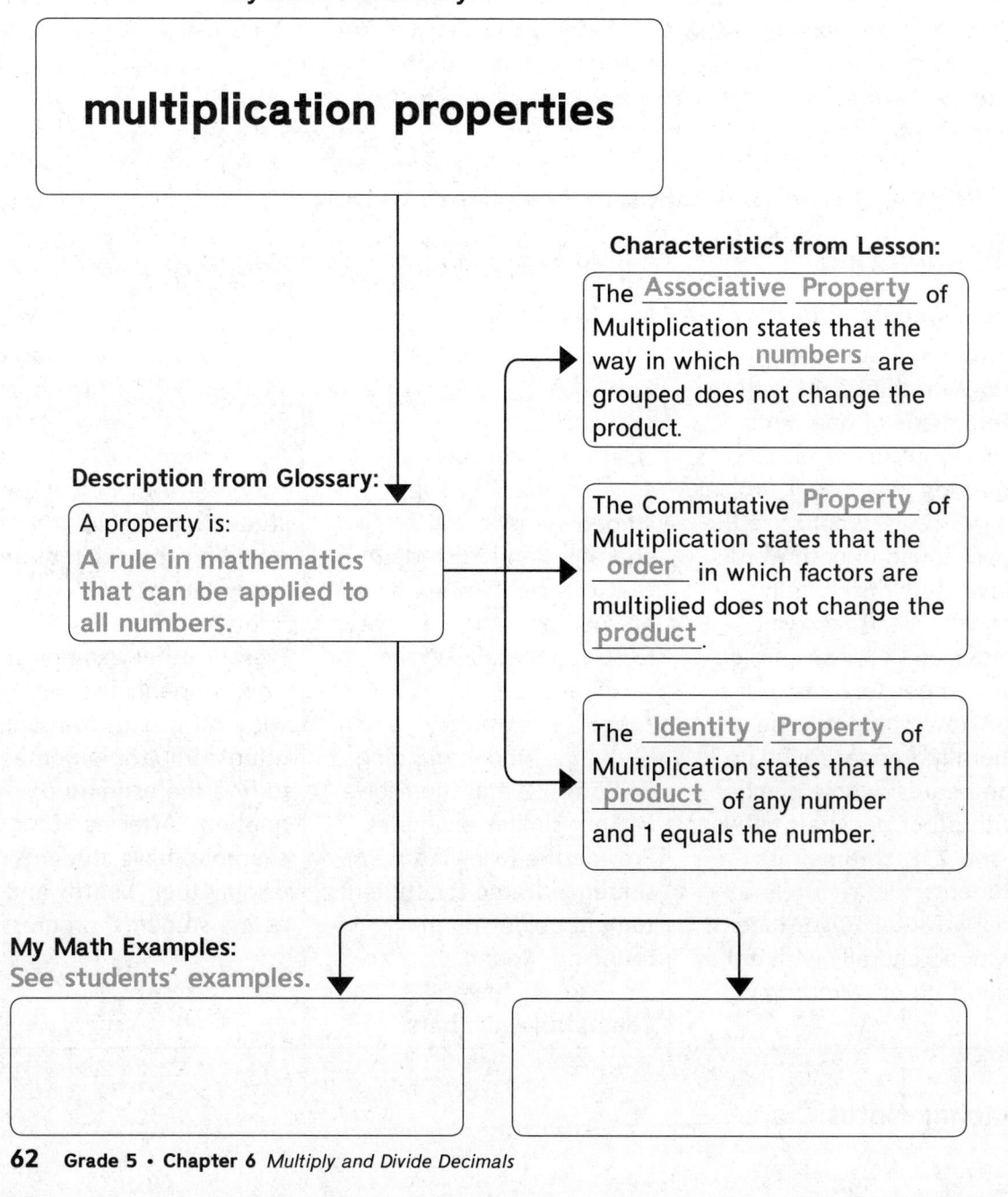

multiplication properties

Description from Glossary:
A property is:
A rule in mathematics that can be applied to all numbers.

Characteristics from Lesson:

The _Associative_ _Property_ of Multiplication states that the way in which _numbers_ are grouped does not change the product.

The Commutative _Property_ of Multiplication states that the _order_ in which factors are multiplied does not change the _product_.

The _Identity_ _Property_ of Multiplication states that the _product_ of any number and 1 equals the number.

My Math Examples:
See students' examples.

62 Grade 5 • Chapter 6 *Multiply and Divide Decimals*

Lesson 9 Estimate Quotients

English Learner Instructional Strategy

Vocabulary Support: Utilize Resources

On the board, create a list of words related to estimating that students can refer to and use during the lesson. Include the following words and add any others that seem appropriate: *about, around, close to, approximately, near, compatible.*

As students work through the Problem-Solving exercises, be sure to remind them that they can refer to the Glossary or the Multilingual eGlossary for help with math vocabulary. Direct students to other translation tools if they are having difficulty with non-math terms in the problems, such as: *purchase, tickets, state fair, drawing pens, canoe, rental, river, exhibit,* or *Egyptian mummies.*

English Language Development Leveled Activities

Emerging Level	Expanding Level	Bridging Level
Word Recognition Draw a number line from 1 to 2 on the board with increments of one tenth. Place your finger on 1.25. As you slide your finger to 1 say, *round down.* Emphasize the /ow/ sound in both words. Have students chorally repeat. Then place your finger on 1.75. As you slide your finger to 2 say, *round up.* Have students repeat chorally. Repeat rounding to the nearest whole number with other decimals between 1 and 2. Each time, ask students, *Do we round up or round down?* Have students respond chorally with either **round up** or **round down**.	**Number Sense** Write 21 ÷ 7 and 23 ÷ 7. Say, *21 ÷ 7 contains compatible numbers. 23 ÷ 7 does not. Which is easier to solve mentally?* Have students answer, **21 ÷ 7** chorally. Remind students that compatible numbers are numbers that are easy to solve mentally. Write 26.85 ÷ 3. Say, *We can use rounding to get compatible numbers.* Model rounding 26.85 to 27 and then solve. Present similar examples. Provide the following sentence frame for students to help guide you in rounding: **Round _____ to _____ so we have compatible numbers.**	**Show What You Know** Write the division problem 14.78 ÷ 3.2 on the board. Say, *Round two compatible numbers to find an estimated quotient.* Have students guide you through rounding the decimal and finding an estimated quotient. 15 ÷ 3 = 5 Present other examples. Have students use write-on/wipe-off boards to round using compatible numbers to find the estimated quotient. After each example, have students display their boards and assess students' progress.

Teacher Notes:

NAME _____ DATE _____

Lesson 9 Vocabulary Concept Web

Estimate Quotients

Use the concept web to identify examples of compatible numbers for each division expression.

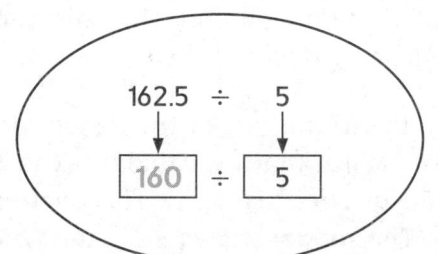

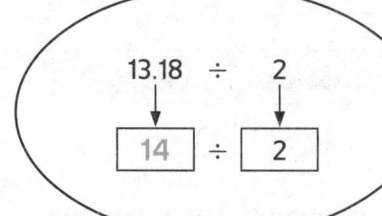

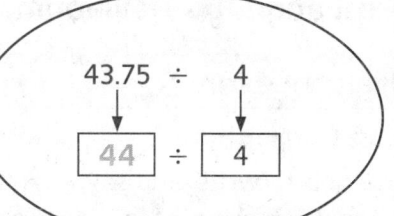

- 162.5 ÷ 5 → 160 ÷ 5
- 13.18 ÷ 2 → 14 ÷ 2
- 43.75 ÷ 4 → 44 ÷ 4

compatible numbers

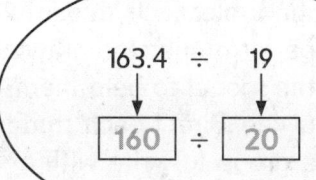

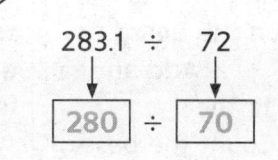

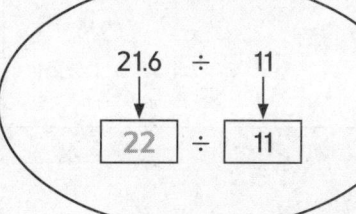

- 163.4 ÷ 19 → 160 ÷ 20
- 283.1 ÷ 72 → 280 ÷ 70
- 21.6 ÷ 11 → 22 ÷ 11

Grade 5 · Chapter 6 *Multiply and Divide Decimals*

Lesson 10 Inquiry/Hands On: Divide Decimals

English Learner Instructional Strategy

Vocabulary Support: Build Background Knowledge

Before the lesson, display the division Anchor Charts students created in Chapter 3 and the decimal Anchor Charts from Chapter 5. Use the charts to review math vocabulary related to division and decimals, such as *divisor, dividend, quotient, remainder, decimal, tenths*, and so on.

As you work through the Model the Math part of the lesson, provide sentence frames to help students utilize math vocabulary: **The divisor is _____. The dividend is _____. The quotient is _____. The decimal point goes between _____ and _____. The model shows _____ tenths.**

English Language Development Leveled Activities

Emerging Level	Expanding Level	Bridging Level
Making Connections	**Listen and Write**	**Round the Table**
On the board, write and say, *3.4*. Model 3.4 using base-ten blocks. Point to the model and then to the number on the board as you say, *3.4*. Then add ÷ 2 to the board. Model dividing the base-ten blocks into two groups of 1.7. Count the groups: One, two. Point to the groups and then to the expression on the board as you say, *3.4 ÷ 2*. Add an equals sign to the expression. Count the base-ten blocks in one group and say, *1.7*. Turn to the board and add 1.7 after the equals sign. Finally, read the entire equation and have students chorally repeat.	Have students work in pairs. Say, *I will read aloud a problem. Write down the problem. Then solve the problem using base-ten blocks.* Read aloud one of Practice Exercises 3–6. Give pairs time to solve, and then call on a volunteer to share his or her answer. Provide a sentence frame for students to use: **We have _____ flats and _____ rods. The answer is _____.** Have students switch partners. Repeat the activity with another Practice It Exercise.	Have students work in small groups. Assign half the groups Apply It Exercise 7 and the other half Exercise 8. Have one student in the group write the problem on a large piece of paper. Have the other students work jointly to solve the problem by passing the paper around the table. Each student will perform one step drawing the model to help them divide. Direct each student in a group to write with a different color to ensure all students participate in solving the problem. Afterward, choose one student to present the solution to the class.

Teacher Notes:

NAME _____ DATE _____

Lesson 10 Note Taking

Inquiry/Hands On: Divide Decimals

Read the question. Write words you need help with and research each word. Use your lesson to write your Cornell notes. Write or draw math examples to explain your thinking. Share your examples with a classmate.

Building on the Essential Question

How do you divide decimals using base-ten blocks?

Words I need help with:

See students' words.

Notes:

The decimal __2.4__ is modeled below using base-ten blocks.

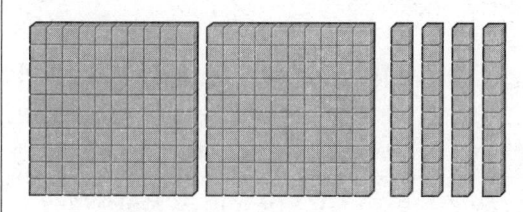

If you divide the whole blocks into 2 equal groups, there will be __1__ whole in each group.

If you divide the tenths into 2 equal groups, there will be __2__ tenths in each group.

Each group has __1__ whole and __2__ tenths.

Each group represents the decimal __1.2__.

You have just modeled the following division equation:

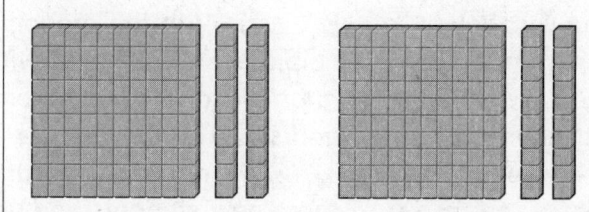

__2.4__ ÷ 2 = __1.2__

My Math Examples:

See students' examples.

64 Grade 5 • Chapter 6 *Multiply and Divide Decimals*

Lesson 11 Divide Decimals by Whole Numbers

English Learner Instructional Strategy

Vocabulary Support: Word List

Before the lesson, write Noun and Verb on the board, and then list the following words appropriately below each term: *division, divisor, dividend, divide, dividing, divided*. As you work through the lesson, try to use each math term at least once. Point to the word on the board as you say it and emphasize the pronunciation. Say the word a second time by itself, and have students repeat chorally. After the lesson, ask volunteers to use the terms in sentences. Monitor usage and pronunciation, and correct students as necessary through modeling.

English Language Development Leveled Activities

Emerging Level	Expanding Level	Bridging Level
Activate Prior Knowledge Randomly distribute the following My Vocabulary Cards to students: *dividend, divisor*, and *quotient*. Choose a problem from the Independent Practice section and then model solving it on the board. Afterward, point to each component of the problem, in random order, as you ask, *Is this the dividend, divisor, or quotient?* Have the students with the corresponding vocabulary card raise their cards. Then say the vocabulary word and have students chorally repeat. Have students exchange cards, and then repeat the activity with another problem.	**Academic Vocabulary** Distribute the following My Vocabulary Cards to three students: *dividend, divisor,* and *quotient*. Draw a division bracket on the board. Have the student with *dividend* write in a four-digit decimal number to the hundredths and say, **The dividend is ____**. Have the student with *divisor* write a one-digit whole number divisor and say, **The divisor is ____**. With the group's help, have the student with *quotient* solve the problem and say, **The quotient is ____**. Distribute the cards to three new students and repeat the activity.	**Act It Out** Have students get into groups of three or four, and distribute a short piece of string or strip of paper (about 10 cm to 20 cm long) to each group. Have students measure their string/paper in centimeters. Have them write a division problem solving for the length once it is divided into four equal parts. Have groups fold the paper in half lengthwise and then fold it in half again. Have them measure to check their division. Provide a sentence frame for students to use in sharing their answer: ____ **divided by four is** ____.

Teacher Notes:

NAME _____ DATE _____

Lesson 11 Multiple Meaning Word
Divide Decimals by Whole Numbers

Complete the four-square chart to review the multiple meaning word or phrase.

Everyday Use	Math Use in a Sentence
Sample answer: Disagree or cause one to disagree. Two friends were divided on the way they should respond to the loss of their favorite team.	Sample sentence: You can divide items equally among friends.

divide

Math Use	Example From This Lesson
An operation on two numbers in which the first number is split into the same number of equal groups as the second number.	Sample answer: 2.4 divided by 2 is 1.2

Write the correct term on the line to complete the sentence.

Division of decimals is similar to division of whole numbers except there is a __decimal point__ appearing in the dividend and the quotient.

Grade 5 · Chapter 6 *Multiply and Divide Decimals* **65**

Lesson 12 Inquiry/Hands On: Use Models to Divide Decimals

English Learner Instructional Strategy

Sensory Support: Models

Before the lesson, create matching sets of index cards with decimal division problems written in standard form and word form. For example, on one card write 0.45 ÷ 0.03, and on a separate card write *forty-five hundredths divided by three hundredths*. Create enough cards for one card per student. Distribute the cards, and then say, *Find the student who has the same problem written on his or her card.* Allow students time to find their partners with matching cards. Once students are paired, say, *Work together to solve the division problem on your cards. Use a model to find the answer.* Then ask each pair, *What decimal does your model show?* Provide this sentence frame for students to respond: **The model shows ____, so the answer is ____.**

English Language Development Leveled Activities

Emerging Level	Expanding Level	Bridging Level
Money Sense Using manipulative coins, make $0.84 with three quarters, one nickel, and four pennies. Write and say, *$0.84* and have students chorally repeat. Invite four students to the board. Say, *I want to divide $0.84 between four people.* Write and say, *$0.84 ÷ 4.* Use a model to solve 0.84 ÷ 4. Afterward, write the answer on the board and say the equation: *$0.84 ÷ 4 = $0.21.* Ask, *How much does each person get?* Have students answer chorally. Point out that the coins cannot be evenly divided, and model creating an equivalent amount (eight dimes and four pennies).	**Turn & Talk** Read aloud Talk About It Exercise 1, then say, *Turn to the student nearest you and discuss your answer.* Give students a chance to discuss their ideas. Then come together again as a group and ask volunteers to share their answers. Repeat the procedure for Talk About It Exercise 2. Afterward, have students spend a few minutes writing in their math journals about how dividing with decimal numbers is similar to dividing with whole numbers.	**Numbered Heads Together** Have students get into groups of four. Ask the students in each group to number off as 1–4. Have the students in each group work together to solve Apply It Exercise 10. Choose numbers from 1–4 to designate which student in each group will use a complete sentence to describe their group's answer. Repeat the activity for Apply It Exercise 8.

Teacher Notes:

NAME _____ DATE _____

Lesson 12 Guided Writing

Inquiry/Hands On: Use Models to Divide Decimals

How do you model dividing decimals using base-ten blocks?

Use the exercises below to help you build on answering the Essential Question. Write the correct word or phrase on the lines provided.

1. Rewrite the question in your own words.
   See students' work.

2. What key words do you see in the question?
   model, dividing, decimals, base-ten blocks

3. What decimal is modeled using the base-ten blocks below?
   2.4

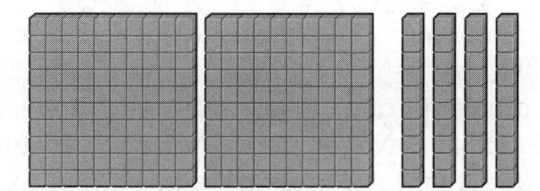

4. If you divide a decimal by tenths, you will regroup the decimal into tenths .

5. The decimal 0.6 written in word form is six   tenths . When you divide 2.4 by 0.6, you are dividing by tenths .

6. So, you will regroup the wholes in 2.4 into tenths. There are 20 tenths in 2 wholes. After regrouping, there are 24 tenths altogether.

7. If you separate the 24 tenths into groups of 6 tenths, you will have 4 groups.

8. 24 ÷ 6 = 4 and 24 tenths ÷ 6 tenths = 4 tenths. So, 2.4 ÷ 6 = 0.4

9. How do you model dividing decimals using base-ten blocks?
   Model the dividend. Regroup the dividend using the place value of the divisor. Separate the model into equal groups each containing the amount of the divisor. The total number of groups represents the quotient using the place value of the divisor.

Lesson 13 Divide Decimals

English Learner Instructional Strategy

Collaborative Support: Show What You Know

During Independent Practice, assign one problem each to student pairs. Have them work together to solve their assigned problem. Afterward, have pairs come to the board and narrate each step as they solve their problem. Provide sentence frames to help students share what they know:

First multiply the divisor by ____.
Then multiply the divided by ____.
Next place the ____ in the quotient.
____ divided by ____ is ____.

Have the other students multiply to check the answer.

English Language Development Leveled Activities

Emerging Level	Expanding Level	Bridging Level
Listen and Respond	**Developing Oral Language**	**Background Knowledge**
Write *next to, inside*, and *above*. Then write a decimal number division problem using a division bracket. Model solving it. Then identify the location of each component of the problem—dividend, divisor, and quotient—using words from the board. Write a new division problem. Have students write the same problem using write-on/wipe-off boards. As you model solving, have students follow along on their boards. Afterward, call out a location or division term and have students circle the appropriate number. Repeat with a new problem.	Divide students into four groups, and then assign each group a problem from Independent Practice. Have students work together to solve the problem, taking notes to describe each step in the process. Afterward, display the following sentence frames and have each group explain the steps: **First** ____. **Next** ____. **Then** ____. **Last** ____.	Have students get into groups of three or four. Have them measure each other's heights in meters, using decimals to represent centimeters. Then have students calculate their group's average height by adding all their heights and dividing the sum by the number of people in the group. Ask each group to display their results and how they found the average. Display the following sentence by adding all their heights and dividing the sum by the number of people in their group. frame to help them: **We found the average by** ____.

Teacher Notes:

NAME _____ DATE _____

Lesson 13 Concept Web

Divide Decimals

Use the concept web to identify the parts of a decimal division sentence.

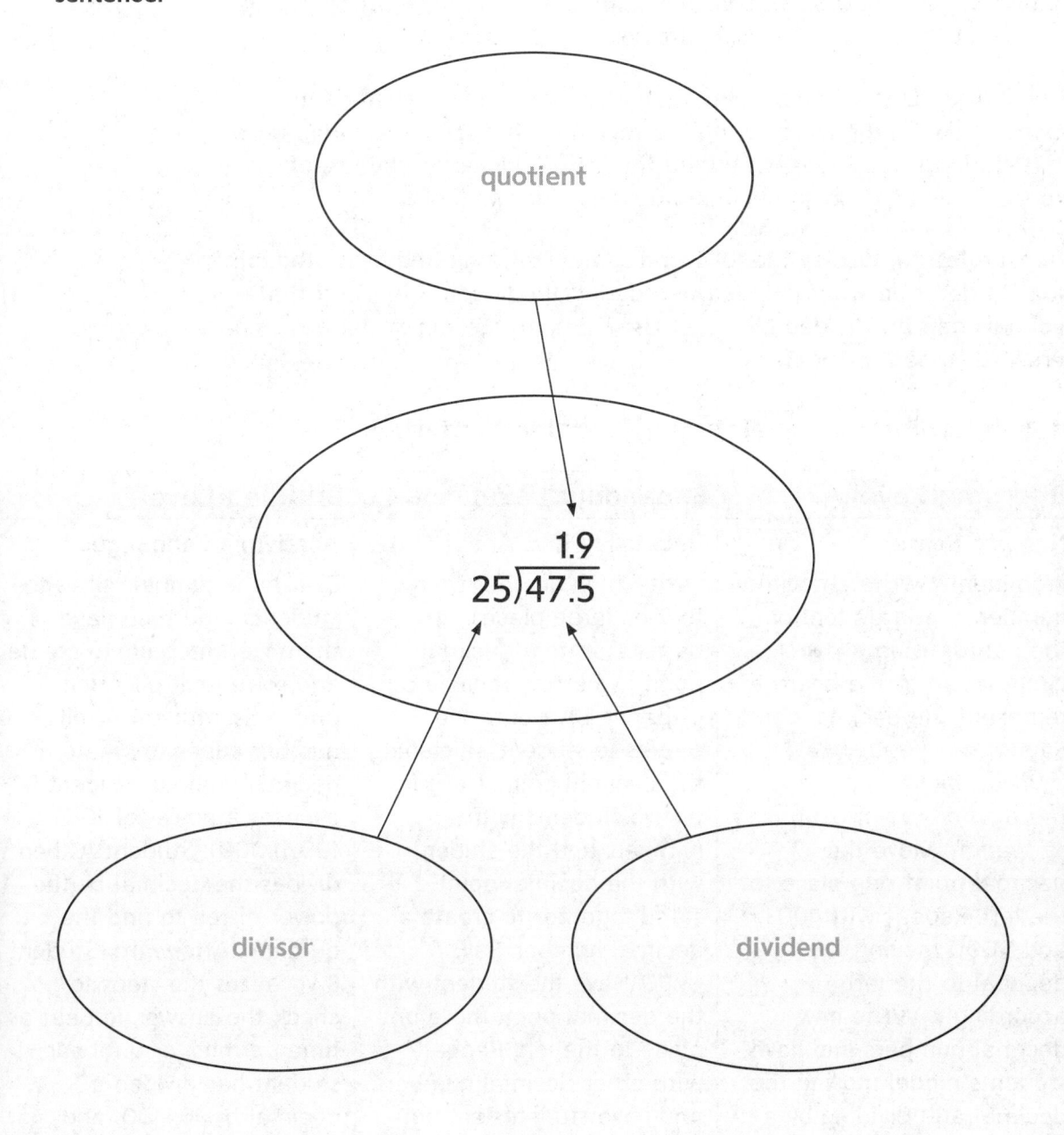

quotient

$$25\overline{)47.5}^{\,1.9}$$

divisor

dividend

Grade 5 • Chapter 6 *Multiply and Divide Decimals* **67**

Lesson 14 Divide Decimals by Powers of Ten

English Learner Instructional Strategy

Vocabulary Support: Activate Prior Knowledge

Display the classroom cognate chart and point out *exponent* and its Spanish cognate, *exponente*. Review exponents with students by asking them what they remember from previous chapters.

Display a KWL chart and, in the first column, record what students recalled about exponents during the review. In the second column, record what students hope to learn during the lesson, including the use of exponents when dividing decimal numbers by powers of ten.

After the lesson, display the following sentence frame and have students use it to describe what they learned during the lesson: **I learned that decimals can be divided by ____.** Use the third column of the KWL chart to record student responses.

English Language Development Leveled Activities

Emerging Level	Expanding Level	Bridging Level
Number Sense Prominently write a decimal number in a large font on the board, using a paper circle taped to the board to represent the decimal point. Say, *When you divide a decimal by 10, you move the decimal point one place to the left.* Move the decimal point one place to the left. Repeat with 100 and 1,000 moving the decimal to the left accordingly. Write new decimal numbers and have students model moving the decimal and dividing by a power of 10 of their choice. Then have the student say: **I divided by ____.**	**Recognize and Act It Out** Write each number from 0 to 9 on large pieces of paper. Create a decimal point by cutting a circle out of paper. Distribute the papers to students, including the decimal point. Have four or five students with numbers and the student with the decimal point stand together to create a decimal number. Say, *Divide by 10.* Have the student with the decimal point move one place to the left. Repeat with other decimal numbers, and have students act out dividing by 100 and 1,000. Have additional papers numbered with 0 if needed.	**Academic Language** Distribute number cubes to students, and then have them work in pairs to create and solve multiplication problems. Student A rolls number cubes to create a decimal number. Student B chooses a power of 10 (10, 100, 1,000). Student A then divides the decimal by the power of ten to find the quotient. Afterward, Student B vocalizes the steps to check the answer. Repeat as time permits, or until each student has divided a decimal by 10, 100, and 1,000.

Teacher Notes:

NAME _____ DATE _____

Lesson 14 Vocabulary Definition Map
Divide Decimals by Powers of Ten

Use the definition map to write a description and list characteristics about the vocabulary word or phrase. Write or draw math examples. Share your examples with a classmate.

My Math Vocabulary:

exponent

Description from Glossary:

In a power, the number of times the base is used as a factor.

Characteristics from Lesson:

<u>Powers</u> of ten can be written with exponents.

When you <u>divide</u> a decimal by a power of ten, move the decimal point to the <u>left</u> the same number of <u>zeros</u> in the power of ten.

$10^3 =$ <u>1,000</u>

My Math Examples:
See students' examples.

68 Grade 5 • Chapter 6 *Multiply and Divide Decimals*

Chapter 7 Expressions and Patterns

What's the Math in This Chapter?

Mathematical Practice 4: Model with mathematics

Write then say the following math problem: *Each day Mrs. Dowler's class uses 10 pieces of paper in math. How many pieces of paper will they use in 4 weeks (5 days in a school week)?*

Allow students time to work on a solution. Encourage students to share various strategies they used to solve the problem. For example, repeated addition, using a table, and writing an expression. Model all possible strategies. Explain that all of these strategies are ways to model math and help make sense of a problem.

Ask, *What patterns do you see?* Have students turn and talk with a peer. Have students share ideas. The discussion goal is for students to identify a table and repeated addition as patterns. Say, *These patterns and models helped us solve the problem.*

Display a chart with Mathematical Practice 4. Restate Mathematical Practice 4 and have students assist in rewriting it as an "I can" statement, for example: **I can model a problem to solve it.** Post the new "I can" statement.

Inquiry of the Essential Question:

How are patterns used to solve problems?

Inquiry Activity Target: **Students come to a conclusion that a problem can be modeled using a pattern.**

As an introduction to the chapter, present the Essential Question to students. The inquiry graphic organizer will offer opportunities for students to observe, make inferences, and apply prior knowledge of patterns representing the Essential Question. As they investigate, encourage students to draw, write, and collaborate with peers to demonstrate their observations and thinking. Then have students present additional questions they may have to a peer to extend discussions.

Regroup students and restate Mathematical Practice 4 and the Essential Question. Pose questions to reflect on what has been learned to guide students in making connections between the Mathematical Practice and the Essential Question.

NAME _____ DATE _____

Chapter 7 Expressions and Patterns

Inquiry of the Essential Question:

How are patterns used to solve problems?

Read the Essential Question. Describe your observations (I see...), inferences (I think...), and prior knowledge (I know...) of each math example. Write additional questions you have below. Then share your ideas and questions with a classmate.

Phrase: subtract 2 from 8, then divide by 3

Expression: $(8 - 2) \div 3$

I see ...

I think...

I know...

72, 67, 62, 57, 52, 47, ?
 −5 −5 −5 −5 −5

The next number in the pattern is 47 − 5, or _____.

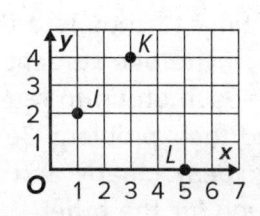

I see ...

I think...

I know...

J (1, 2)
K (3, 4)
L (5, 0)

I see ...

I think...

I know...

Questions I have...

_____

_____

_____

Grade 5 • **Chapter 7** *Expressions and Patterns* **69**

Lesson 1 Inquiry/Hands On: Numerical Expressions

English Learner Instructional Strategy

Vocabulary Support: Cognates

Write *numerical expression* and *evaluate* and the Spanish cognates, *expresión numérica* and *evaluar*, on a classroom cognate chart. Provide a simple concrete model of the words' meanings. For example, write 4 × 5 on the board. Underline the numbers and circle the multiplication sign as you describe the elements of a numerical expression. Underline *valu* in both e*valu*ate and e*valu*ar as you say, *When we evaluate the expression, we find its value.* During the lesson, have students refer to the classroom cognate chart or the Multilingual eGlossary if they need to review math vocabulary related to numerical expressions or evaluating expressions: *sum, product, multiply, parentheses, multiplication, addition.*

English Language Development Leveled Activities

Emerging Level	Expanding Level	Bridging Level
Sentence Frames Display the following sentence frames: ___ **plus** ___ **equals** ___. ___ **times** ___ **equals**. On the board, draw a bar diagram with 3s in the first three spaces and 5 in the last space. Say, *I will find the total.* Write 3 + 3 + 3 + 5. Read the expression aloud. Have students chorally repeat. Say, *I can write the expression another way.* Write (3 × 3) + 5. Read the expression aloud. Have students chorally repeat. Say, *I will evaluate the expression.* Use the sentence frames as you model evaluating. Repeat with a new problem and have students guide you.	**Partners Work** Have students work in pairs to complete Practice It Exercises 3–4. Afterward, discuss the differences between the two problems, including the number of operations needed in each expression in order to solve. Display a sentence frame to help students: **The first problem needed ___ operation(s) because ___. The second problem needed ___ operation(s) because ___.** Preview Apply It Exercise 5, and based on the discussion, have students predict the number of operations needed in the expression that will solve it.	**Public Speaking Norms** On the board, write (___ × ___) + ___. Divide students into groups of four and say, *Write a real-world problem that can be solved using an expression similar to the one written on the board.* After groups have had time to complete the task, invite one student from each group to read aloud their problem. Be sure they speak clearly and loud enough for the other students to hear. Have students work together to solve each group's problem.

Teacher Notes:

NAME _____ DATE _____

Lesson 1 Vocabulary Cognates

Inquiry/Hands On: Numerical Expressions

Use the Glossary to define the math word in English and in Spanish in the word boxes. Write a sentence using your math word.

evaluate	evaluar
Definition To find the value of an expression by replacing variables with numbers.	**Definición** Hallar el valor de una expressión reemplazando las variables por númerous.

My math word sentence:

Sample answer: The expression 2 + 2 + 2 can be evaluated using multiplication as 2 × 3 = 6.

numerical expression	expressión numérica
Definition A combination of numbers and operations.	**Definición** Combinación de númerous y operaciones.

My math word sentence:

Sample answer: The numerical expression for adding two and two and two is 2 + 2 + 2.

Lesson 2 Order of Operations
English Learner Instructional Strategy

Graphic Support: Word Web

Write *order of operations* and its Spanish cognate, *orden de las operaciones*, on a classroom cognate chart. Display a word web and write *order* in the center. Discuss with students the different meanings and uses of *order* and record them in the word web. For example, ordering in a restaurant, calling a classroom or courtroom to order, putting things in order, or following orders. Display a second word web and use it to discuss meanings and uses of *operation*. Be sure to include the math meaning of *operation*.

Model solving an expression that requires multiple operations. Then say, *Doing the operations in the right order gives me a correct answer.*

English Language Development Leveled Activities

Emerging Level	Expanding Level	Bridging Level
Word Knowledge Write the following list: *1. Get a cup. 2. Pour water. 3. Drink.* Say, *This is the right order.* Rewrite the list as: 2. Pour water. 1. Get a cup. 3. Drink. Say, *This is the wrong order.* If possible, have a towel ready and demonstrate each list using water and a cup. Then list the order of operations on the board and solve a problem from the lesson. First solve by following the order of operations correctly. Say, *This answer is correct.* Then follow the order incorrectly to solve, point to the answer, and say, *The order was incorrect. Our answer is wrong.*	**Academic Vocabulary** List the following operations on a large piece of paper: *addition, subtraction, multiplication, division.* Write a single operation expression on the board. Have students identify the operation using the sentence frame, **The operation is ___.** Have students evaluate the expression. Write a multiple operation expression on the board. Have students identify the operations using the sentence frame, **The operations are ___ and ___.** Have students identify the order to perform the operations and evaluate the expression.	**Academic Language** Provide several multi-operation expressions and solutions for student pairs. Make some of the solutions correct and others incorrect. Have pairs use the order of operations to determine which expressions were evaluated correctly. For those expressions that were not evaluated correctly, have pairs determine the correct order for evaluating the expression and solve for the correct answer.

Teacher Notes:

NAME _____ DATE _____

Lesson 2 Vocabulary Definition Map
Order of Operations

Use the definition map to write a description and list characteristics about the vocabulary word or phrase. Write or draw math examples. Share your examples with a classmate.

My Math Vocabulary:

order of operations

Description from Glossary:

A set of rules to follow when more than one operation is used in an expression.

Characteristics from Lesson:

The first rule is to perform operations in <u>parentheses</u>. Then perform operations inside brackets, and finally inside <u>braces</u>.

The <u>second</u> rule is to find the value of <u>exponents</u>.

The <u>third</u> rule is to <u>multiply</u> and <u>divide</u> in order from left to right.

The <u>fourth</u> rule is to <u>add</u> and <u>subtract</u> in order from left to right.

My Math Examples:
See students' examples.

Lesson 3 Write Numerical Expressions
English Learner Instructional Strategy

Language Structure Support: Tiered Questions

Throughout the lesson, ask questions according to EL students' level of English comprehension for formative assessment. For example, ask emerging students simple questions that elicit one-word answers: *What do we do first? Do we add or subtract? Which number do we divide?*

For expanding students, ask questions that elicit simple phrases or short sentences: *Which operations do we use to solve the problem? Which operation comes first and which comes second?*

For bridging students, ask questions that require more complex answers: *Why do we use that operation? How can we check our answer?*

English Language Development Leveled Activities

Emerging Level	Expanding Level	Bridging Level
Word Recognition Give a volunteer two connecting cubes. Say, *You have two cubes. I will give you three more cubes.* Write the numerical expression: 2 + 3 and below it the following phrase: *two plus three.* Model saying the phrase as you hand the student three more cubes. Have students chorally repeat. Then say, *I will take away one cube.* Write: − 1 next to the expression on the board and add *take away one* to the phrase. Model saying the phrase as you take away one cube. Have students chorally repeat. Repeat the activity with a new volunteer and expression.	**Recognize and Act It Out** Display the following phrase frames: ___ **divided by** ___, ___ **times** ___, ___ **plus** ___, ___ **take away** ___ Show ten connected cubes. Say, *I have ten cubes. I will divide the cubes and put half on the table.* Model and have students identify the phrase that represents the action. **ten divided by two** Write the numerical expression: 10 ÷ 2. Using connecting cubes, have pairs act out one model for each expression (division, multiplication, addition, subtraction) and then use the phrase frames to describe their models.	**Academic Language** Discuss with students how they recognize language indicating an operation in word problems. Model writing numerical expressions derived from this language and list on an anchor chart. Ask volunteers to read aloud word problems from the lesson and then list the words and phrases from the problems that signify an operation. Discuss the list as a group.

Teacher Notes:

NAME _____ DATE _____

Lesson 3 Note Taking

Write Numerical Expressions

Read the question. Write words you need help with and research each word. Use your lesson to write your Cornell notes. Write or draw math examples to explain your thinking. Share your examples with a classmate.

Building on the Essential Question

How do you write numerical expressions?

Notes:

A numerical expression is a combination of numbers and <u>operations</u>.

The numerical expression <u>3</u> + <u>4</u> represents the phrase *add three and four*. When you evaluate the expression, you get <u>7</u>.

The numerical expression <u>7</u> × <u>2</u> represents the phrase *multiply seven by two*. When you evaluate the expression, you get <u>14</u>.

The order of operations is a set of rules to follow when more than one <u>operation</u> is used in an expression.

1. Perform operations in <u>parentheses</u>.

2. Find the value of <u>exponents</u>.

3. <u>Multiply</u> and <u>divide</u> in order from left to right.

4. <u>Add</u> and <u>subtract</u> in order from left to right.

The numerical expression (<u>3</u> + <u>4</u>) × <u>2</u> represents the phrase *add three and four, then multiply by two*.

The first operation to perform in this expression is <u>addition</u> because it is enclosed in the <u>parentheses</u>.

The second operation to perform in this expression is <u>multiplication</u>.

When you evaluate the expression, you get <u>14</u>.

Words I need help with:
See students' words.

My Math Examples:
See students' examples.

Grade 5 • Chapter 7 *Expressions and Patterns*

Lesson 4 Problem-Solving Investigation Strategy: Work Backward

English Learner Instructional Strategy

Language Structure Support: Report Back

Distribute two number cubes each to pairs of students. Say, *You will count backward from 100.* Have students roll one cube to determine how much they will count back at a time and roll the other cube to determine how many times they will count. For example, pairs that roll a five and four will count back by fives four times, ending on 80.

Display the following sentence frames: **We rolled ____ and ____. We counted back by ____. We counted ____ times. We stopped on ____.** Have students use the sentence frames to report back. Be sure they are correctly differentiating between the /d/, /ed/, and /t/ sounds when indicating past tense.

English Language Development Leveled Activities

Emerging Level	Expanding Level	Bridging Level
Word Recognition	**Recognize and Act It Out**	**Academic Language**
Have students help you to describe putting on a sock and shoe. List the steps on the board: *1. Put on the sock. 2. Put on the shoe. 3. Tie the shoe.* Say, *To take off a shoe and sock we work backward.* Say *backward* again and have students chorally repeat. List the steps: *1. Untie the shoe. 2. Take off the shoe. 3. Take off the sock.* Say, *We can also work backward to solve a math problem.* Model by reviewing a problem from the lesson.	Provide a volunteer with an unknown number of connecting cubes. Say, *You have some cubes. I will give you five more cubes.* Hand the student five cubes. Have the student count aloud the total number of cubes. Display the known information: **5 cubes were added, the total number of cubes is ____.** Say, *We can work backwards to find the number of cubes we started with. What operation can we use?* **subtraction** Write the subtraction expression and have students solve and check the answer.	Have pairs work together on problems from the lesson. Student A will identify the facts that are known and what they need to find out to solve the problem. Student B will describe the steps needed to work backwards to solve the problem and record the steps on index cards. Have pairs check their work for reasonableness by working forward. Ask volunteers to share their solutions and read aloud the steps written on their index cards.

Teacher Notes:

NAME _____ DATE _____

Lesson 4 Problem-Solving Investigation
STRATEGY: Work Backward

Work backward to solve each problem.

1. **Seth** bought a movie ticket, popcorn, and a drink. After the movie, **he** played **4** video **games** that **each cost** the **same**. He spent a **total** of **$19**. How much did it **cost** to play **each** video **game**?

Movie	Costs
Popcorn	$4
Drink	$3
Ticket	$8

Understand	Solve
I know: I need to find:	

Plan	Check
Total spent: $19 I need to work _backward_ to solve.	

2. Students sold raffle tickets to raise money for a field trip. The first **20 tickets** sold cost **$4 each**. To sell more tickets, they **lowered** the **price** to **$2 each**. If they **raised $216**, how many **tickets** did they **sell in all**?

Understand	Solve
I know: I need to find:	

Plan	Check
Total raised: $216 I need to _work_ backward to solve.	

Grade 5 • Chapter 7 Expressions and Patterns **73**

Lesson 5 Inquiry/Hands On: Generate Patterns

English Learner Instructional Strategy

Language Structure Support: Modeled Talk

Write *pattern* and its Spanish cognate, *patrón*, on a classroom cognate chart. Provide concrete examples of meaning by having students locate and describe nonmath patterns within the classroom.

Display the following sentence frames to help students describe the patterns in the lesson: **Add _____ toothpicks. The _____ figure will have _____ toothpicks. The expression _____ plus _____ shows the number of toothpicks needed.**

As you work through the Build It lesson, model using each sentence frame so students can hear standard English pronunciation. After saying each sentence, ask a question that will allow a volunteer to respond using the same sentence frame. For example, *How many toothpicks do we add? How many toothpicks will the figure have?* and so on.

English Language Development Leveled Activities

Emerging Level	Expanding Level	Bridging Level
Act It Out	**Pairs Check**	**Show What You Know**
On the board, write and say, *Add 2.* Model the pattern using connecting cubes: 1, 3, 5, 7. Invite a student to use connecting cubes to create the next number in the pattern. Provide the student with seven connected cubes and say, *Add 2.* Have the student add two cubes. Invite another student forward to create the next number. Provide him or her with nine connected cubes and ask the other students, *What do we add?* Have them respond chorally, **Add 2.** Continue for a few more numbers, and then start over with another simple pattern.	Have students work in pairs to solve Apply It Exercises 7 and 8. Have one student complete Exercise 7 as the other student provides support and suggestions. Then have students switch roles to complete Exercise 8. Have each pair meet with another pair to compare answers. Once students have agreed on the correct answers, have them share with class. Provide sentence frames to help students share their answers: **On day 6, Tammi swam _____ laps than Kelly swam. The candles at Store _____ cost _____ each.**	Have pairs of students use toothpicks to create their patterns. Say, *Create the first three figures of your pattern.* Once pairs have completed their three figures, have them get up and move to sit by another pair's figures. Say, *Identify the pattern and create the next two figures.* Have students describe the patterns and why the figures they created followed the same pattern. If the original three figures did not follow a regular pattern, discuss why it was not possible to create two new figures.

Multicultural Teacher Tip

Word problems are an important part of the math curriculum, but they can be particularly challenging for ELs, and not just because of language issues. Allow students to share examples from their own cultures, including popular national sports or physical activities they participated in, foods and drinks from their culture, traditional clothing worn in their home countries, and so on. When appropriate, help ELs reword an exercise to include a familiar cultural reference.

NAME _____ DATE _____

Lesson 5 Guided Writing

Inquiry/Hands On: Generate Patterns

How do you identify a generated pattern?

Use the exercises below to help you build on answering the Essential Question. Write the correct word or phrase on the lines provided.

1. Rewrite the question in your own words.
   See students' work.

2. What key words do you see in the question?
   identify, generated, pattern

3. The shape below is made with __4__ lines.

4. The shape below is made with __7__ lines. Which is __3__ more lines than the previous shape.

5. The shape below is made with __10__ lines. Which is __3__ more lines than the previous shape.

6. If the pattern continues, the next shape will have __3__ more lines than the previous shape. The next shape in the pattern will be made with __13__ lines.

7. The shape below is made with __13__ lines.

8. How do you identify a pattern?
   Sample answer: Find the first few numbers in the pattern and see how each one changes from the previous number. The pattern should be easily identified by the change in each number in the pattern.

74 Grade 5 • Chapter 7 *Expressions and Patterns*

Lesson 6 Patterns
English Learner Instructional Strategy

Vocabulary Support: Frontload Academic Vocabulary

Write *term* and its Spanish cognate, *término*, on a classroom cognate chart. Provide a simple concrete example, such as: 2, 4, 6, 8, 10. Ask students to identify the pattern. **add two** Then use the sequence on the board to discuss the meanings of *term* and *sequence*.

Display the following sentence frames: **The pattern sequence is ___. The next three terms are ___.** Assign Independent Practice exercises to student pairs, and distribute either a *term* or *sequence* My Vocabulary Card to each student in a pair. After solving, have the student with the *sequence* card describe the pattern. Have the student with the *term* card identify the next three terms in the sequence. Pairs exchange My Vocabulary Cards and roles after each exercise.

English Language Development Leveled Activities

Emerging Level	Expanding Level	Bridging Level
Word Knowledge Write each term for a numerical pattern, such as: 3, 6, 9, 12, 15, 18, on individual pieces of paper. Have six students, each holding one of the numbers, stand in sequential order. Read aloud the pattern and say, *This is a numerical pattern. Each number is a term in the sequence.* Say *term* again and have students repeat chorally. Then ask each student in order, *Which term are you holding?* Display the following sentence frame, and have each student step forward to answer: **My term is ___. We are a sequence.** Repeat with a new sequence and group of students.	**Recognize and Act It Out** Display the following sentence frames: **The pattern is ___. The fourth/fifth term in the sequence is ___.** Have students get into pairs or small groups. Read aloud the following sequence: 4, 8, 12. Have each group use counters to represent the numerical pattern. Ask a volunteer to identify the pattern using the sentence frame. Then have students use counters to display the next two terms in the sequence. **16, 20** Have volunteers use the sentence frame to report back the fourth and fifth terms. Repeat the activity with a new sequence.	**Building Oral Language** Distribute counters to student pairs and say, *Create a numerical pattern using counters. Your sequence will have four terms.* After students have created their patterns, tell pairs to switch places so they are sitting by another pair's sequence. Ask, *What is the pattern? What is the next term in the sequence?* Display the following sentence frames for students to use: **The pattern is ___. The next term in the sequence is ___.** If there is no pattern, have students explain why the counters do not show a sequence.

Teacher Notes:

NAME _____ DATE _____

Lesson 6 Concept Web
Patterns

Write the next term in each sequence on the concept web.

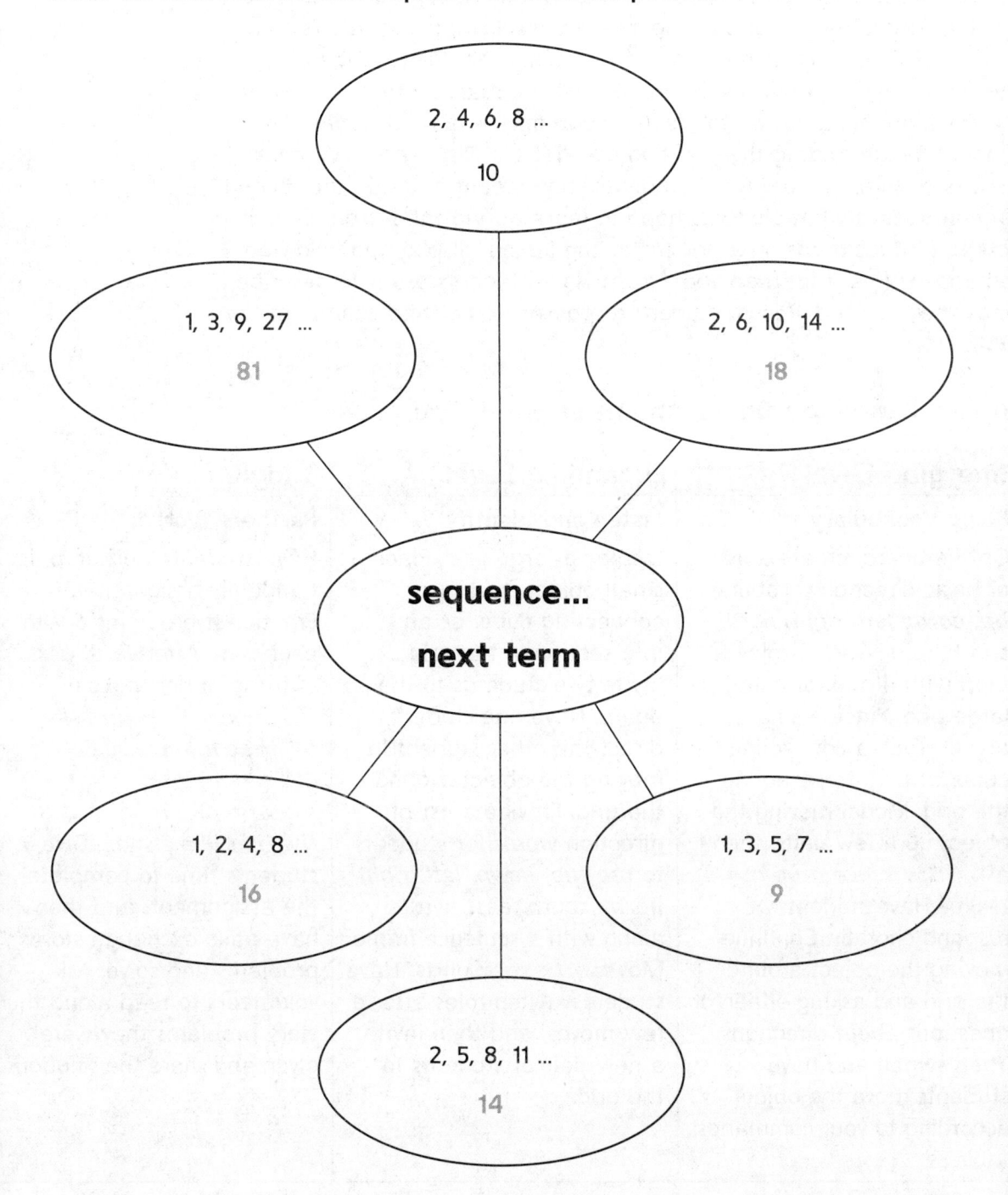

- 2, 4, 6, 8 …
  10

- 1, 3, 9, 27 …
  81

- 2, 6, 10, 14 …
  18

sequence… next term

- 1, 2, 4, 8 …
  16

- 1, 3, 5, 7 …
  9

- 2, 5, 8, 11 …
  14

Grade 5 • Chapter 7 *Expressions and Patterns* 75

Lesson 7 Inquiry/Hands On: Map Locations

English Learner Instructional Strategy

Sensory Support: Maps

Write *map* and its Spanish cognate, *mapa*, on a classroom cognate chart. Provide concrete examples of meaning by displaying a variety of maps. Be sure to include at least one map of the local area. Invite students to locate the school and/or their neighborhoods on the map. Discuss how letters and/or numbers are used to create a grid on the map. Describe the school's location using the grid numbers/letters. Display a KWL chart. In the first column, record what students know about maps. In the second column, record what students hope to learn during the lesson, including how to plot locations on a grid. After the lesson, display the following sentence frame: **I learned that _____.** Have students use it to describe what they learned. Record student responses in the third column of the KWL chart.

English Language Development Leveled Activities

Emerging Level	Expanding Level	Bridging Level
Basic Vocabulary	**Listen and Identify**	**Partners Work**
On the board, create a list of basic direction vocabulary: *up, down, left, right, north, south, east, west*. Display a map with a grid, or create a large grid. Place a small object, such a connecting cube, at an intersection in the grid. Model moving the object up a few units, then ask, *Did I move it up or down?* Have students respond chorally. Continue moving the object around the grid and asking either/or questions about direction. Then switch and have students move the object according to your commands: *Move it ___ ___ units.*	Display a large grid. Place a small object, such a connecting cube, at an intersection in the grid. Invite two students to the board. Have one student direct the other student in moving the object around the grid. Provide a list of direction words for students to use—*up, down, left, right, north, south, east, west*—along with a sentence frame: **Move ___ ___ units.** Have students switch roles after a few moves, and then invite a new pair of students to the grid.	Have students work in pairs. Randomly assign one of Practice It problems 6–9 to each pair. Afterward, pass out grid paper and say, *Create a 4-by-4 grid similar to those in the Practice It exercises. Write your own story problem that uses the grid like a map.* Give students time to complete the assignment, and then have pairs exchange story problems and solve. Ask volunteers to read aloud the story problems they were given and share the solutions.

Teacher Notes:

NAME _____ DATE _____

Lesson 7 Note Taking

Inquiry/Hands On: Map Locations

Read the question. Write words you need help with and research each word. Use your lesson to write your Cornell notes. Write or draw math examples to explain your thinking. Share your examples with a classmate.

Building on the Essential Question	Notes:
How do you plot the location of an item on a map?	This grid represents locations on a map.
	When you move <u>up</u> on the grid you move **north** on the map.
	When you move <u>down</u> on the grid you move **south** on the map.
	When you move <u>left</u> on the grid you move **west** on the map.
Words I need help with: See students' words.	When you move <u>right</u> on the grid you move **east** on the map.
	Each square on the grid represents a **block** on the map.
	The library is <u>three</u> blocks <u>north</u> of the school.
	The park is <u>two</u> blocks <u>east</u> of the school.
	Marcia's home is one block south of the park.
	The dot that represents Marcia's home will be <u>one</u> square <u>below</u> the dot that represents the park. Draw and label a dot for Marcia's home.

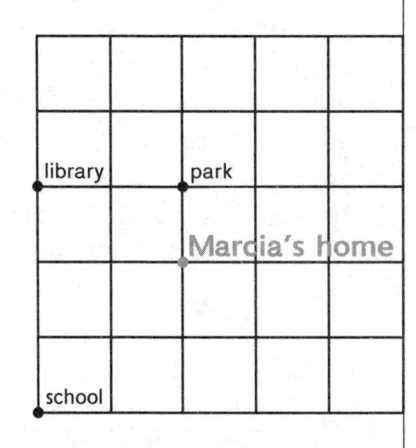

My Math Examples:
See students' examples.

Lesson 8 Ordered Pairs

English Learner Instructional Strategy

Collaborative Support: Pairs Frontload Vocabulary

Write *coordinates* and *origin* and the Spanish cognates, *coordinadas* and *origen*, on a classroom cognate chart. Provide concrete examples by displaying a coordinate plane and identifying the origin, along with related vocabulary introduced in this lesson.

To frontload new vocabulary for this lesson, pair more proficient English speakers with emerging or expanding students. Distribute one of the following My Vocabulary Cards to each pair: *coordinate plane, origin, ordered pairs, x-coordinate, y-coordinate*. Display a large demonstration coordinate plane. After pairs read and discuss their term, invite each pair to the board. Have the more proficient English speaker read the definition on his or her card as the other student labels and gestures to the corresponding aspect of the coordinate plane.

English Language Development Leveled Activities

Emerging Level	Expanding Level	Bridging Level
Word Recognition	**Recognize and Act It Out**	**Academic Vocabulary**
Use tape to create a large coordinate grid on the floor. Have a student volunteer toss a beanbag onto the grid. If necessary, move the beanbag onto the nearest point in the coordinate plane. Say, *The beanbag landed on a point*. Emphasize *point* and have students chorally repeat. Identify the ordered pair and write it on the board. Say, *We plotted (___, ___)*. Emphasize *plotted* and have students chorally repeat. Continue until all students have had a turn tossing the beanbag and identifying the ordered pairs for its location.	Have each student create a 10 by 10 coordinate grid on graph paper and then draw a shape, such as a star, triangle, or square, on their grid. Display the sentence frame: **Plot the ordered pair (___, ___).** Have students work in pairs while keeping their grids out of their partner's view. Students will take turns naming ordered pairs. After one student names an ordered pair, the other student will mark it on his or her grid. Marks that land inside the drawn shape are "hits" that score a point. The first student to earn five points wins.	Post a coordinate plane and write a list of ordered pairs. Model the terms *x-coordinate, y-coordinate, ordered pair,* and *origin* while plotting the first ordered pair in the list. Ask volunteers to plot the remaining ordered pairs in the list. Then have student pairs take turns providing coordinates. Student A will provide an x-coordinate and student B will provide a y-coordinate. Have them come to the grid, plot the coordinates, and use academic terms to describe the point's location. Repeat for five plotted points.

Teacher Notes:

NAME _____ DATE _____

Lesson 8 Vocabulary Chart
Ordered Pairs

Use the three-column chart to organize the vocabulary in this lesson. Write the word in Spanish. Then write the correct terms to complete each definition.

English	Spanish	Definition
coordinate	coordenada	One of two numbers in an <u>ordered</u> pair.
coordinate plane	plano de coordenadas	A plane that is formed when <u>two</u> number lines <u>intersect</u>.
ordered pair	par ordenado	A <u>pair</u> of numbers that is used to name a <u>point</u> on the coordinate plane.
origin	origen	The point (0, 0) on a coordinate plane where the <u>vertical</u> axis meets the <u>horizontal</u> axis.
x-coordinate	coordenada x	The <u>first</u> part of an ordered pair that indicates how far to the <u>right</u> of the y-axis the corresponding point is.
y-coordinate	coordenada y	The <u>second</u> part of an ordered pair that indicates how far <u>above</u> the x-axis the corresponding point is.

Grade 5 • Chapter 7 *Expressions and Patterns*

Lesson 9 Graph Patterns
English Learner Instructional Strategy

Vocabulary Support: Utilize Resources

As students work through the problem-solving exercises, be sure to remind them that they can refer to the Glossary or the Multilingual eGlossary for help, or direct students to other translation tools if they are having difficulty with non-math language in the problems.

Point out signal words and phrases that frequently appear in story problems, such as: *what is the difference, how much more, how many more than*, and help students understand that these phrases will show them which kinds of equations are needed for solving. Remind students to look in their math journals for guidance or to record additional signal words and phrases they encounter in new story problems.

English Language Development Leveled Activities

Emerging Level	Expanding Level	Bridging Level
Word Knowledge	**Recognize and Act It Out**	**Academic Vocabulary**
Draw a four-column chart. Label the columns *x, add 2, y,* and *(x, y)*. In the first row, write 1, add two, 3, and (1, 3). Say, *The pattern is to add two. We can use a table to make ordered pairs. Then we can graph the pattern.* After you write a number in the *x* column, have students say chorally **add two** and provide the *y*-coordinate. Complete the chart up to the ordered pair (6, 8). Then model using the coordinates to graph the pattern. Repeat the activity with a new rule for Column 2.	Say, *Train A picks up two passengers at each stop. Train B picks up three passengers at each stop.* Write and say, *x* = stop number and *y* = total number of passengers. Work with students to create a two-column table for each train, with columns labeled *x* and *y*. Model using the information in the tables to create two sets of ordered pairs. Create one coordinate grid and work with students to plot each set of ordered pairs using a different color. Discuss the difference in how each pattern appears in the grid.	Read a problem from the lesson aloud to students. Have pairs discuss the pattern and terms and then use them to create tables to represent the situation. Student A creates the first table and student B creates the second table. Each student in the pair generates ordered pairs from the table and plots the ordered pairs in a different color using the same coordinate grid. Discuss as a group the solution each pair found. Have pairs compare tables, ordered pairs, and graphs.

Teacher Notes:

NAME _____ DATE _____

Lesson 9 Multiple Meaning Word
Graph Patterns

Complete the four-square chart to review the multiple meaning word or phrase.

Everyday Use	Math Use in a Sentence
Sample answer: A repeated decorative design on a fabric.	Sample sentence: A pattern creates a predictable set of numbers.
Math Use	**Example From This Lesson**
A sequence of numbers, figures, or symbols that follows a rule or design.	Sample answer: The pattern in the following ordered pairs is that every *y*-coordinate is two more than the *x*-coordinate. (1, 3), (2, 4), (3, 5), (4, 6), (5, 7), (6, 8)

(center: **pattern**)

Write the correct term on the line to complete the sentence.

When you graph a pattern of __addition__, the points are plotted from the bottom left to the top right on the graph.

78 Grade 5 • Chapter 7 *Expressions and Patterns*

Chapter 8 Fractions and Decimals

What's the Math in This Chapter?

Mathematical Practice 3: Construct viable arguments and critique the reasoning of others

Display 2 circle fraction models. One showing $\frac{1}{2}$ and the other showing $\frac{2}{8}$. Say, *These fractions are the same amount. They are equivalent. Do you agree?* Allow time for students to think. Then have them turn and talk with a peer. Regroup and discuss their observations. The discussion goal should be for students to "construct viable arguments" proving that your fractions are not equivalent.

Ask, *How can I change my fraction models to show equivalent fractions?* change $\frac{2}{8}$ to $\frac{4}{8}$

Discuss with students how they critiqued your reasoning (the fraction models) and provided viable arguments as to why the fractions were not equivalent. Say, *All of you were applying Mathematical Practice 3.*

Display a chart with Mathematical Practice 3. Restate Mathematical Practice 3 and have students assist in rewriting it as an "I can" statement, for example: **I can use strategies I know to respond to the problem solving techniques of others.** Post the new "I can" statement.

Inquiry of the Essential Question:

How are factors and multiples helpful in solving problems?

Inquiry Activity Target: **Students come to a conclusion that they can use strategies they know to solve problems.**

As an introduction to the chapter, present the Essential Question to students. The inquiry graphic organizer will offer opportunities for students to observe, make inferences, and apply prior knowledge of problem-solving representing the Essential Question. As they investigate, encourage students to draw, write, and collaborate with peers to demonstrate their observations and thinking. Then have students present additional questions they may have to a peer to extend discussions.

Regroup students and restate Mathematical Practice 3 and the Essential Question. Pose questions to reflect on what has been learned to guide students in making connections between the Mathematical Practice and the Essential Question.

NAME _____ DATE _____

Chapter 8 Fractions and Decimals

Inquiry of the Essential Question:

How are factors and multiples helpful in solving problems?

Read the Essential Question. Describe your observations (I see..), inferences (I think...), and prior knowledge (I know...) of each math example. Write additional questions you have below. Then share your ideas and questions with a classmate.

$\frac{5}{9} = 5 \div 9$

I see ...

I think...

I know...

Two pounds of grapes are divided equally among five fruit baskets. How many pounds of grapes are in each basket?

| 1 | 2 | 3 | 4 | 5 | | 1 | 2 | 3 | 4 | 5 |

Each basket has $\frac{2}{5}$ pounds of grapes.

I see ...

I think...

I know...

$\frac{1}{5} = \frac{1 \times 2}{5 \times 2} = \frac{2}{10}$

Write the fraction with a denominator of 10.

I see ...

I think...

I know...

Questions I have...

_____

_____

Grade 5 • Chapter 8 *Fractions and Decimals* **79**

Lesson 1 Fractions and Division
English Learner Instructional Strategy

Vocabulary Support: Cognates

Write the words *numerator* and *denominator* and the Spanish cognates, *numerador* and *denominador* on a classroom cognate chart. Have students practice pronouncing the multisyllabic terms aloud. Write a fraction on the board, identify, and label each term.

Then write *improper* on the board. Have students brainstorm other words with the *im-* prefix, such as: *impossible, impolite, immature, immobile, imperfect.* Point out that *im-* has the same meaning in English as Spanish, "not." Display the following sentence frame and have students use it to identify proper and improper fractions: **This is a(n) proper/improper fraction because the numerator is greater/less than the denominator.**

English Language Development Leveled Activities

Emerging Level	Expanding Level	Bridging Level
Look and Identify Write the fraction $\frac{4}{3}$ on the board. Point to the numerator. Say, *The top number in a fraction is the numerator.* Emphasize *numerator.* Point to the denominator. Say, *The bottom number is the denominator.* Emphasize *denominator.* Have students practice saying **numerator** and **denominator** correspondingly as you point to each. Provide additional fractions. Invite students to the board to identify each numerator and denominator. Ask, *Which is the numerator/denominator?* Have students answer with a gesture.	**Memory Device** Write *numerator* and *up*. Underline the *u* in each word. Say, *The numerator has a u for up. The numerator is up on top.* Repeat with *denominator* and *down*, underlining the *d* in each word. Say, *The denominator has a d for down. The denominator is down below.* Display the following sentence frames: **The numerator is ____. The denominator is ____.** Write fractions on the board and have students take turns using the sentence frames to identify the numerators and denominators.	**Internalize Language** Display the following sentence frame: **In this fraction, ____ is the numerator, and ____ is the denominator.** Give one number cube apiece to student pairs. Have each pair roll their number cube twice to create a fraction. The first number is the numerator and the second number is the denominator. Have them write the fraction and identify it as either proper or improper. Have pairs work together to change any improper fractions into mixed numbers. Encourage pairs use the sentence frame to describe their fractions.

Teacher Notes:

T80 Grade 5 • Chapter 8 *Fractions and Decimals*

NAME _____ DATE _____

Lesson 1 Vocabulary Cognates

Fractions and Division

Use the Glossary to define the math word in English and in Spanish in the word boxes. Write a sentence using your math word.

fraction	fracción
Definition A number that represents part of a whole or part of a set.	**Definición** Número que representa partes iguales de un entero o de un conjunto.
My math word sentence: Sample answer: The fraction that represents two parts of a set of three parts is $\frac{2}{3}$.	

numerator	numerador
Definition The top number in a fraction; the part of the fraction that tells the number of parts you have.	**Definición** Número que se encuentra en la parte superior de una fracción; indica cuántas de las partes iguales se usan.
My math word sentence: Sample answer: In the fraction $\frac{2}{3}$, the numerator is 2.	

denominator	denominador
Definition The bottom number in a fraction. It represents the number of parts in the whole.	**Definición** Número que se encuentra en la parte inferior de una fracción. Representa la cantidad de partes en que se divide un entero.
My math word sentence: Sample answer: In the fraction $\frac{2}{3}$, the denominator is 3.	

Grade 5 • Chapter 8 *Fractions and Decimals*

Lesson 2 Greatest Common Factor

English Learner Instructional Strategy

Graphic Support: KWL Chart

Write *common factor* and the Spanish cognate, *factor común*, on a classroom cognate chart. Have Spanish-speaking students refer to the Glossary for the Spanish definition. Discuss with students the nonmath meanings of *common*, and then introduce the vocabulary term as it is used in a math context.

Display a KWL chart. In the first column, record what students recall about *common* factors from previous math lessons. In the second column, write and say aloud, *How can common factors help us solve problems?* Record student ideas and other questions, including how to identify greatest common factors. After the lesson, display the following sentence frame and have students use it to report back: **I learned that common factors ___.** Record student responses in the third column of the KWL chart.

English Language Development Leveled Activities

Emerging Level	Expanding Level	Bridging Level
Word Knowledge Use counters to model the number 32. Divide the counters into two groups of 16. Write 2 × 16 = 32. Say, *A fact is something true. Two times sixteen equals thirty-two is a math fact.* Stress the word *fact*. Underline 2 and 16. Say, *Two and sixteen are factors of thirty-two.* Stress the word factors and have students chorally repeat. Divide 16 counters into two groups of eight. Write 2 × 8 = 16. Say, *Two times eight equals sixteen is a fact. Which two numbers are factors of 16?* Allow students to answer verbally or by pointing. Repeat with other multiplication facts and factors.	**Number Sense** Write 30. With students' help, create a list of factors of 30: 1, 2, 3, 5, 6, 10, 15, 30. Write 54 and create a list of its factors: 1, 2, 3, 6, 9, 18, 27, 54. Have a student circle all factors that appear in both lists. **1, 2, 3, 6** Say, *The numbers that appear in both lists are the common factors of 30 and 54.* Ask students which common factor is greatest. **6** Say, *Six is the greatest common factor.* Provide more examples and display sentence frames for students to use: **___ are the common factors of ___ and ___. The greatest common factor is ___.**	**Building Oral Language** Have students work in pairs. Have each student write a two-digit number on an index card and then exchange cards with his or her partner. On the back of the card, have students list the factors of their number. Have students work together to find the greatest common factor for their two numbers and circle it. Display this sentence frame for students to use when identifying the greatest common factor as they report back to you or another pair of students: **The greatest common factor of ___ and ___ is ___.**

Multicultural Teacher Tip

Some ELs may have been taught a different method for making factor trees. For example, in Mexico, students draw a vertical line to use in determining factors. On the left side, they write the number to be factored, and then the first factor is written on the right side. The number divided by the factor is then written on the left side, below the original number. The next factor is written on the right, and the process continues until there are no more factors. In factoring 18, for example, the result would be *18, 9, 3, 1* listed on the left side, and the prime factors *2, 3, 3* listed on the right.

NAME _____ DATE _____

Lesson 2 Vocabulary Definition Map
Greatest Common Factor

Use the definition map to write a description and list characteristics about the vocabulary word or phrase. Write or draw math examples. Share your examples with a classmate.

My Math Vocabulary:

> **greatest common factor**

Description from Glossary:

> The greatest of the common factors of two or more numbers.

Characteristics from Lesson:

> A <u>factor</u> is a number that is multiplied by another number.

> A common factor is a number that is a factor of <u>two</u> or more numbers.

> Prime factorization is a way of expressing a <u>composite</u> number as a product of its <u>prime</u> factors.

My Math Examples:
See students' examples.

Grade 5 • Chapter 8 *Fractions and Decimals* **81**

Lesson 3 Simplest Form
English Learner Instructional Strategy

Vocabulary Support: Frontload Academic Vocabulary

Write *equivalent* and the Spanish cognate, *equivalente* on a classroom cognate chart. Display a meterstick and say, *This is one meter long. It is also 100 centimeters long. One meter is* **equivalent** *to 100 centimeters. One meter and 100 centimeters are two ways to describe the* **same** *or* **equal** *lengths.* Review what students know about equivalent fractions.

Display a word web. Write *simple* in the center. Work with students to fill the web with synonyms for *simple*, such as: *easy, basic, not hard, plain*. Remind students that the suffix *-est* means "most," so *simplest* means "most simple." Write the fractions: $\frac{17}{68}, \frac{13}{52}, \frac{1}{4}$. Ask, *Which fraction is the easiest to understand?* Have students answer, and then say, *These fractions are equivalent, but $\frac{1}{4}$ is written in the* **simplest form**.

English Language Development Leveled Activities

Emerging Level	Expanding Level	Bridging Level
Word Recognition Cut out two large circles. Cut one in half and adhere the two halves next to each other on the board. Below them write $\frac{1}{2} = \frac{1}{2}$. Say, *One half is* **equal** *to the other half.* Cut the second circle into quarters. Replace one of the halves on the board with two quarters. Rewrite the equation as $\frac{1}{2} = \frac{2}{4}$. Point to each side of the equation and the circle parts as you say, *These fractions look different, but they name the same number. One half is* **equivalent** *to two fourths.* Stress *equivalent* and have students chorally repeat. Repeat the activity modeling with a square.	**Recognize and Act It Out** Display a one-half fraction tile next to 2 onefourth fraction tiles. Say, *One half and two fourths are equivalent fractions.* Guide students to identify other fractions equivalent to one half. Then have pairs work together using fraction tiles or circles to model these fractions in simplest form: $\frac{2}{8}, \frac{4}{48}, \frac{15}{30}$, and $\frac{20}{50}$. Display the following sentence frames and have students report back: ___ **and** ___ **are equivalent fractions. The fraction** ___ **is in simplest form.**	**Academic Language** Have students work in pairs. Write the fraction $\frac{8}{32}$. Say, *You have two minutes to write equivalent fractions for $\frac{8}{32}$. The pair with the most equivalent fractions after two minutes will earn two points. All pairs with the fraction written in simplest form will earn one point.* After two minutes, have each pair share the fractions they have written. Award points as described, and then write a new fraction. Repeat the exercise, and continue writing new fractions and awarding points until one pair has earned six points.

Teacher Notes:

NAME _____ DATE _____

Lesson 3 Note Taking
Simplest Form

Read the question. Write words you need help with and research each word. Use your lesson to write your Cornell notes. Write or draw math examples to explain your thinking. Share your examples with a classmate.

Building on the Essential Question	Notes:
How do you write a fraction in simplest form?	$\frac{9}{12}$ is the fraction to simplify. The numerator of the fraction is __9__ and the denominator is __12__. The factors of 9 are __1__, __3__, __9__. The factors of 12 are __1__, __2__, __3__, __4__, __6__, __12__. The common factors of 9 and 12 are __1__ and __3__. The **greatest** common factor (GCF) of 9 and 12 is __3__. Divide both the numerator and denominator by the GCF. 9 ÷ __3__ = __3__ 12 ÷ __3__ = __4__ A fraction is in __simplest__ form when the greatest common factor (GCF) of the numerator and the denominator is 1. The factors of 3 are __1__ and __3__. The factors of 4 are __1__, __2__, __4__. The common factor of 3 and 4 is __1__. The greatest common factor (GCF) of 3 and 4 is __1__. Equivalent fractions are fractions that have the __same__ value. $\frac{9}{12}$ and $\frac{3}{4}$ are equivalent fractions.
Words I need help with: See students' words.	

My Math Examples:

See students' examples.

Lesson 4 Problem-Solving Investigation Strategy: Guess, Check, and Revise

English Learner Instructional Strategy

Language Structure Support: Report Back

During the lesson, display the following communication guide to aid students in reporting back on the problem solving process and the specific strategy of Guess, Check, Revise.

I understand ____.
I need to find out ____.
My plan is to ____.
My guess is ____.
I need to revise my guess to ____.
The answer is ____.
I know my answer is reasonable because ____.

As students report back, be sure they are differentiating between the /s/ and /z/ sounds as they are enunciating words with s, such as: *revise, is, guess,* and *answer.* If necessary, model correct pronunciation and have students repeat.

English Language Development Leveled Activities

Emerging Level	Expanding Level	Bridging Level
Word Knowledge Write a whole number between 1 and 100 on a piece of paper. Conceal the paper from students. Display a sentence frame for students to use: **My guess is ____.** Ask a volunteer to guess the number. Write the guess and say, *Let me check your guess.* Stress the word *check.* Look at the number on the paper, and then provide feedback, such as: *too high, too low,* or *correct.* Then ask, *Would you like to revise your guess?* Stress the word *revise.* If the student wishes to guess again, have them answer, **Yes, I will revise.** Repeat until the number is identified.	**Recognize and Act It Out** Write then say, *A racing video game costs $10 more than a soccer video game. The total for both games is $50. How much does each game cost?* Ask students to name two numbers whose sum is 50. Say, *This will be our first guess.* Write the addition expression and say, *Now let's check our guess.* Write a subtraction expression to subtract the lesser addend from the greater one. Solve, and then ask, *How does this compare to what we know? Should we revise our guess?* Repeat the guess, check, revise process until you arrive at the correct solution. $30 + $20 = $50.	**Internalize Language** Have pairs work together using the guess, check, and revise strategy to solve an Apply the Strategy exercise from the lesson. Display the following sentence frames for students to use as they report back: **Our first guess was ____. When we checked our guess, it was ____. We revised our guess by ____. The answer is ____.** Afterward, discuss how many times students had to revise their guesses to find the correct solution.

Teacher Notes:

NAME _____ DATE _____

Lesson 4 Problem-Solving Investigation

STRATEGY: Guess, Check, and Revise

Guess, check, and revise to solve each problem.

1. Bike path A is **4 miles** long.
   Bike path B is **7 miles** long.
   If **April** biked a **total** of **37 miles**,
   how many **times** did **she** bike **each** path?

Understand	Solve
I know:	
I need to find:	
Plan	**Check**
I will ___guess___, check, and revise to solve.	

2. Ruben sees **14 wheels** on a total of **6** bicycles **and** tricycles. How **many** bicycles **and** tricycles are there?

Understand	Solve
I know:	
I need to find:	
Plan	**Check**
A bicycle has ___2___ wheels. A tricycle has ___3___ wheels. I will guess, ___check___, and revise to solve.	

Grade 5 • Chapter 8 *Fractions and Decimals* **83**

Lesson 5 Least Common Multiple

English Learner Instructional Strategy

Collaborative Support: Peers/Mentor

Write the words *multiple* and *multiples* and the Spanish cognates, *múltiplo* and *múltiplos*, on a classroom cognate chart. Provide a visual math example by listing multiples of a number such as 2 or 3.

Pair emerging students with expanding or bridging students who share the same native language. Have pairs work collaboratively on the Independent Practice exercises finding the least common multiple. Have the more proficient English speaker identify the least common multiple using the following sentence frame: **The least common multiple is ____.** Then have the emerging student repeat the same sentence.

English Language Development Leveled Activities

Emerging Level	Expanding Level	Bridging Level
Word Recognition	**Read and Respond**	**Academic Language**
Write 3 × 1. Say, *I will multiply three times one.* Say *multiply* again and have students repeat chorally. Write the answer. Then write 3 × 2 = 6, 3 × 3 = 9, and so on through 3 × 10 = 30. Circle each product. Point to the products as you say, *These are multiples of three.* Stress *multiples* as you say it again and have students repeat chorally. Model multiples of four. Circle all the products and point to them as you ask, *What are these?* Prompt students to answer, **multiples of four.** Be sure they are differentiating between *multiply* and *multiples*.	Have students work in pairs. Ask one student to write a single digit number on an index card. Have the other student write the first four multiples of the number on a second index card. Collect and shuffle the cards. Redistribute one random card back to each student. Have each student find the classmate with the card that corresponds to theirs. Display the following sentence for students to use when naming the multiples listed on their cards: **The first four multiples of ____ are ____, ____, ____, ____.**	Have pairs take turns rolling a number cube. Each student in the pair will write the first 8-10 multiples of the number he or she rolled. Together the pair will identify the least common multiple of the two numbers. Display the following sentence frames to help them: **____ is the least common multiple ____ of and ____.** Afterward, turn the activity into a game. Have pairs compete to see which can be the first to find the least common multiple using two numbers you select.

Teacher Notes:

NAME _____ DATE _____

Lesson 5 Vocabulary Chart

Least Common Multiple

Use the three-column chart to organize the vocabulary in this lesson. Write the word in Spanish. Then write the correct terms to complete each definition.

English	Spanish	Definition
multiple	múltiplo	A multiple of a number is the product of __that__ number and any __whole__ number.
common multiple	múltiplo común	A __whole__ number that is a __multiple__ of two or more numbers.
least common multiple (LCM)	minimo común múltiplo (m.c.m.)	The __smallest__ whole number greater than 0 that is a __common__ multiple of each of two or more numbers.
product	producto	The __answer__ to a multiplication problem.

Lesson 6 Compare Fractions
English Learner Instructional Strategy

Language Structure Support: Sentence Frames

Write the term *common denominator* and the Spanish cognate, *denominador común* on a classroom cognate chart. Have Spanish speaking students refer to the Glossary for the Spanish definition. Write two fractions with common denominators, label them as such, and refer to the examples as you discuss the new vocabulary. During the lesson, display tiered sentence frames to help students with different levels of English proficiency participate. For example:

The fractions are ____ and ____.
The denominators are ____ and ____.
The least common multiple of the denominators is ____.
The least common denominator is ____.
I can find equivalent fractions by ____.
The fractions are equivalent because ____.

English Language Development Leveled Activities

Emerging Level	Expanding Level	Bridging Level
Word Knowledge Write $\frac{2}{6}$ and $\frac{3}{6}$. Say, *Circle the numbers that are the same.* Have a volunteer circle both denominators. Say, *The denominators are the same. The fractions have a common denominator.* Emphasize common denominator and have students repeat chorally. Write the fractions $\frac{2}{3}$ and $\frac{2}{4}$. Circle the denominators and say, *The denominators are different. These fractions do not have a common denominator.* Stress do not. Repeat with other pairs of fractions. Have students clap when they identify common denominators and remain silent when they do not.	**Recognize and Act It Out** Write $\frac{1}{3}$ and $\frac{1}{4}$. Model finding equivalent fractions with a common denominator and compare the fractions. Say, $\frac{4}{12}$ is greater than $\frac{3}{12}$, so $\frac{1}{3}$ is greater than $\frac{1}{4}$. Write $\frac{3}{4}$ and $\frac{4}{5}$. Ask, *Which fraction is greater?* Have pairs work together to find equivalent fractions with common denominators. Display the following sentence frames so students can share their answers: ____ is the least common multiple of 4 and 5. ____ is an equivalent fraction of ____. The fraction ____ is greater than the fraction ____.	**Academic Language** Have students work in pairs to create and compare fractions. Each student in the pair creates a fraction by rolling a number cube twice. The lower number is the numerator and the larger number is the denominator in the fraction. Pairs work together to compare their fractions. Afterward, have students describe the steps taken to compare their fractions. If a least common denominator must be found, students should describe the steps needed to find equivalent fractions with common denominators.

Teacher Notes:

T85 Grade 5 • Chapter 8 *Fractions and Decimals*

NAME _____ DATE _____

Lesson 6 Concept Web

Compare Fractions

Use the concept web to identify the least common multiple of each set of numbers.

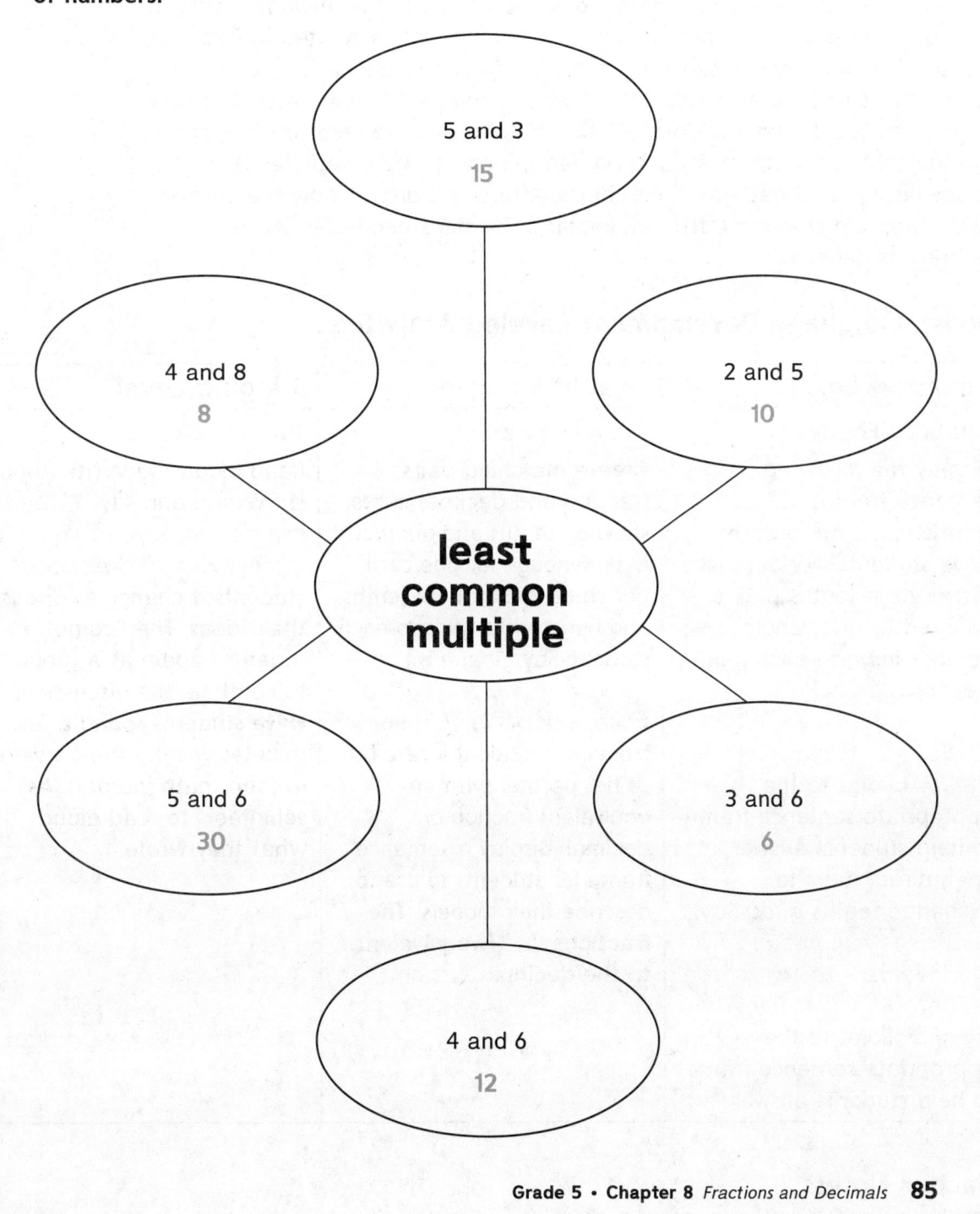

- 5 and 3 — 15
- 4 and 8 — 8
- 2 and 5 — 10

least common multiple

- 5 and 6 — 30
- 3 and 6 — 6
- 4 and 6 — 12

Lesson 7 Inquiry/Hands On: Use Models to Write Fractions as Decimals

English Learner Instructional Strategy

Vocabulary Support: Anchor Chart

Divide students into four groups. Direct students to make an anchor chart showing what they know about fractions. Each chart should include a title at the top of the poster and definitions for math vocabulary related to fractions, such as *denominator, numerator, equivalent fractions,* and so on. Suggest that students include a shaded model labeled with a fraction describing how much is shaded. Direct students to label the different elements in their charts with appropriate math vocabulary. When the charts are completed, have groups display and describe their charts. Afterward, discuss how the anchor charts can help students better understand the steps needed to write fractions as decimals.

English Language Development Leveled Activities

Emerging Level	Expanding Level	Bridging Level
Sentence Frames	**Number Sense**	**Turn & Talk**
Display the following sentence frames: ____ **tenths.** ____ **hundredths.** Have students work in pairs. Distribute a tenths grid, a hundredths grid, and a number cube to each pair. Say, *Roll the cube. Shade that many columns in the tenths grid. What fraction is shaded?* Point to the appropriate sentence frame to help students answer. Then direct pairs to exchange tenths girds. Say, *Shade the hundredths grid to model the fraction of the tenths grid. What fraction is shaded?* Point to the appropriate sentence frame to help students answer.	Create matching pairs of fraction and decimal cards showing tenths and hundreds. Make enough for one card per student. Distribute tenths and hundredths grids to each student. Say, *Shade one of the grids to model your fraction or decimal.* Then have each student locate his or her partner with an equivalent fraction or decimal. Display a sentence frame for students to use to describe their models: **The fraction ____ is equivalent to the decimal ____.**	Read aloud the Write About It Exercise and say, *Turn to the student nearest you and discuss your answer.* Give students a chance to discuss their ideas. Then come together again as a group to continue the discussion. Have students spend a few minutes writing the answers in their math journals. Ask volunteers to read aloud what they wrote.

Teacher Notes:

NAME _____ DATE _____

Lesson 7 Guided Writing

Inquiry/Hands On: Use Models to Write Fractions as Decimals

How do you use models to write fractions as decimals?

Use the exercises below to help you build on answering the Essential Question. Write the correct word or phrase on the lines provided.

1. Rewrite the question in your own words.
   See students' work.

2. What key words do you see in the question?
   fractions, decimals, models

3. A decimal can be modeled by shaded squares on a 10 by 10 grid. The grid contains __100__ squares.

   One shaded square represents __0.01__.

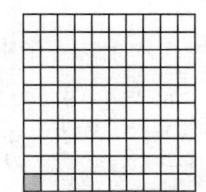

   Ten shaded squares represents __0.10__.

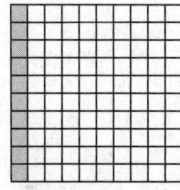

4. When modeling a fraction on a 10 by 10 grid, the first step is to find an __equivalent__ fraction with a denominator of 10 or 100.

5. Equivalent fractions are fractions that have the __same__ value. If you multiply the __numerator__ and __denominator__ by the same number, you will find an equivalent fraction.

6. The following fractions are equivalent: $\frac{2}{5} = \frac{40}{100}$.

7. How many squares will be shaded to represent this equivalent fraction? __40__

8. The fraction $\frac{40}{100}$ and the decimal __0.40__ is modeled on the grid. The fraction $\frac{2}{5}$ written as a decimal is __0.40__.

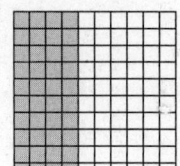

9. How do you use models to write fractions as decimals?
   Find an equivalent fraction with a denominator of 10 or 100. Model the equivalent fraction on a 10 by 10 grid. Identify the decimal represented by the shaded squares in the grid.

86 Grade 5 • Chapter 8 *Fractions and Decimals*

Lesson 8 Write Fractions as Decimals
English Learner Instructional Strategy

Collaborative Support: Round the Table

Place students into multilingual groups of 4 or 5. Assign even or odd numbered Independent Practice exercises to each group and have students work jointly to solve each exercise by passing a write-on/wipe-off board around the table. Each student will perform one step in rewriting the fraction as a decimal. Provide a step-by-step list for groups to follow, such as:

1) Write the fraction.
2) Determine if an equivalent fraction will be written with a denominator of 10 or 100.
3) Determine the number to multiply the numerator and denominator by to find an equivalent fraction.
4) Find the equivalent fraction.
5) Use place-value to write the fraction as a decimal.

English Language Development Leveled Activities

Emerging Level	Expanding Level	Bridging Level
Number Sense Draw a number line from 0 to 1. Mark and label the following decimal numbers: 0.25, 0.5, 0.75. Below the number line, draw a second number line from 0 to 1. Mark and label the following fractions: $\frac{1}{4}$, $\frac{1}{2}$, $\frac{3}{4}$. Review with students that 0.5 is equivalent to $\frac{1}{2}$. Gesture to the number lines and say, *The fraction $\frac{1}{2}$ and the decimal 0.5 appear in the same place along the number lines. $\frac{1}{2}$ and 0.5 are **equivalent**. They are the same number.* Repeat with $\frac{1}{4}$ and 0.25 and $\frac{3}{4}$ and 0.75.	**Listen and Write** Say, *Write the fraction eight tenths.* Have students write it on a write-on/wipe-off board. Say, *Write the decimal number eight tenths.* Have students write it on their boards. Use a Tenths Model from Work Mat 5, to visually verify the equivalency of $\frac{8}{10}$ and 0.8. Say, *The fraction $\frac{8}{10}$ and the decimal 0.8 are different ways to write the same number. They are equivalent.* Repeat with $\frac{7}{10}$ and 0.7. Then have pairs find the decimal equivalent of $\frac{9}{50}$ and explain their answer using the following sentence frame: ____ is equivalent to $\frac{9}{50}$ because ____.	**Building Oral Language** Have students work in groups of three. Student A will choose a numerator, student B will choose a denominator, and student C will find the decimal equivalent of the fraction, rounding to the nearest hundredth. Have groups repeat the process three times, switching roles each time. Afterward, have a volunteer in each group share one of the equivalencies using the following sentence frame: **I know ____ is equivalent to ____ because ____.**

Teacher Notes:

NAME _____ DATE _____

Lesson 8 Vocabulary Cognates

Write Fractions as Decimals

Use the Glossary to define the math word in English and in Spanish in the word boxes. Write a sentence using your math word.

equivalent fractions	fracciones equivalentes
Definition Fractions that have the same value.	**Definición** Fracciones que tienen el mismo valor.

My math word sentence:
Sample answer: $\frac{1}{2}$ of a pizza is equal to $\frac{4}{8}$ of a pizza. So, $\frac{1}{2}$ and $\frac{4}{8}$ are equivalent fractions.

equivalent decimals	decimales equivalentes
Definition A number that has a digit in the tenths place, hundredths place, and beyond.	**Definición** Decimales que tienen el mismo valor.

My math word sentence:
Sample answer: 0.5 and 0.50 are equivalent decimals.

Chapter 9 Add and Subtract Fractions

What's the Math in This Chapter?

Mathematical Practice 5: Use appropriate tools strategically

Distribute connecting cubes, fraction tiles, and base-ten blocks to each student. Group students. Distribute to each group a piece of paper with the following addition and subtraction problems:

1. $378 - 281$
2. $\frac{1}{2} + \frac{1}{4}$
3. $24 - 8$
4. $16 + 5$
5. $159 + 188$
6. $\frac{2}{3} - \frac{1}{2}$

Ask each group of students to not only solve the problems, but also write down the best manipulative to use for that problem. After students have finished compare answers, but more importantly compare the manipulatives they used.

Which model is best for fractions? **fractions tiles** *Which model is better for large numbers?* **base-ten blocks** *Which model is best for smaller numbers?* **connecting cubes**

Display a chart with Mathematical Practice 5. Restate Mathematical Practice 5 and have students assist in rewriting it as an "I can" statement, for example: **I can decide which model is the best to use as a tool to solve problems.** Post the new "I can" statement.

Inquiry of the Essential Question:

How can equivalent fractions help me add and subtract fractions?

Inquiry Activity Target: **Students come to a conclusion that using models can be a tool in problem-solving.**

As an introduction to the chapter, present the Essential Question to students. The inquiry graphic organizer will offer opportunities for students to observe, make inferences, and apply prior knowledge of using models representing the Essential Question. As they investigate, encourage students to draw, write, and collaborate with peers to demonstrate their observations and thinking. Then have students present additional questions they may have to a peer to extend discussions.

Regroup students and restate Mathematical Practice 5 and the Essential Question. Pose questions to reflect on what has been learned to guide students in making connections between the Mathematical Practice and the Essential Question.

NAME _____ DATE _____

Chapter 9 Add and Subtract Fractions

Inquiry of the Essential Question:

How can equivalent fractions help me add and subtraction fractions?

Read the Essential Question. Describe your observations (I see...), inferences (I think...), and prior knowledge (I know...) of each math example. Write additional questions you have below. Then share your ideas and questions with a classmate.

Use models to show sums like $\frac{3}{5} + \frac{4}{5}$.

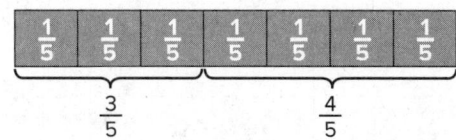

$\underbrace{\boxed{\tfrac{1}{5}}\boxed{\tfrac{1}{5}}\boxed{\tfrac{1}{5}}}_{\frac{3}{5}}\underbrace{\boxed{\tfrac{1}{5}}\boxed{\tfrac{1}{5}}\boxed{\tfrac{1}{5}}\boxed{\tfrac{1}{5}}}_{\frac{4}{5}}$

There are seven $\frac{1}{5}$-tiles. So, the sum is $\frac{7}{5}$ or $1\frac{2}{5}$.

I see ...

I think...

I know...

$\frac{7}{9} - \frac{4}{9} = \frac{7-4}{9}$

$\phantom{\frac{7}{9} - \frac{4}{9}} = \frac{3}{9}$ or $\frac{1}{3}$

I see ...

I think...

I know...

Solve subtraction problems like $9\frac{4}{5} - 1\frac{3}{10}$.

$9\frac{4}{5} \rightarrow 9\frac{8}{10}$ Write $9\frac{4}{5}$ as $9\frac{8}{10}$.
$-1\frac{3}{10} \rightarrow -1\frac{3}{10}$
$\phantom{-1\frac{3}{10} \rightarrow}\; 8\frac{5}{10}$ or $8\frac{1}{2}$

I see ...

I think...

I know...

Questions I have...

_____

_____

_____

88 Grade 5 • Chapter 9 *Add and Subtract Fractions*

Lesson 1 Round Fractions

English Learner Instructional Strategy

Vocabulary Support: Sentence Frames

Display tiered sentence frames, corresponding to differing levels of English language proficiency, for students to use throughout the lesson.

For emerging students: Encourage silent students to gesture to or write the answer. **The fraction is ___. The numerator/denominator is ___.**

For expanding students: **The fraction is closer to ___. The fraction rounds to ___.**

For bridging students: **Round to 0 if ___. Round to $\frac{1}{2}$ if ___. Round to 1 if ___.**

English Language Development Leveled Activities

Emerging Level	Expanding Level	Bridging Level
Developing Oral Language Draw a number line from 0 to 1 marked with tenths. Model using the line to round three fractions: one that rounds to 0, one that rounds to $\frac{1}{2}$, and one that rounds to 1. Display the following sentences for students to use: **Round to zero. Round to one half. Round to one.** Provide several fractions between 0 and 1. Have students use the sentences on the board to guide you in rounding each fraction. If necessary, model saying the correct sentence and have students repeat chorally.	**Number Game** Draw a number line from 0 to 1 marked with eighths. Label the whole numbers. Write the following fractions on sticky notes: $\frac{1}{8}, \frac{2}{8}, \frac{3}{8}, \frac{4}{8}, \frac{5}{8}, \frac{6}{8}, \frac{7}{8},$ and $\frac{8}{8}$. Have each student pick a sticky note, round their fraction to the nearest benchmark fraction, and place the note accordingly above the number line. **Display the following sentence frame for students to use when describing their answer: The fraction ___ rounds to ___.**	**Building Oral Language** Direct groups of 3 or 4 students to draw a number line from 0 to 1 on a write-on/wipeoff board. Ask them to mark the number line in sixths and label $\frac{1}{2}$. Have students in each group alternate rolling a number cube to generate a numerator for a fraction with a denominator of 6. Each student will find his or her fraction on the number line, round it to the nearest benchmark (0, $\frac{1}{2}$, 1), and then use the following sentence frame to explain how they know the answer: **The fraction ___ rounds to ___ because ___.**

Multicultural Teacher Tip

Many cultures emphasize the use of decimal numbers over fractions. For this reason, ELs may be unfamiliar with fractions and how they describe the relationship between a part and the whole. You may want to create a chart showing common fractions, their decimal equivalents, and a visual example, such as a shaded circle or rectangle.

Student page

NAME _____ DATE _____

Lesson 1 Vocabulary Definition Map
Round Fractions

Use the definition map to write a description and list characteristics about the vocabulary word or phrase. Write or draw math examples. Share your examples with a classmate.

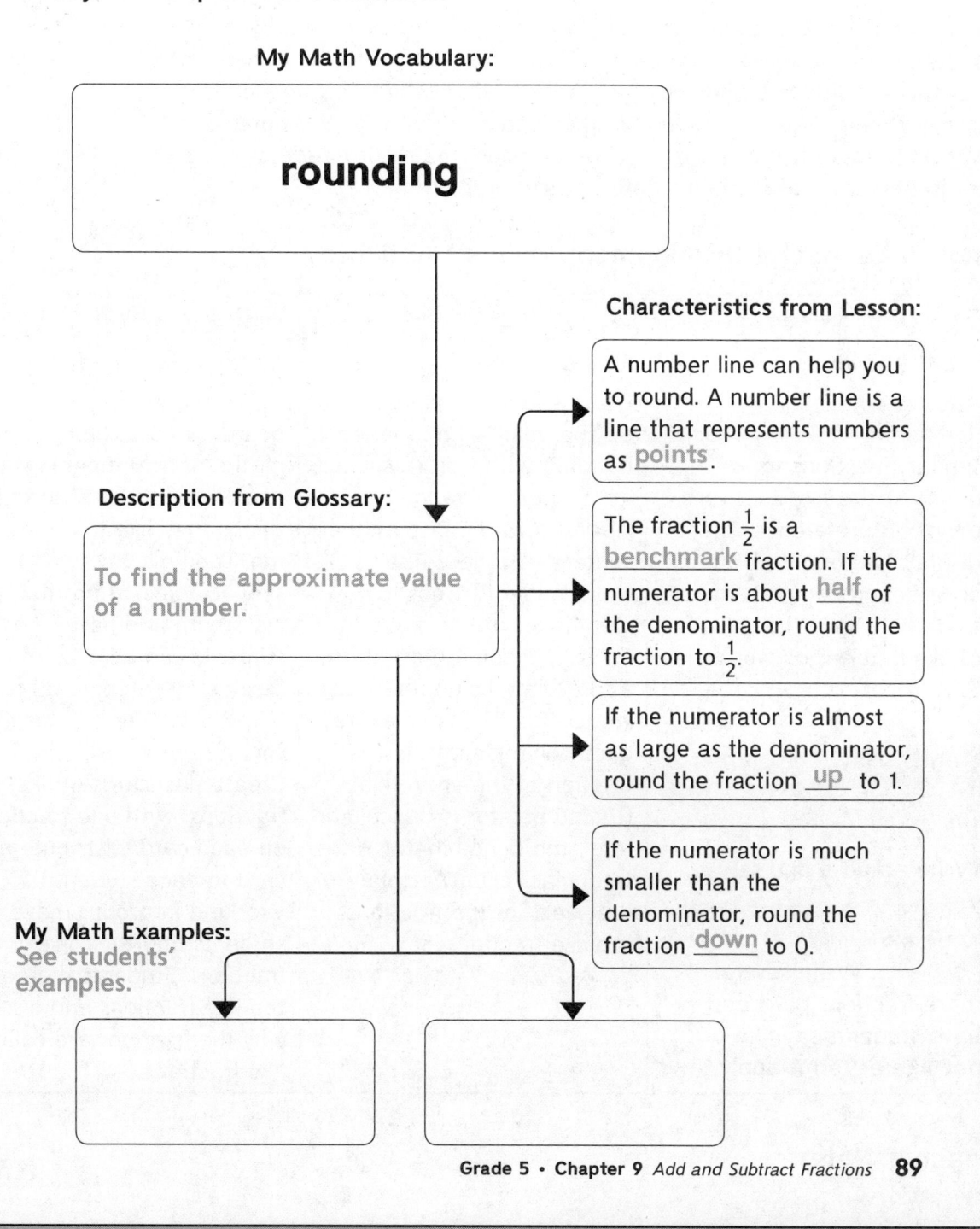

My Math Vocabulary:

rounding

Characteristics from Lesson:

A number line can help you to round. A number line is a line that represents numbers as <u>points</u>.

The fraction $\frac{1}{2}$ is a <u>benchmark</u> fraction. If the numerator is about <u>half</u> of the denominator, round the fraction to $\frac{1}{2}$.

If the numerator is almost as large as the denominator, round the fraction <u>up</u> to 1.

If the numerator is much smaller than the denominator, round the fraction <u>down</u> to 0.

Description from Glossary:

To find the approximate value of a number.

My Math Examples:
See students' examples.

Grade 5 • Chapter 9 *Add and Subtract Fractions* 89

Lesson 2 Add Like Fractions
English Learner Instructional Strategy

Vocabulary Support: Utilize Resources

As students work through lesson exercises, be sure to remind them that they can refer to the Glossary or the Multilingual eGlossary for help with math vocabulary. Direct students to other translation tools if they need to clarify non-math language in the problems.

Point out signal words and phrases that frequently appear in story problems, such as *altogether* and *total amount*. Help students understand that these words and phrases often indicate that addition is needed for solving. Remind students to look in their math journals for the signal words/phrases list they have already compiled and record additional signal words/phrases that they encounter in new problems.

English Language Development Leveled Activities

Emerging Level	Expanding Level	Bridging Level
Word Knowledge	**Memory Device**	**Multiple Meanings**
Ask, *Do you like apples? Do you like funny movies?* or similar questions to demonstrate *like* as a verb denoting preference. Then display two books or other objects and compare them to demonstrate the meaning of like in terms of similarities. Say, *The ___ is like the ___ because ___.* Write $\frac{2}{6}$ and $\frac{1}{6}$. Say, *These are like fractions because they have the same denominator.* Write $\frac{5}{6}$ and $\frac{3}{7}$ and say, *These are not like fractions. Their denominators are not the same.* Write several more fractions pairs and have students say **like** or **not like** as terms apply.	Draw a four-column chart on the board and label each heading with a fruit, such as: *Bananas, Oranges, Apples, Grapes*. Display the sentence frame, **I like ___.** Have students use it to identify his or her preference for fruit. Place a mark in the appropriate column as each student replies. Then express each column's total as a fraction of the entire class (the numerator is the column total, the denominator is the total class count). Explain that each of the fractions are like fractions. Say, *Each of these like fractions is a part of the same whole.*	Write the word *like* on the board. Discuss common and mathematical meanings of *like*. Have students describe things they like to do for fun. Then display two similar objects from the classroom and have students compare them. Display a sentence frame to help them: **The ___ is like the ___ because ___.** Create flashcards of like fractions, with one fraction on each card. Distribute a card to each student. Working in groups, have students find their like fractions. Students then add the like fractions and discuss why the fractions are called like fractions.

Teacher Notes:

NAME _____ DATE _____

Lesson 2 Multiple Meaning Word
Add Like Fractions

Complete the four-square chart to review the multiple meaning word.

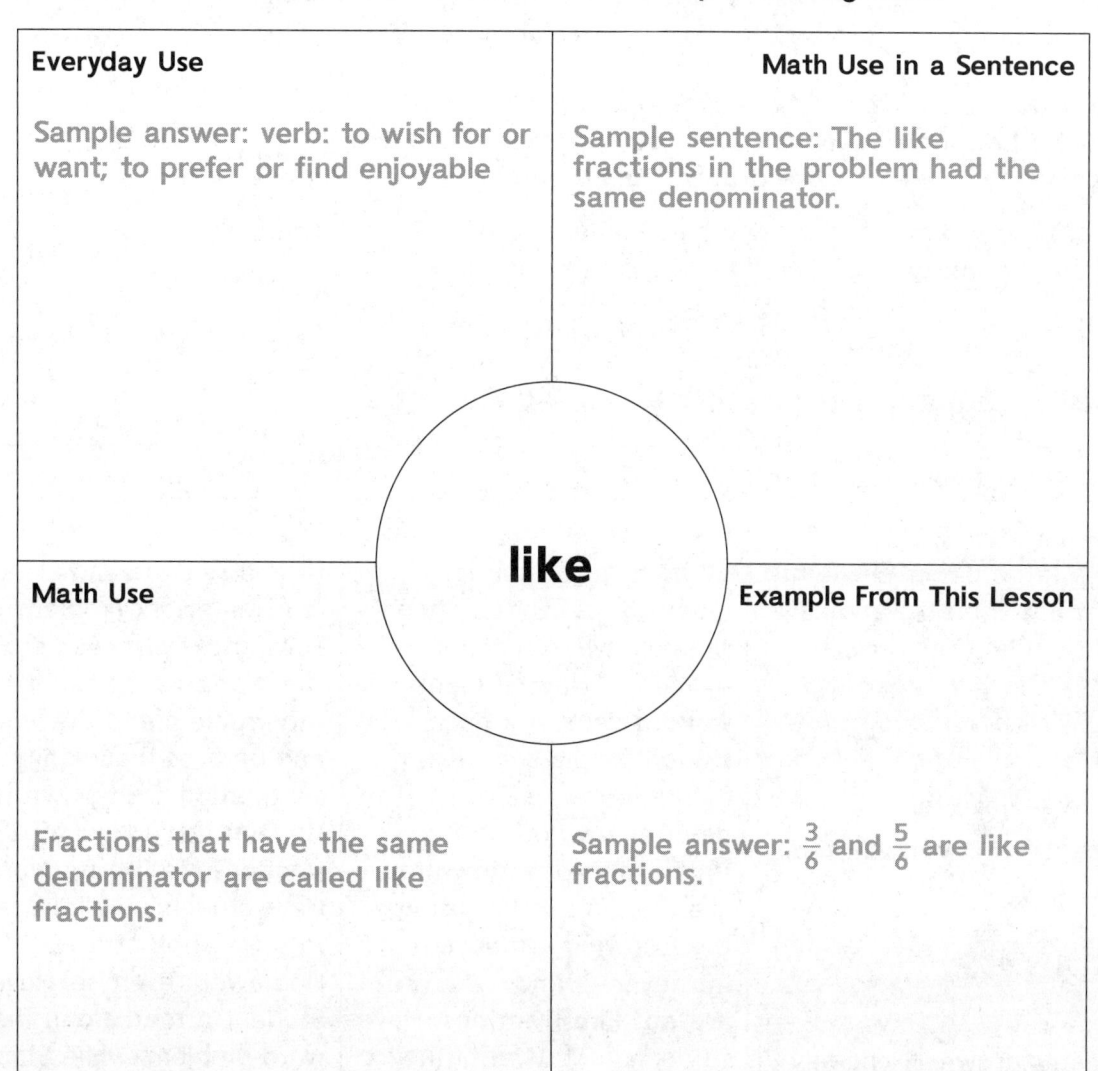

Everyday Use	Math Use in a Sentence
Sample answer: verb: to wish for or want; to prefer or find enjoyable	Sample sentence: The like fractions in the problem had the same denominator. |// like //
Math Use	**Example From This Lesson**
Fractions that have the same denominator are called like fractions.	Sample answer: $\frac{3}{6}$ and $\frac{5}{6}$ are like fractions.

Write the correct terms on the blank lines to complete the sentence.

To add like fractions, you add the <u>numerators</u> and keep the denominator the <u>same</u>.

90 Grade 5 • Chapter 9 *Add and Subtract Fractions*

Lesson 3 Subtract Like Fractions
English Learner Instructional Strategy

Language Structure Support: Report Back

Divide students into three small groups and assign each group one problem from Problem Solving Exercises 16–18. Direct students to solve their assigned problem and write the answer in simplest form. Display the following sentence frames to aid groups as they report back:

Vicki used ____ of the water in the bucket.
Roshanda bought ____ pound more of roast beef than ham.
Chris spent ____ hour more drawing than reading.

Also provide a sentence frame to help students explain how they changed their original answer so it would be in simplest form:

The fraction ____ in simplest form is ____.

English Language Development Leveled Activities

Emerging Level	Expanding Level	Bridging Level
Number Sense	**Number Game**	**Pair Work**
Cut a large paper circle into eight equal pieces and have eight students hold a piece. Say, *Eight pieces in the whole circle. Each piece is one-eighth of the whole.* Write $\frac{8}{8}$. Have three students sit down with their pieces. Say, *Three pieces are taken away.* Write $-\frac{3}{8}$ after the $\frac{8}{8}$. Ask, *How many pieces are left?* Have students answer **5** chorally. Model solving the subtraction problem on the board and then gesture to the remaining five students. Say, *Five pieces of the whole remain.* Write $=\frac{5}{8}$ on the board. Repeat with a new shape cut into 9 pieces.	Write a set of fractions on index cards. Each fraction should have a denominator of 9 or 10. Have students work in pairs. Ask each student to pick one index card and have pairs determine whether the fractions they've drawn are like fractions or not using the following sentence frame: ____ **and** ____ **are/are not like fractions.** Have students redraw until they have like fractions. Then have them model subtraction of the lesser fraction from the greater fraction.	Have a student read aloud a My Homework Problem Solving exercise. As he or she reads, list fractions from the problem and the words and phrases that signify subtraction. Discuss with students how to identify if the fractions are like fractions. Have students work in pairs to solve another My Homework exercise. Have Student A read aloud the word problem, while Student B lists the fractions and words or phrases that signify subtraction. The pair will then determine if the fractions are like fractions and subtract to solve the problem.

Teacher Notes:

T91 Grade 5 • Chapter 9 *Add and Subtract Fractions*

NAME _____ DATE _____

Lesson 3 Concept Web

Subtract Like Fractions

Use the concept web to identify examples of like fractions. Write *true* or *false*.

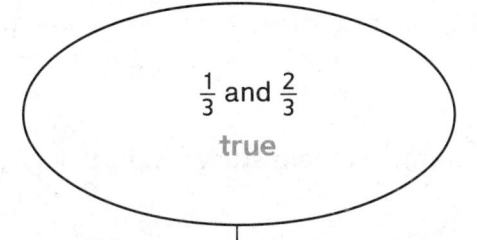

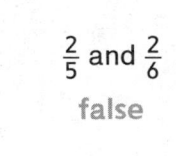

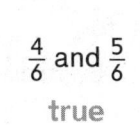

$\frac{1}{3}$ and $\frac{2}{3}$
true

$\frac{2}{5}$ and $\frac{2}{6}$
false

$\frac{4}{6}$ and $\frac{5}{6}$
true

These are like fractions. True or False?

$\frac{7}{8}$ and $\frac{5}{8}$
true

$\frac{1}{4}$ and $\frac{3}{4}$
true

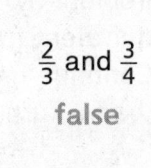

$\frac{2}{3}$ and $\frac{3}{4}$
false

Lesson 4 Inquiry/Hands On: Use Models to Add Unlike Fractions

English Learner Instructional Strategy

Vocabulary Support: Activate Prior Knowledge

Display anchor charts, word webs, KWL charts, or any other graphic organizers from previous lessons related to fractions. Be sure to include the classroom cognate chart as well. Have students take turns coming up to the graphic organizers and sharing a piece of information about fractions with the other students.

Display the following sentence frames to help students participate during the lesson:

____ tiles

____ plus ____

____ is equivalent to ____.

The families ate ____ strawberry pies altogether.

English Language Development Leveled Activities

Emerging Level	Expanding Level	Bridging Level
Choral Responses Write $\frac{3}{5}$ and $\frac{6}{10}$ on the board. Model the fractions with fraction tiles. Say each fraction aloud. Have students repeat chorally. Circle the denominators and say, *The denominators are unlike. These are unlike fractions.* Stress the prefix *un-*. Have students chorally repeat, **unlike fractions.** Arrange the fraction tiles to make it visually apparent that $\frac{3}{5}$ and $\frac{6}{10}$ are equivalent fractions. Say, *They are unlike, but they show the same amount. They are equivalent fractions.* Have students chorally repeat: **equivalent fractions.** Repeat with a new example.	**Act It Out** Divide students into pairs and distribute $\frac{1}{4}$, $\frac{1}{3}$, or $\frac{1}{12}$ fractions tiles to each pair. Say, *Model a fraction using your tiles. Write the fraction on a piece of paper.* Collect the pieces of paper and use them to create an addition problem of unlike fractions. Invite the pairs of students with the tiles corresponding to the problem to come forward. Have the four students work together to solve the problem by creating equivalent fractions with tiles. Continue until all pairs have had a turn.	**Basic Vocabulary** Display a list of sequence words, such as *first, next, then,* and so on. Distribute $\frac{1}{4}$, $\frac{1}{3}$, or $\frac{1}{12}$ fractions tiles to students. Say, *Model a fraction using your tiles.* Invite two students with unlike fractions to come forward with their models. Have the other students direct them in how to add the unlike fractions by creating equivalent fraction models. Ensure they use the sequence words in their descriptions. Invite another pair of students to come forward and repeat the activity.

Teacher Notes:

NAME _____ DATE _____

Lesson 4 Note Taking

Inquiry/Hands On: Use Models to Add Unlike Fractions

Read the question. Write words you need help with and research each word. Use your lesson to write your Cornell notes. Write or draw math examples to explain your thinking. Share your examples with a classmate.

Building on the Essential Question

How do you use models to add unlike fractions?

Words I need help with:

See students' words.

Notes:

The fraction modeled below is $\frac{2}{3}$. The denominator is __3__.

| $\frac{1}{3}$ | $\frac{1}{3}$ |

Fractions that have __different__ denominators are called unlike fractions.

The fraction $\frac{1}{6}$ and the fraction $\frac{2}{3}$ are __unlike__ fractions.

To add unlike fractions, the fractions must have a __common__ denominator.

__Two__ $\frac{1}{6}$-tiles will match the length of $\frac{1}{3}$.

__Four__ $\frac{1}{6}$-tiles will match the length of $\frac{2}{3}$.

| $\frac{1}{6}$ | $\frac{1}{6}$ | $\frac{1}{6}$ | $\frac{1}{6}$ | $\frac{1}{6}$ |

| $\frac{1}{3}$ | $\frac{1}{3}$ |

$\frac{1}{6} + \frac{4}{6} = \frac{5}{6}$ So, $\frac{1}{6} + \frac{2}{3} = \frac{5}{6}$

There are __5__ $\frac{1}{6}$-tiles altogether.

My Math Examples:

See students' examples.

Lesson 5 Add Unlike Fractions
English Learner Instructional Strategy

Graphic Support: Word Web

Display a word web and write *un-* in the center oval. Say, *Un- is a prefix that means "not." When it is added to the beginning of a word, it changes the word's meaning to the opposite.* Work with students to generate a list of words with the *un-* prefix, such as: *unkind, unwell, unhappy,* and so on. Record student responses on the word web.

Write *unlike,* and ask students to tell what it means based on their understanding of the prefix.

During the lesson, display sentence frames to help students participate: ____ and ____ are unlike fractions. The least common denominator is ____. ____ and ____ are equivalent fractions.

English Language Development Leveled Activities

Emerging Level	Expanding Level	Bridging Level
Phonemics	**Academic Vocabulary**	**Academic Language**
Distribute a piece of string or yarn to students. Write the word *tied*. Tie a loose knot around a pencil. Say, **tied**. Have students tie a loose knot around their own pencils and chorally say, **tied**. Add the prefix *un-* to *tied*. Say, *Un- means "not."* Untie the knot and say, **untied**. Have students untie their knots and chorally say, **untied**. Write *like* and *unlike*. Display *like* and *unlike* pairs of objects and have students chorally identify them as **like** or **unlike**. Repeat with fraction pairs, having students chorally identify them as **like** or **unlike**.	Write $\frac{1}{4}$ and $\frac{3}{4}$. Have students identify the fractions as like or unlike. Repeat with $\frac{2}{3}$ and $\frac{3}{4}$. Have a student volunteer explain how to determine if fractions are like or unlike. Say, *When you have unlike fractions, you can rename the fractions using the least common denominator.* Have students help you find a least common denominator of the two unlike fractions. 12 Once the fractions have been renamed ($\frac{8}{12}$ and $\frac{9}{12}$) using the least common denominator, have students identify the renamed fractions as *like fractions*.	Have students work in pairs. Give each pair a 10-part spinner numbered 1–10 to create fractions. Have them spin the spinner two times. The lowest number spun is the numerator and the largest number spun is the denominator for a fraction. Direct pairs to use the spinner to create two unlike fractions. Have one student create like fractions by determining a least common denominator and then add them. Have the other student verbally describe the steps to follow as the first student finds the sum. Have pairs switch roles and repeat the activity.

Teacher Notes:

NAME _____ DATE _____

Lesson 5 Vocabulary Cognates
Add Unlike Fractions

Use the Glossary to define the math word in English and in Spanish in the word boxes. Write a sentence using your math word.

like fractions	fracciones semejantes
Definition Fractions that have the same denominator.	**Definición** Fracciones que tienen el mismo denominador.

My math word sentence:
Sample answer: The fractions $\frac{3}{4}$ and $\frac{1}{4}$ are like fractions.

unlike fractions	fracciones no semejantes
Definition Fractions that have different denominators.	**Definición** Fracciones que tienen denominadores diferentes.

My math word sentence:
Sample answer: The fractions $\frac{3}{4}$ and $\frac{1}{3}$ are unlike fractions.

Lesson 6 Inquiry/Hands On: Use Models to Subtract Unlike Fractions

English Learner Instructional Strategy

Vocabulary Support: Anchor Chart

Divide students into four groups. Say, *Make an anchor chart showing what you know about adding unlike fractions.* Each chart should include a title at the top of the poster and definitions for math vocabulary related to adding unlike fractions, such as *denominator, equivalent fractions, least common denominator,* and so on. Suggest that students include a model showing how to create equivalent fractions in order to add unlike fractions. Direct students to label different elements in their charts with appropriate math vocabulary. When the charts are completed, have groups display and describe their charts. Afterward, discuss how the anchor charts can help students better understand the steps they will need to follow in order to subtract unlike fractions.

English Language Development Leveled Activities

Emerging Level	Expanding Level	Bridging Level
Sentence Frames Display the following sentence frames: **We need ____ tiles. The denominators are ____. ____ and ____ are unlike fractions. ____ and ____ are equivalent fractions.** As you work through the Build It and Try It parts of the lesson, have students use the sentence frames to repond to your questions. As needed, model the answers using the sentence frames and have students chorally repeat. Keep the sentence frames displayed throughout the lesson so students can refer to them as needed.	**Numbered Heads Together** Have students get into groups of four. Randomly assign one of the Practice It exercises to each group. Ask the students in each group to number off as 1–4. Have the students in each group work together to solve their assigned problem. Afterward, display the following sentence frames: **We used ____ tiles to represent ____. The difference is ____.** Choose numbers from 1–4 to designate which student in each group will use the sentence frames to describe their group's answer.	**Turn & Talk** Read aloud the Write About It exercise and say, *Turn to the student nearest you and discuss your answer.* Give students a chance to discuss their ideas. Then come together again as a group to continue the discussion. Have students spend a few minutes writing the answers in their math journals. Ask volunteers to read aloud what they wrote.

Teacher Notes:

Grade 5 • Chapter 9 *Add and Subtract Fractions*

NAME _____ DATE _____

Lesson 6 Guided Writing

Inquiry/Hands On: Use Models to Subtract Unlike Fractions

How do you subtract unlike fractions using models?

Use the exercises below to help you build on answering the Essential Question. Write the correct word or phrase on the lines provided.

1. Rewrite the question in your own words.
   See students' work.

2. What key words do you see in the question?
   subtract, unlike fractions, models

3. The fractions modeled right are $\frac{2}{3}$ and $\frac{1}{2}$.

4. The length of the fraction tile for $\frac{1}{2}$ is <u>less</u> than the fraction tiles for $\frac{2}{3}$.

5. The subtraction expression $\frac{2}{3} - \frac{1}{2}$ is modeled below.

6. You can find the <u>difference</u> of a modeled subtraction expression by finding which fraction tiles will <u>fill in</u> the area of the dashed box.

7. Which of the following tiles will fill in the dashed box: a $\frac{1}{2}$-tile, a $\frac{1}{3}$-tile, or a $\frac{1}{6}$-tile?
   $\frac{1}{6}$-tile

8. Find the difference. $\frac{2}{3} - \frac{1}{2} = \frac{1}{6}$

9. How do you subtract unlike fractions using models?
   Model each fraction, with the greater fraction (minuend) above the lesser fraction (subtrahend). Create a dashed box below the first fraction that shows how much longer the top fraction is than the bottom fraction. Find the fraction tiles that will fill the dashed box.

Lesson 7 Subtract Unlike Fractions

English Learner Instructional Strategy

Collaborative Support: Think-Pair-Share

Before beginning the lesson, pair emerging students with expanding or bridging students. As you work through the lesson and seek student responses, direct your questions or prompts to student pairs instead of individual students. Give pairs time to think about and discuss their response. Allow the more proficient English speaker to answer. Record his or her answer on the board. Model saying it again and have the class chorally repeat. Be sure to prompt a response from each pair at least once during the lesson.

English Language Development Leveled Activities

Emerging Level	Expanding Level	Bridging Level
Word Knowledge Write a fraction and ask, *What is the denominator in this fraction?* Have students chorally answer. Repeat with a second fraction that has a different denominator. Circle the denominator of each fraction and say, *The denominators are not the same number. They are unlike. These are unlike fractions.* Repeat the activity, but with a pair of like fractions. Continue with additional fractions pairs, and have students identify if each pair are like fractions or unlike fractions.	**Act It Out** Use fraction tiles to model the fractions $\frac{3}{6}$ and $\frac{2}{6}$. Have students identify these as like or unlike fractions. Then have students model the fractions $\frac{2}{3}$ and $\frac{1}{2}$. Ask a volunteer to identify the fractions as like or unlike fractions, and to explain how to determine the answer. Say, *When you have unlike fractions you can rename the fractions using the least common denominator.* Have students help you find a least common denominator of the two unlike fractions. 6 Model the renamed fractions using the least common denominator ($\frac{4}{6}$ and $\frac{3}{6}$).	**Academic Language** Give student pairs a 10-part spinner numbered 1–10 to use for creating fractions. Have them spin the spinner two times. The lowest number spun is the numerator and the largest number spun is the denominator. Direct pairs to create two unlike fractions. One student will rename the unlike fractions to like fractions using a common denominator and then find the difference between them. Have the other student verbally describe the steps to follow as the first student finds the difference. Have pairs switch roles and repeat the activity.

Teacher Notes:

T95 Grade 5 • Chapter 9 *Add and Subtract Fractions*

NAME _____ DATE _____

Lesson 7 Concept Web

Subtract Unlike Fractions

Use the concept web to identify the least common denominator used to subtract unlike fractions.

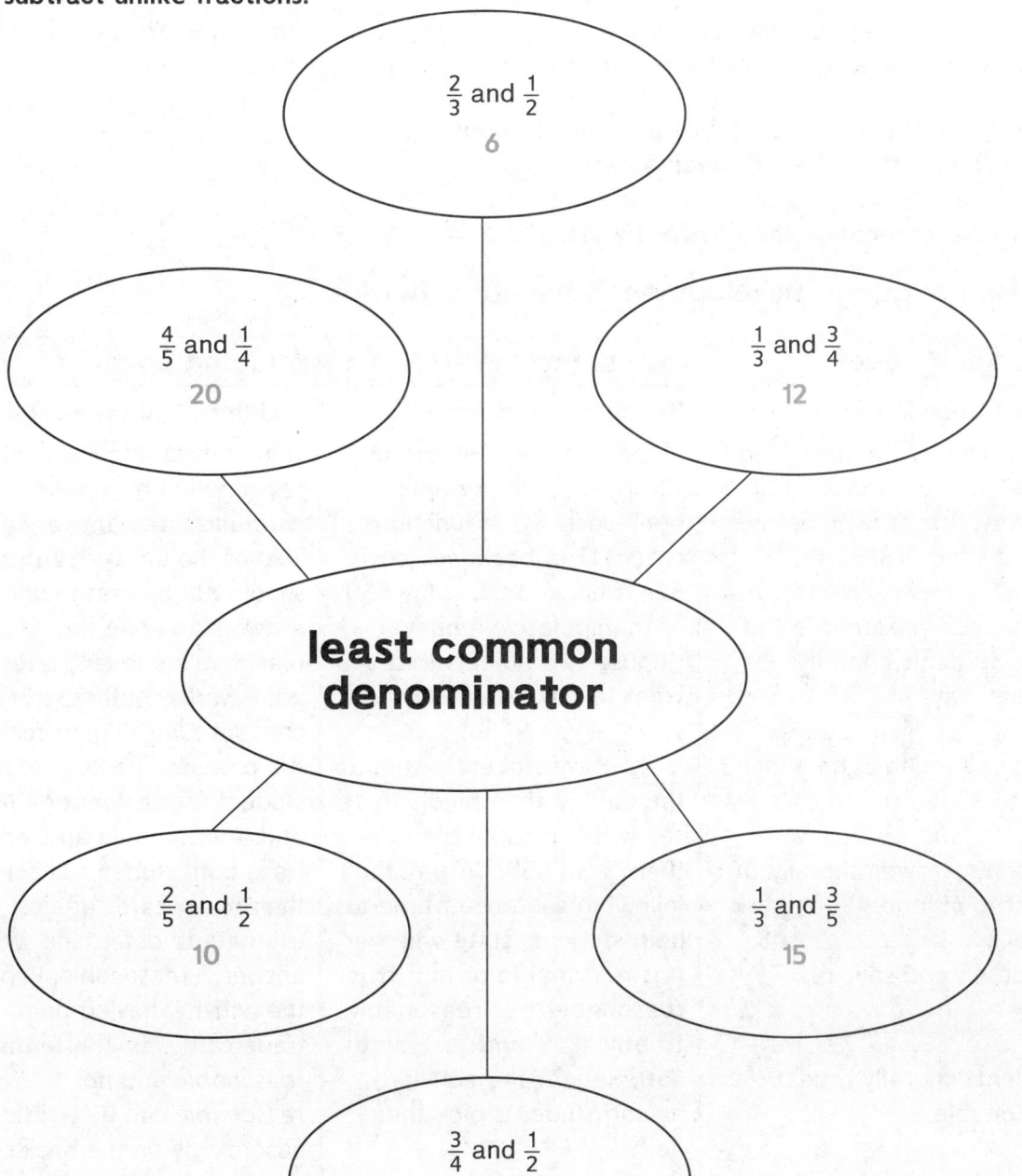

$\frac{2}{3}$ and $\frac{1}{2}$
6

$\frac{4}{5}$ and $\frac{1}{4}$
20

$\frac{1}{3}$ and $\frac{3}{4}$
12

least common denominator

$\frac{2}{5}$ and $\frac{1}{2}$
10

$\frac{1}{3}$ and $\frac{3}{5}$
15

$\frac{3}{4}$ and $\frac{1}{2}$
4

Grade 5 • Chapter 9 *Add and Subtract Fractions* **95**

Lesson 8 Problem-Solving Investigation Strategy: Determine Reasonable Answers

English Learner Instructional Strategy

Graphic Support: Four-Column Chart

On the board, draw a large four-column chart labeled: *Understand, Plan, Solve, Check*. Write the following sentence frames in the designated columns:

Understand: **We know _____. We need to find out _____.**
Plan: **We will to find a reasonable answer _____.**
Solve: **The answer is _____.**
Check: **We can check the answer by using _____.**

English Language Development Leveled Activities

Emerging Level	Expanding Level	Bridging Level
Word Knowledge	**Recognize and Act It**	**Building Oral Language**
Write the addition problem 27 + 18 = 65 and 27 + 18 = 45. Say, *My answer was 65. My friend's answer was 45. Which answer is reasonable?* Emphasize *reasonable* and have students chorally repeat. Say, *I will estimate to find out.* Have students help you round each addend. Write 30 + 20 and ask, *What is the sum?* Have students answer chorally. **50** Point to 65 and 45, and ask, *Which is closer to 50?* **45** Circle 45 and say, Yes, *45 is closer to the estimate. 45 is a reasonable answer.* Have students chorally repeat reasonable.	Place price tags on about 10 objects. For example, a book costs $11, a lunchbox costs $17, a backpack costs $24, and so on. Display $50 of manipulative money. Choose two objects that you want to buy. Say, *I have $50. I want to buy _____ and _____.* Have students estimate the cost of the objects to see if it is reasonable to buy them with $50. Display the following sentence frame to help students state whether it is reasonable or not: **It is reasonable/not reasonable to buy _____ and _____ with $50.** Repeat the activity having students pick the items.	Create card sets. Have each set contain an expression card and three answer cards. Two of the answer cards should not be reasonable answers and one should be a reasonable answer. Have pairs work together with a card set. One student reads the problem aloud. The other student uses estimation to find a reasonable answer. Have both students check the answer cards against the estimate to determine which answer is reasonable. Repeat the activity, having pairs trade card sets. The terms *reasonable* and *not reasonable* can be written accordingly on the back of the cards to make the activity self-checking.

Teacher Notes:

NAME _____ DATE _____

Lesson 8 Problem-Solving Investigation

STRATEGY: Determine Reasonable Answers

Solve each problem by determining a reasonable answer.

1. Use the **table** to determine whether **245** pounds, **260** pounds, or **263** pounds is the most **reasonable estimate** for how much **more** the ostrich weighs **than** the flamingo. Explain.

Bird	Weight (lb)
Flamingo	$9\frac{1}{10}$
Ostrich	$253\frac{1}{2}$

Understand	Solve
I know: I need to find:	

Plan	Check
1. Round the weight of a flamingo. 2. Round the weight of an ostrich. 3. Find the difference of the rounded weights.	My answer is reasonable because...

2. A grocer sells **12** pounds of apples. Of those, $5\frac{3}{4}$ pounds are **green** and $3\frac{1}{4}$ pounds are **golden**. The **rest** are **red**. Which is a more **reasonable** estimate for how many pounds of **red apples** the grocer **sold**? **3** pounds or **5** pounds? Explain.

Understand	Solve
I know: I need to find:	

Plan	Check
1. Round the weight of green apples sold. 2. Round the weight of golden apples sold. 3. Subtract the rounded weight of green apples sold. 4. Subtract the rounded weight of golden apples sold.	My answer is reasonable because...

96 Grade 5 • Chapter 9 *Add and Subtract Fractions*

Lesson 9 Estimate Sums and Differences

English Learner Instructional Strategy

Collaborative Support: Round the Table

Place students into multilingual groups of 4 or 5. Assign three Independent Practice problems to each group. Have one student write the first problem on a large piece of paper. Then have students work jointly to solve the problem by passing the paper around the table. Each student will perform one step in solving the equation. Direct each member of the group to write with a different color to ensure all students participate in solving the problem. Once the first problem is solved, have the next student in turn write the second problem, pass it to the next student to begin solving, and so on until all three problems are solved. Afterward, choose one student to present his or her group's solutions to the class.

English Language Development Leveled Activities

Emerging Level	Expanding Level	Bridging Level
Word Recognition	**Memory Devices**	**Recognize and Act It Out**
Write *some* on the board. Display a bottle of water and say, *Here is some water.* Place a handful of paper clips on a desk and say, *Here are some paper clips.* Repeat with other items. Write *sum* on the board. Return to the paper clips and divide them into two piles. Count each pile and write the corresponding addition problem. Have students assist you in solving the problem. Display a sentence frame for students to use to chorally say the answer: **The sum is ___.** Repeat with another group of objects, such as some pencils.	Display the following rhyme: *Sums total and show what you add. Sums will give you more than you had. With subtraction you end up having less. The minus sign shows you'll have a difference.* Model *plus* displaying crossed index fingers while you read aloud the first two lines. Have students chorally repeat with their fingers crossed. Model *minus* displaying a parallel index finger while you read aloud the last two lines. Have students chorally repeat as they model *plus* and *minus* with their fingers.	Organize students in groups ranging from 2 to 8 students in a group. Give each group one paper plate to cut into equal-sized pieces, one piece for each member. Have them label each piece as a fraction of the whole. For example, a group for three students will label each of their three pieces as $\frac{1}{3}$. Each student will hold a piece of the plate. Reorganize students into new groups of two unlike fractional parts for example thirds and eighths. Have them combine their fractional pieces, and then estimate the sum of the pieces. Remind students to estimate to the nearest whole number.

Teacher Notes:

NAME _____ DATE _____

Lesson 9 Vocabulary Chart
Estimate Sums and Differences

Use the three-column chart to organize the review vocabulary in this lesson. Write the word in Spanish. Then write the correct terms to complete each definition.

English	Spanish	Definition
estimate	estimación	A number close to an <u>exact</u> value. An <u>estimate</u> indicates about how much.
sum	suma	The answer to an <u>addition</u> problem.
fraction	fracción	A number that represents part of a <u>whole</u> or part of a <u>set</u>.
like fractions	fracciones semejantes	Fractions that have the <u>same</u> denominator.
unlike fractions	fracciones no semejantes	Fractions that have <u>different</u> denominators.

Lesson 10 Inquiry/Hands On: Use Models to Add Mixed Numbers

English Learner Instructional Strategy

Vocabulary Support: Modeled Talk

During the Model the Math part of the lesson, point to each part of the fraction to reinforce the connection between the written fraction and its model. For example, point to each $\frac{1}{8}$ piece as you say, *One-eighth and one-eighth is two-eighths.* Point to the 2 and then the 8 in the written fraction as you repeat, *Two-eighths.* Have students chorally repeat. Continue in this manner throughout Model the Math.

During the Draw It part of the lesson, display sentence frames to help students participate in the discussion: _____ **whole.** _____ **thirds.** _____ **and** _____ **thirds.** Have students create a word problem that could be represented by the Draw It model. Allow emerging students to create the word problem in their native language, and ask a bridging student volunteer to help translate.

English Language Development Leveled Activities

Emerging Level	Expanding Level	Bridging Level
Academic Vocabulary Write the following math vocabulary on the board: *mixed number, equivalent fraction,* and *whole fraction.* Say each word and have students chorally repeat. Model solving a simple mixed number addition problem using fraction circles. As you solve, pause to ask students to identify elements of the problem using the words on the board. For example, point to a mixed number and ask, *Is this a mixed number, equivalent fraction, or whole fraction?* Have students respond verbally or by pointing to the correct term on the board.	**Word Recognition** Display the following sentence frames: **Shade** _____ **parts. Write** _____ **on the board. The answer is** _____. Write $1\frac{1}{3} + 2\frac{2}{3}$ on the board. Draw empty 3-part fraction circles to help in solving, but do not shade them. Ask a volunteer to come to the board. Have the other students use the sentence frames to guide him or her in shading the circles and solving the mixed number addition problem. Write a new problem and draw new fraction circles. Invite another volunteer to come forward and repeat the activity.	**Listen and Write** Have students work in pairs. On the board, write $2\frac{5}{8} + 1\frac{5}{8}$. Distribute fraction circles to each pair to use in solving the problem. Have one student explain each step needed to shade the circles and find the solution as the other student follows the steps. Ask volunteers to share their answers. Then write a new problem and have the students in each pair switch roles.

Teacher Notes:

NAME _____ DATE _____

Lesson 10 Note Taking

Inquiry/Hands On: Use Models to Add Mixed Numbers

Read the question. Write words you need help with and research each word. Use your lesson to write your Cornell notes. Write or draw math examples to explain your thinking. Share your examples with a classmate.

Building on the Essential Question How do you use models to add mixed numbers?	**Notes:** The mixed number modeled below is $1\frac{3}{4}$. The mixed number modeled below is $1\frac{2}{4}$. To find the sum of two mixed numbers, <u>combine</u> the models of the fractions. A whole fraction circle is equal to <u>4</u> quarter fraction circles. Find the sum of $1\frac{3}{4} + 1\frac{2}{4}$. There are a total of <u>2</u> whole fraction circles. There are a total of <u>5</u> quarter fraction circles. $1\frac{3}{4} + 1\frac{2}{4} = 1 + \frac{1}{4} + \frac{1}{4} + \frac{1}{4} + 1 + \frac{1}{4} + \frac{1}{4} = \underline{2} + \frac{5}{4}$ $= \underline{2} + 1\frac{1}{\underline{4}} = \underline{3}\frac{1}{\underline{4}}$ $1\frac{3}{4} + 1\frac{2}{4} = \underline{3}\frac{1}{\underline{4}}$
Words I need help with: See students' words.	
My Math Examples: See students' examples.	

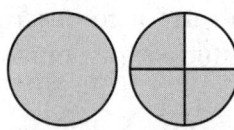

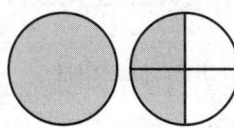

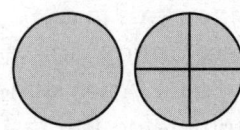

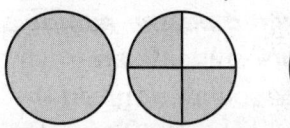

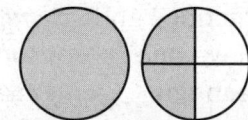

Lesson 11 Add Mixed Numbers
English Learner Instructional Strategy

Graphic Support: KWL Chart

Write the word *mixed numbers* and the Spanish cognate, *el número mixto,* on a classroom cognate chart. Write a mixed number on the board and use it as a concrete example of the phrase's meaning.

Display a KWL chart. In the first column, record what students already know about mixed numbers from previous lessons. In the second column, record what students hope to learn during the lesson, including how to add mixed numbers and estimate their sums. After the lesson, display the following sentence frame. Have students use it to describe what they learned: **I learned that** _____. Record student responses in the third column of the KWL chart.

English Language Development Leveled Activities

Emerging Level	Expanding Level	Bridging Level
Number Sense	**Recognize and Act It Out**	**Multiple Meanings**
Write whole numbers and fractions on pieces of paper. Have students hold the papers. Line up all students with whole numbers in one line and those with fractions in another line. Point to the line of whole numbers and say, *whole numbers.* Point to the line of fractions and say, *fractions.* Have one student from each line stand together in a pair, with the whole number to the left of the fraction. Say, *A whole number with a fraction. This is a mixed number.* Emphasize *mixed number* as you say it again and have students chorally repeat.	Have student pairs create number cards by writing whole numbers 1–12 on pieces of paper and then place the cards in a container or bag. Say, *Mix up the cards.* After students have completed the task, direct them to say, **We mixed up the numbers.** Be sure students are saying the /t/ sound in *mixed* to indicate past tense. Have students take turns drawing three cards to form a mixed number. For example, the first number drawn is the whole number, and the second and third numbers drawn create the fraction. Once the pair has created four mixed numbers, direct them to say, **We made four mixed numbers.**	Have students work in groups to create a two column chart labeled *Name* and *Rename.* Groups will select three objects and write the name of each object in the *Name* column. Students will then rename each object with a synonym and write it in the *Rename* column. For example, *box* can be renamed as *container.* Then have groups write the name of three fractions or mixed numbers in the *Name* column, and then write equivalent fractions or mixed numbers in the *Rename* column.

Teacher Notes:

NAME _____ DATE _____

Lesson 11 Vocabulary Definition Map
Add Mixed Numbers

Use the definition map to write a description and list characteristics about the vocabulary word or phrase. Write or draw math examples. Share your examples with a classmate.

My Math Vocabulary:

mixed number

Description from Glossary:

A number that has a whole number part and a fraction part.

Characteristics from Lesson:

In the mixed number $2\frac{1}{4}$ the whole number part is <u>2</u> and the fraction part is $\frac{1}{4}$.

You can write a mixed number as a sum of the <u>wholes</u> and the <u>fractions</u>.

A fraction with a <u>numerator</u> that is greater than the <u>denominator</u> can be written as a mixed number.

My Math Examples:
See students' examples.

Grade 5 • Chapter 9 *Add and Subtract Fractions* **99**

Lesson 12 Subtract Mixed Numbers

English Learner Instructional Strategy

Language Structure Support: Tiered Questions

During the lesson, be sure to ask questions according to students' level of English language proficiency. Ask emerging students simple questions that elicit one-word answers or gestures: *Is this a mixed number? Do we round to 2 or 3? Are we finding a sum or difference?*

For expanding level students, ask questions that elicit simple phrases or short sentences: *What does the fraction round to? What do we need to do next?*

For bridging level students, ask questions that require more complex answers: *How do you know? What do we do if? What steps do we need to take to solve the problem?*

English Language Development Leveled Activities

Emerging Level	Expanding Level	Bridging Level
Build Background Knowledge Show students images of single pieces of fruit, such as an orange and a banana. Say, *These are different kinds of fruit.* Then show students an image of a fruit salad. Say, *The different fruits are mixed together.* Show students a whole number and a fraction written on separate sheets of paper. Say, *These are different kinds of numbers: a whole number and a fraction.* Then show a mixed number written on paper. Say, *The numbers are mixed together. This is a mixed number.* Have students chorally say **mixed number**. Repeat with other examples.	**Recognize and Act It Out** Create pairs of mixed numbers by writing mixed numbers from problems in the lesson on pieces of paper. Distribute one pair of mixed numbers to pairs of students. Display a sentence frame for pairs to use to identify the elements of each mixed number: **The whole number is ____. The fraction is ____.** Have each pair model their mixed numbers using fraction tiles. If the fractions are unlike, have students rename them using a least common denominator and model the renamed mixed numbers.	**Academic Language** Create sets of mixed numbers by writing pairs of mixed numbers from problems in the lesson on pieces of paper. Distribute one pair of mixed numbers to each pair of students. Say, *Use subtraction to find the difference between your numbers. Record the steps you take to find the answer.* Display the following order words: *First, Next, Then, Last.* Have students refer to the order words as they record the steps needed to solve. Afterward, have each pair share their answer, along with the steps followed to solve it.

Teacher Notes:

NAME _____ DATE _____

Lesson 12 Four-Square Vocabulary
Subtract Mixed Numbers

Write the definition for each math word. Write what each word means in your own words. Draw or write examples that show each math word meaning. Then write your own sentences using the words.

Definition	My Own Words
A fraction in which the greatest common factor of the numerator and the denominator is 1.	See students' examples.

simplest form

My Examples	My Sentence
Sample answer: The fraction $\frac{5}{4}$ written in simplest form is $1\frac{1}{4}$.	Sample sentence: When the fractional part of a mixed number is not greater than 1.

Definition	My Own Words
Fractions that have the same value.	See students' examples.

equivalent fractions

My Examples	My Sentence
Sample answer: $\frac{3}{4} = \frac{6}{8}$	Sample sentence: After multiplying the numerator and denominator of $\frac{3}{4}$ by 3, you find equivalent fractions $\frac{3}{4}$ and $\frac{9}{12}$.

Lesson 13 Subtract with Renaming
English Learner Instructional Strategy

Collaborative Support: Partners Work/Pairs Check

Assign Exercises 2–7 in Independent Practice. Have students work in pairs. For the first problem, have one student coach the other in finding the estimate and the actual difference. For the second problem, have students switch roles. When pairs have finished the second problem, have them get together with another pair and check answers.

Provide the following sentence frames:

What is your estimate for _____? Our estimate is _____.

What is your exact answer for _____? Our answer is _____.

Afterward, ask them to shake hands and continue working in their original pairs for the next two problems, exchanging roles as before.

English Language Development Leveled Activities

Emerging Level	Expanding Level	Bridging Level
Word Knowledge Display a bag and ask, *What is this called?* Guide students to provide a list of possible names for the object, such as: *bag, sack,* and *container.* Include any words from their native languages on the list as well. Say, *This is a bag, but I can rename it as _____.* Write a fraction on the board, such as $\frac{1}{2}$. Say, *This fraction is one half. I can* **rename** *the fraction.* Write $\frac{3}{6}$. Say, *I renamed the fraction. I used an* **equivalent fraction.** Work with students to help rename the fraction in several different ways. Repeat with other fractions.	**Recognize and Act It Out** Write $2\frac{1}{3}$ and model it using fraction tiles. Say, *We will rename this mixed number.* Display the fraction tile representing one whole and say, *I will regroup one whole as three thirds.* Perform the task, Write $2\frac{1}{3} = 1\frac{4}{3}$, and then gesture to the model of $1\frac{4}{3}$. Say, *I have renamed the original mixed number to an equivalent mixed number.* Provide a mixed number and have students use fraction tiles to find an equivalent mixed number. Display a sentence frame for students to use in explaining the result: **I renamed the mixed number _____ as _____.**	**Academic Language** Have groups of students roll a number cube six times, record the numbers, and then use them to create two mixed numbers. One student from the group will identify which mixed number is greater. Have students work together to subtract the lesser number from the greater number, renaming if needed. Have one student in each group list the steps followed to find the difference. Repeat until each student has had a turn listing the steps. If necessary, encourage groups to use fraction tiles to help them with renaming and subtraction.

Teacher Notes:

NAME _____ DATE _____

Lesson 13 Guided Writing

Subtract with Renaming

How do you subtract mixed numbers with renaming?

Use the exercises below to help you build on answering the Essential Question. Write the correct word or phrase on the lines provided.

1. Rewrite the question in your own words.
   <u>See students' work.</u>

2. What key words do you see in the question?
   <u>subtract, renaming, mixed numbers</u>

3. The fractional part of the mixed number $2\frac{2}{5}$ is $\frac{2}{5}$ and the fractional part of $1\frac{1}{2}$ is $\frac{1}{2}$. The fractions $\frac{2}{5}$ and $\frac{1}{2}$ are <u>unlike</u> fractions.

4. Before subtracting mixed numbers, find an equivalent fraction so the fractional parts have the <u>same</u> denominator.

5. The least common denominator for $\frac{2}{5}$ and $\frac{1}{2}$ is <u>10</u>.
   $2\frac{2}{5} = 2\frac{4}{10}$ and $1\frac{1}{2} = 1\frac{5}{10}$

6. The mixed number $2\frac{4}{10}$ is <u>greater</u> than the mixed number $1\frac{5}{10}$. The fraction $\frac{4}{10}$ is <u>less</u> than $\frac{5}{10}$.

7. Since the fractional part of the minuend ($2\frac{4}{10}$) is <u>less</u> than the fractional part of the subtrahend ($1\frac{5}{10}$), rename one whole into a fraction with a denominator of 10.

8. $2\frac{4}{10} = 1 + 1 + \frac{4}{10} = 1 + \frac{10}{10} + \frac{4}{10} = 1 + \frac{14}{10} = 1\frac{14}{10}$

9. Now, you can subtract the mixed numbers: $2\frac{2}{5}$ and $1\frac{1}{2} = 2\frac{4}{10} - 1\frac{5}{10} = 1\frac{14}{10} - 1\frac{5}{10} = \frac{9}{10}$

10. How do you subtract mixed numbers with renaming?
    <u>Sample answer: Rewrite the fractional parts of the mixed numbers so they have the same denominator. If the fractional part of the minuend is less than the fractional part of the subtrahend, rewrite the mixed number so one of the wholes is written as an improper fraction. Subtract the fractional parts and subtract the wholes to find the difference.</u>

Chapter 10 Multiply and Divide Fractions

What's the Math in This Chapter?

Mathematical Practice 5: Use appropriate tools strategically

Prior to the lesson, cut four paper plates into fourths. Say, *We are going to use these models to show how to divide fractions.* Have four volunteers come to the front of the room and distribute 1 paper plate to each student. On the board write $4 \div \frac{1}{4}$. Explain that you have 4 students up front to represent the 4 in the division problem.

Say, *To divide by one-fourth, we need to know how many fourths are in 4. How can we do that?* **Count each student's pieces.** As a group count how many fourths the students are holding up front. **16** On the board complete the division problem, $4 \div \frac{1}{4} = 16$.

Ask students, *What strategies did we use to help us solve the division problem?* The goal of the discussion is to have students understand that using a model is one strategy they can use to help solve division problems. Dividing fractions can be confusing. Seeing concrete models is helpful for students to understand the process.

Display a chart with Mathematical Practice 5. Restate Mathematical Practice 5 and have students assist in rewriting it as an "I can" statement, for example: **I can use models to better understand fractions.** Post the new "I can" statement.

Inquiry of the Essential Question:

What strategies can be used to multiply and divide fractions?

Inquiry Activity Target: **Students come to a conclusion that using concrete models can help when multiplying and dividing fractions.**

As an introduction to the chapter, present the Essential Question to students. The inquiry graphic organizer will offer opportunities for students to observe, make inferences, and apply prior knowledge of concrete models representing the Essential Question. As they investigate, encourage students to draw, write, and collaborate with peers to demonstrate their observations and thinking. Then have students present additional questions they may have to a peer to extend discussions.

Regroup students and restate Mathematical Practice 5 and the Essential Question. Pose questions to reflect on what has been learned to guide students in making connections between the Mathematical Practice and the Essential Question.

NAME _____ DATE _____

Chapter 10 Multiply and Divide Fractions

Inquiry of the Essential Question:

What strategies can be used to multiply and divide fractions?

Read the Essential Question. Describe your observations (I see..), inferences (I think...), and prior knowledge (I know...) of each math example. Write additional questions you have below. Then share your ideas and questions with a classmate.

Use models to find products like $\frac{2}{3} \times 21$.

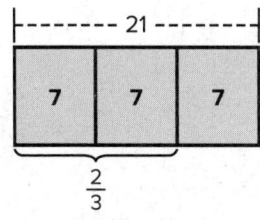

|―――― 21 ――――|
| 7 | 7 | 7 |
  $\underbrace{}_{\frac{2}{3}}$

$\frac{2}{3} \times 21 = 7 + 7 = 14$

I see…

I think…

I know…

Use models to find products like $\frac{2}{3} \times \frac{1}{5}$.

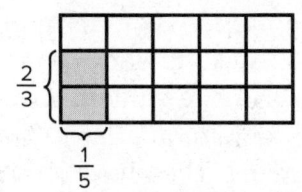

$\frac{2}{3}\{$ [grid with 2 of 15 sections shaded]
  $\underbrace{}_{\frac{1}{5}}$

2 out of 15 sections are shaded. So, $\frac{2}{3} \times \frac{1}{5} = \frac{2}{15}$.

I see…

I think…

I know…

Use models to find quotients like $5 \div \frac{1}{2}$.

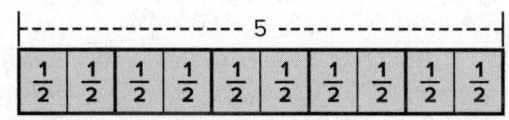

|―――――――― 5 ――――――――|
| $\frac{1}{2}$ | $\frac{1}{2}$ | $\frac{1}{2}$ | $\frac{1}{2}$ | $\frac{1}{2}$ | $\frac{1}{2}$ | $\frac{1}{2}$ | $\frac{1}{2}$ | $\frac{1}{2}$ | $\frac{1}{2}$ |

There are 10 halves in the model.
So, $5 \div \frac{1}{2} = 10$.

I see…

I think…

I know…

Questions I have…

_____
_____
_____

Lesson 1 Inquiry/Hands On: Part of a Number

English Learner Instructional Strategy

Vocabulary Support: Activate Prior Knowledge

Throughout the lesson, display and fill-in a KWL chart. Start by completing the first column during a review of what students have learned previously about multiplication, including math vocabulary such as *factor, product, multiply,* and how they have used models to find products. Fill in the column with what students already know. Ask students what they expect to learn about multiplying fractions during the lesson, including how they think it will differ from multiplying whole numbers. Record student responses in the second column. After the lesson, display sentence frames to help students describe what they learned about multiplying fractions and records their responses in the chart's third column: **I learned that _____.**

English Language Development Leveled Activities

Emerging Level	Expanding Level	Bridging Level
Report Back	**Word Lists**	**Round the Table**
Display the following sentence frames: **The denominator is _____. The numerator is _____. _____ parts.** Model solving $18 \times \frac{2}{3}$ using a bar diagram. As you work through each step of the problem, prompt student responses that can be answered using the sentence frames. For example, *What is the denominator? What is the numerator? How many parts do we add together?* If necessary, model the answer and have students report back chorally. Repeat with a new problem.	Have students brainstorm words related to multiplication and fractions, such as *denominator, numerator, product, factor, equivalent fraction, multiplication sign,* and so on. List their responses on the board. Write a fraction multiplication problem on the board, and then have students guide you in solving it using a bar diagram. Challenge students to use as many of the words from the list as possible. Cross off each word as it is used. If any words remain after the problem is solved, model using the word in a sentence.	Divide students into small groups. Assign an Apply It exercise to each group. Have one student write the problem on a large piece of paper. Then have students work jointly to solve the problem by passing the paper around the table. Each student will perform one step in solving. Direct each member of the group to write with a different color to ensure all students participate in solving the problem. Afterward, choose one student to present his or her group's solution to the class.

Teacher Notes:

T103 Grade 5 • Chapter 10 *Multiply and Divide Fractions*

NAME _____ DATE _____

Lesson 1 Note Taking

Inquiry/Hands On: Part of a Number

Read the question. Write words you need help with and research each word. Use your lesson to write your Cornell notes. Write or draw math examples to explain your thinking. Share your examples with a classmate.

Building on the Essential Question How can you use parts of a number to multiply and divide?	**Notes:** ⊢------------ 12 ------------⊣ [| | |] The bar diagram represents the number __12__. The bar diagram is divided into __4__ equal sections. Each section of the bar represents $\frac{1}{4}$ of the whole. $12 \div 4$ is the same as $\frac{1}{4}$ of 12 $12 \div 4$ is the same as $\frac{1}{4} \times 12$ $\frac{1}{4}$ of 12 is the same as $\frac{1}{4} \times 12$
Words I need help with: See students' words.	Each section of the bar represents the number __3__. $12 \div 4 =$ __3__ $\frac{1}{4} \times 12 =$ __3__ $\frac{1}{4}$ of $12 =$ __3__

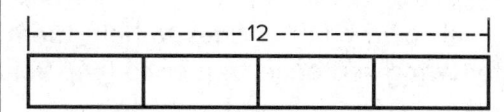

My Math Examples:

See students' examples.

Grade 5 • Chapter 10 *Multiply and Divide Fractions*

Lesson 2 Estimate Products of Fractions

English Learner Instructional Strategy

Collaborate Support: Partner Work

Write *number line* and the Spanish cognate, *línea numérica*, on a classroom cognate chart. Provide a concrete example by drawing a number line that can be used during the lesson to model estimating.

During the Talk Math part of the lesson, have students work in pairs. Direct them to draw their own number lines to help them answer the prompt. Display the following sentence frames to help volunteers share their responses: **Round _____ to _____ and _____ to _____. _____ multiplied by _____ is _____. The answer is about _____.**

English Language Development Leveled Activities

Emerging Level	Expanding Level	Bridging Level
Word Knowledge Write 12 on the board. Have twelve students, each holding a full sheet of blank paper, stand side-by-side in a line. Be sure students are holding the papers linked together end-to-end. Say, *This is a bar diagram. It has 12 equal pieces.* Next to 12 on the board, write: $\times \frac{1}{2}$. Model the expression using the student bar diagram. Hold a yardstick vertically to divide the bar diagram into six students/papers on either side. Say, *We multiplied the bar diagram by a half.* Next to $12 \times \frac{1}{2}$, write: $= 6$. Say, *12 multiplied by $\frac{1}{2}$ is 6.* Repeat with a similar problem.	**Recognize and Act It Out** Use tape to create a large number line from 0 to 1 on the floor, with 0, $\frac{1}{2}$, and 1 marked and labeled. Write fractions between 0 and 1 on index cards. Distribute cards to 5 students and have them stand on the number line to model their fraction. Once students are in position, ask them to round their fraction to 0, $\frac{1}{2}$, or 1 by moving to the appropriate mark, lining up perpendicularly to the number line as necessary. Display a sentence frame for each student to use: **My fraction is _____. It rounds to _____.** Repeat so all students get a turn.	**Building Oral Language** Have students get into pairs. Assign an Independent Practice exercise to each pair. Say, *Solve your problem using either a bar diagram or a number line.* Afterward, display the following sentence frames and have students use them to share their answers and explain how they arrived at their solution: **We estimated that _____ times _____ is _____. We used a number line/bar diagram to find the answer because _____.**

Teacher Notes:

NAME _____ DATE _____

Lesson 2 Four-Square Vocabulary
Estimate Products of Fractions

Complete the four-square chart to review the multiple meaning word or phrase.

Everyday Use Sample answer: to pass and go around something; as in rounding the corner to drive left down the street.	**Math Use in a Sentence** Sample sentence: Rounding fractions can help you estimate products.
Math Use To find the approximate value of a number.	**Example From This Lesson** Sample answer: To estimate $\frac{1}{3} \times 17$, round 17 to 18. $\frac{1}{3} \times 18 = 6$.

rounding

Write the correct terms on the blank lines to complete the sentence.

You can ___estimate___ products of fractions using rounding and ___compatible___ numbers.

Lesson 3 Inquiry/Hands On: Model Fraction Multiplication

English Learner Instructional Strategy

Collaborative Support: Partner Reading

Have students work in pairs during Practice It lesson. Have partners take turns reading aloud information from the problems. Tell students to listen closely to what their partners say and politely suggest corrections for any mistakes in pronunciation or usage as necessary.

Display the following sentence frames to help students share and discuss their answers: **Shade** ____ **sections.** ____ **times** ____ **is** ____. **The model shows** ____. **We can add** ____ **to find the answer.**

For students who need additional help, display the following sentence frames to help them communicate: **I don't understand** ____. **What does** ____ **mean? I need help with** ____.

English Language Development Leveled Activities

Emerging Level	Expanding Level	Bridging Level
Choral Responses Display the following sentence frame: ____ **is equal to** ____. Model solving $\frac{1}{4} \times 5$ using shaded fraction models and repeated addition. Ask, *What is $\frac{5}{4}$ equal to?* Have students respond chorally using the sentence frame: $\frac{5}{4}$ *is equal to* $1\frac{1}{4}$. Then ask, *What is $\frac{1}{4}$ times 5?* Have students answer chorally again using the sentence frame. Listen for the /z/ sound in *is* and the /kw/ sound in *equal,* which are non-transferrable sounds for native Spanish speakers. Model correct pronunciation as needed. Repeat the activity with a new problem.	**Listen and Write** Have students work in pairs. Say, *I am going to tell you a fraction multiplication problem. Write it down, and then solve it using a model.* Say, $\frac{3}{8}$ *times 5.* Give pairs time to solve using shaded fraction models. Ask one pair to share their answer. Provide a sentence frame if necessary: ____ **times** ____ **equals** ____. Repeat with a new problem.	**Basic Vocabulary** Have students work in pairs. Distribute index cards to each pair and display a list of sequence words on the board, such as *first, second, then, next,* and *last.* Say, *Use the sequence words to describe how you would use shaded fraction models to solve this multiplication problem.* Write $\frac{2}{3} \times 5$ on the board. Have pairs write their descriptions on the index cards. Then have pairs exchange cards and follow the steps they receive to solve the problem. If any pair has the wrong answer, discuss which steps were missing or incorrect in the description they received.

Teacher Notes:

NAME _____ DATE _____

Lesson 3 Guided Writing

Inquiry/Hands On: Model Fraction Multiplication

How do you model fraction multiplication?

Use the exercises below to help you build on answering the Essential Question. Write the correct word or phrase on the lines provided.

1. Rewrite the question in your own words.
   See students' work.

2. What key words do you see in the question?
   model, fraction, multiplication

3. Multiplication can be represented as __repeated__ addition.

4. $\underline{3} \times \frac{1}{4} = \frac{1}{4} + \frac{1}{4} + \frac{1}{4}$

5. To model the fraction $\frac{1}{4}$, the model will have __4__ equal sections with __1__ section(s) shaded.

6. The following repeated addition expression is modeled below: $\frac{1}{4} + \frac{1}{4} + \frac{1}{4}$

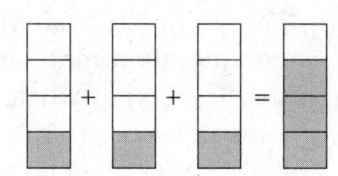

7. There are __3__ total sections shaded in the repeated addition expression. The sum will have __3__ sections shaded. The numerator in the sum will be __3__. The denominator will be the same, which is __4__.

8. $\frac{1}{4} + \frac{1}{4} + \frac{1}{4} = \frac{3}{4}$

9. $3 \times \frac{1}{4} = \frac{3}{4}$

10. How do you model fraction multiplication?
    Sample answer: Identify the repeated addition expression that represents the multiplication. Model the fractions in the repeated addition expression. Identify the number of shaded sections in the repeated addition and that will be the numerator in the product. The denominator remains the same.

Grade 5 • Chapter 10 *Multiply and Divide Fractions* **105**

Lesson 4 Multiply Whole Numbers and Fractions

English Learner Instructional Strategy

Language Structure Support: Word Web

Write the word *simplify* and the Spanish cognate, *simplificar*, on a classroom cognate chart. Underline *simpl* in the word and ask, *What word does this make you think of?* Prompt students to answer, **simple.** Say, *Simplify means "to make something simpler."* Display a word web with *simple* written in the center. Work with students to fill in the surrounding ovals with words that use *simple* as a base, such as: *simpler, simplest, simplify, simplifying, simplified*. Discuss the meanings of each word, focusing on their math meanings. Display the following sentence frames for students to use during the lesson: **I can simplify the fraction ___ to ___. I simplified ___ to ___. I am simplifying ___ to ___. The fraction ___ is in simplest form.**

English Language Development Leveled Activities

Emerging Level	Expanding Level	Bridging Level
Number Sense Write $\frac{1}{2}$ and $\frac{3}{6}$ and say, *These fractions are equivalent.* Use fraction tiles to model the equivalency. Write 2 and ask, *What fraction is equivalent to two?* Display a fraction tile that represents one whole. Say, *This is one whole.* Write $\frac{1}{1}$. Then display two fraction tiles representing one whole each. Say, *Two wholes make the number two.* Write $\frac{2}{1}$. Say, *The numerator is two. The denominator is one.* Model writing other whole numbers as fractions. Check comprehension by asking, *Is this a whole number or fraction?* Have students answer accordingly.	**Word Recognition** Have a volunteer stand on one side of the room, and then guide the student across the room by providing complicated, roundabout directions. Afterward ask, *Is there a simpler way to get across the room?* Guide the student back across the room via a simple, direct route. Say, *I simplified the directions.* Stress *simplified* and have students chorally repeat. Write $\frac{13}{14}$ and model the fraction using only $\frac{1}{4}$ tiles. Then model simplifying the fraction to $3\frac{1}{4}$. Say, *I simplified the fraction $\frac{13}{14}$.* Provide another fraction for pairs to simplify using tiles.	**Academic Language** Create whole number cards and corresponding equivalent fraction cards, enough for one per student. Distribute cards to students. Direct them to find a peer with the equivalent match to their fraction or whole number. Each pair should explain their match using the sentence frame: **The whole number ___ can be written as the fraction ___.** Then have pairs multiply their whole number by a fraction you name, such as $\frac{2}{3}$, and simplify as needed. Have each pair announce the product using the sentence frame: **The product of ___ and ___ is ___.**

Teacher Notes:

NAME _____ DATE _____

Lesson 4 Vocabulary Cognates
Multiply Whole Numbers and Fractions

Use the Glossary to define the math word in English and in Spanish in the word boxes. Write a sentence using your math word.

Commutative Property	**propiedad conmutativa**
Definition Property that states that the order in which numbers are added does not change the sum and that the order in which factors are multiplied does not change the product.	Definición Propiedad que establece que el orden en que se suman los números no altera la suma y que el orden en que se multiplicand los factores no altera el producto.

My math word sentence:

Sample answer: $\frac{2}{3} \times 4 = 4 \times \frac{2}{3}$

unknown	**incógnita**
Definition A missing value in a number sentence or equation.	Definición Valor que falta en una oración numérica o una ecuación.

My math word sentence:
Sample answer: The unknown "□" in the equation $5 \times \frac{1}{4} = \square$ is $1\frac{1}{4}$.

106 Grade 5 • Chapter 10 *Multiply and Divide Fractions*

Lesson 5 Inquiry/Hands On: Use Models to Multiply Fractions

English Learner Instructional Strategy

Vocabulary Support: Modeled Talk

As you work through the Draw It part of the lesson, run your finger along a row or a column to emphasize the difference in meaning between *row* and *column* when you use the words. Then point to one section of the model and say, *This is one section.* Display this sentence frame: ____ section(s). Ask, *How many sections altogether?* Gesture to indicate the entire rectangle, and count the sections aloud if necessary. Have students answer chorally: **12 sections.** After shading one square to model $\frac{1}{3} \times \frac{1}{4}$, point to the one square and ask, *How many sections are shaded?* Have students answer chorally: **One section.** As you work through the remainder of the lesson, use similar gestures to clarify and reinforce math and nonmath vocabulary relevant to the lesson.

English Language Development Leveled Activities

Emerging Level	Expanding Level	Bridging Level
Exploring Language Structure Review the *-er* and *-est* endings used to form comparatives and superlatives with common examples such as *fast/faster/fastest* or *small/smaller/smallest.* Use the words in sentences to demonstrate and compare their meanings. Model solving Practice It Exercise 6. Write the answer as $\frac{3}{24}$, and then ask, *Is this the simplest form?* Give students a chance to answer. Model simplifying the fraction to $\frac{1}{8}$, and say, *One-eighth is the simplest form.* Have students practice simplifying other fractions using a sentence frame: ____ is the simplest form.	**Look, Listen, and Identify** On the board, write out Apply It Exercise 13. Read the problem aloud. Ask, *What do we need to find out?* Invite a volunteer to underline the relevant sentence in the problem. Ask, *What fraction of the day did Tom spend preparing for the party, including decorating the cake?* Have another volunteer come forward to circle the answer. Continue in this manner as you work with students to solve the problem. Repeat the activity with Apply It Exercise 12.	**Partners Work** Have students work with a partner to write a real-world math problem using Practice It Exercise 11. Allow students time to complete the task, and then have partners exchange word problems with another pair of students. Direct partners to read the math problem they received and correct any mistakes in spelling or grammar. Then have them solve the problem and write a complete sentence describing the answer. Ask volunteers to read their answers aloud.

Teacher Notes:

NAME _____ DATE _____

Lesson 5 Note Taking

Inquiry/Hands On: Use Models to Multiply Fractions

Read the question. Write words you need help with and research each word. Use your lesson to write your Cornell notes. Write or draw math examples to explain your thinking. Share your examples with a classmate.

Building on the Essential Question How do you use models to multiply fractions?	**Notes:** The fraction modeled below is $\frac{1}{2}$. The fraction modeled below is $\frac{3}{4}$. To model the multiplication of these two fractions, divide a square into __2__ equal rows since the denominator of the first fraction is 2. Divide the square into __4__ equal columns since the denominator of the second fraction is 4.
Words I need help with: See students' words.	The multiplication expression below is $\frac{1}{2} \times \frac{3}{4}$. The shaded portion is where $\frac{1}{2}$ and $\frac{3}{4}$ **intersect**. There are __3__ sections **shaded** in the multiplication model. This represents the ___numerator___ in the product. There are __8__ **total** sections in the multiplication model. This represents the ___denominator___ in the product. $\frac{1}{2} \times \frac{3}{4} = \frac{3}{8}$
My Math Examples: See students' examples.	

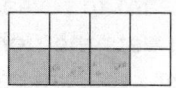

Lesson 6 Multiply Fractions
English Learner Instructional Strategy

Collaborative Support: Think-Pair-Share

Assign Independent Practice Exercise 5–10 to pairs of students. Direct students to first solve the exercises independently by either multiplying the fractions first and then simplifying or simplifying the fractions first then multiplying. Have partners then pair up to share their solutions and discuss which way of solving they prefer.

English Language Development Leveled Activities

Emerging Level	Expanding Level	Bridging Level
Word Knowledge Write *simplify, simplifying,* and *simplified*. Then write $\frac{9}{12}$ and say, *I am going to simplify this fraction.* Point to *simplify*, say it again, and have students chorally repeat. As you simplify $\frac{9}{12}$ say, *I am simplifying the fraction.* Point to *simplifying,* say it again, and have students chorally repeat. After you write the solution $\frac{3}{4}$, say, *I simplified the fraction.* Point to *simplified,* say it again, and have students chorally repeat. Display another fraction that can be simplified, and repeat the activity.	**Math Game** Use a deck of cards with face cards removed. Divide students into two teams. Assign one student to be a referee. One team draws two pairs of cards and arranges them into two fractions. The first card of each pair is the numerator. The second card is the denominator. Have both teams multiply the fractions. If the first team to reveal the solution gets the right answer, they keep the cards. If their answer is wrong, the second team gets the cards only if they have the right answer. If both teams are wrong, the cards go to the bottom of the pile. Teams exchange roles until all cards are gone. The team with the most cards wins.	**Building Oral Language** Divide students into pairs and distribute five-part spinners with the numbers 1, 2, 3, 4, and 5 to each pair. Say, *You will be using the spinner to create two fractions. You will multiply the fractions to find the product.* Explain that the first number spun will be the numerator, and the second number spun will be the denominator. Student A will spin the spinner. Student B will write the fractions on a write-on/wipe-off board. Next, have students work together to find the product, simplifying if necessary. Repeat, asking student pairs to switch roles.

Multicultural Teacher Tip

Mathematical notation varies, so you may find EL students using unfamiliar symbols in place of standard US symbols. For example, students from Latin American countries may use a point in place of × to indicate multiplication. Although the point is also commonly used in the US, the placement and size may vary depending on the student's native culture. In Mexico, the point is larger and set higher between the numbers than in the US. In some Latin American countries, the point is set low and can be confused with a decimal point.

NAME _____ DATE _____

Lesson 6 Vocabulary Chart
Multiply Fractions

Use the three-column chart to organize the review vocabulary in this lesson. Write the word in Spanish. Then write the correct terms to complete each definition.

English	Spanish	Definition
multiplication	multiplicación	An operation on two numbers to find their _product_. It can be thought of as repeated _addition_.
denominator	denominador	The _bottom_ number in a fraction. It represents the number of _parts_ in the whole.
numerator	numerador	The _top_ number in a fraction; the part of the fraction that tells the number of _parts_ you have.
fraction	fracción	A number that represents part of a _whole_ or part of a _set_.
simplest form	minima expresión	A _fraction_ in which the greatest common factor of the numerator and the denominator is _1_.

108 Grade 5 • Chapter 10 *Multiply and Divide Fractions*

Lesson 7 Multiply Mixed Numbers
English Learner Instructional Strategy

Graphic Support: Charts

To help students during the Talk Math part of the lesson, create a 3-column chart labeled *Before*, *During*, and *After*. Model solving a mixed number multiplication problem from the lesson. During each step of the solving process, use the chart to record the steps needed to solve. Afterward, write the following sentence frames for students to use as they work in pairs completing Independent Practice exercises:

Before multiplying, we write the mixed numbers as ____.

During solving, we multiply the ____ and the ____.

After multiplying, we ____ the improper fraction.

English Language Development Leveled Activities

Emerging Level	Expanding Level	Bridging Level
Number Sense Write a mixed number. Say, *A whole number with a fraction is a **mixed number**.* Have students chorally say, **mixed number**. Have students use write-on/wipe-off boards to write examples of mixed numbers. Write an improper fraction. Say, *The numerator is greater than the denominator. This is an **improper fraction**. We can rewrite an improper fraction as a mixed number.* Model finding an equivalent mixed number. Have pairs take turns writing mixed numbers and improper fractions and having their partners identify the type of number.	**Recognize and Act It Out** Write $2\frac{5}{6}$. Ask students to identify the whole number and the fraction. Use fraction tiles, including two whole tiles, to model $2\frac{5}{6}$. Say, *We will find an equivalent fraction.* Model exchanging each whole fraction tile for the equivalent $\frac{6}{6}$ so you have seventeen $\frac{1}{6}$ tiles. Write $= \frac{17}{6}$ next to $2\frac{5}{6}$. Say, *The mixed number and the improper fraction are equivalent.* Provide mixed numbers for students to convert to improper fractions using fractions tiles to model their work. Display a sentence frame for students to use: ____ **is equivalent to** ____.	**Number Game** Divide students into two teams. Distribute write-on/wipe-off boards to each student. Say, *I am going to solve mixed number multiplication problems. Sometimes I will make a mistake. Sometimes I will not.* Challenge students to try and catch your mistakes. Then model solving mixed number multiplication problems. For most of the problems, make a mistake in the solving process. The first team to identify your mistake or to confirm that your solving was correct will score a point. Continue until one team has scored five points.

Teacher Notes:

NAME _____ DATE _____

Lesson 7 Concept Web
Multiply Mixed Numbers

Use the concept web to identify whether mixed numbers are in simplest form. Write *true* or *false*.

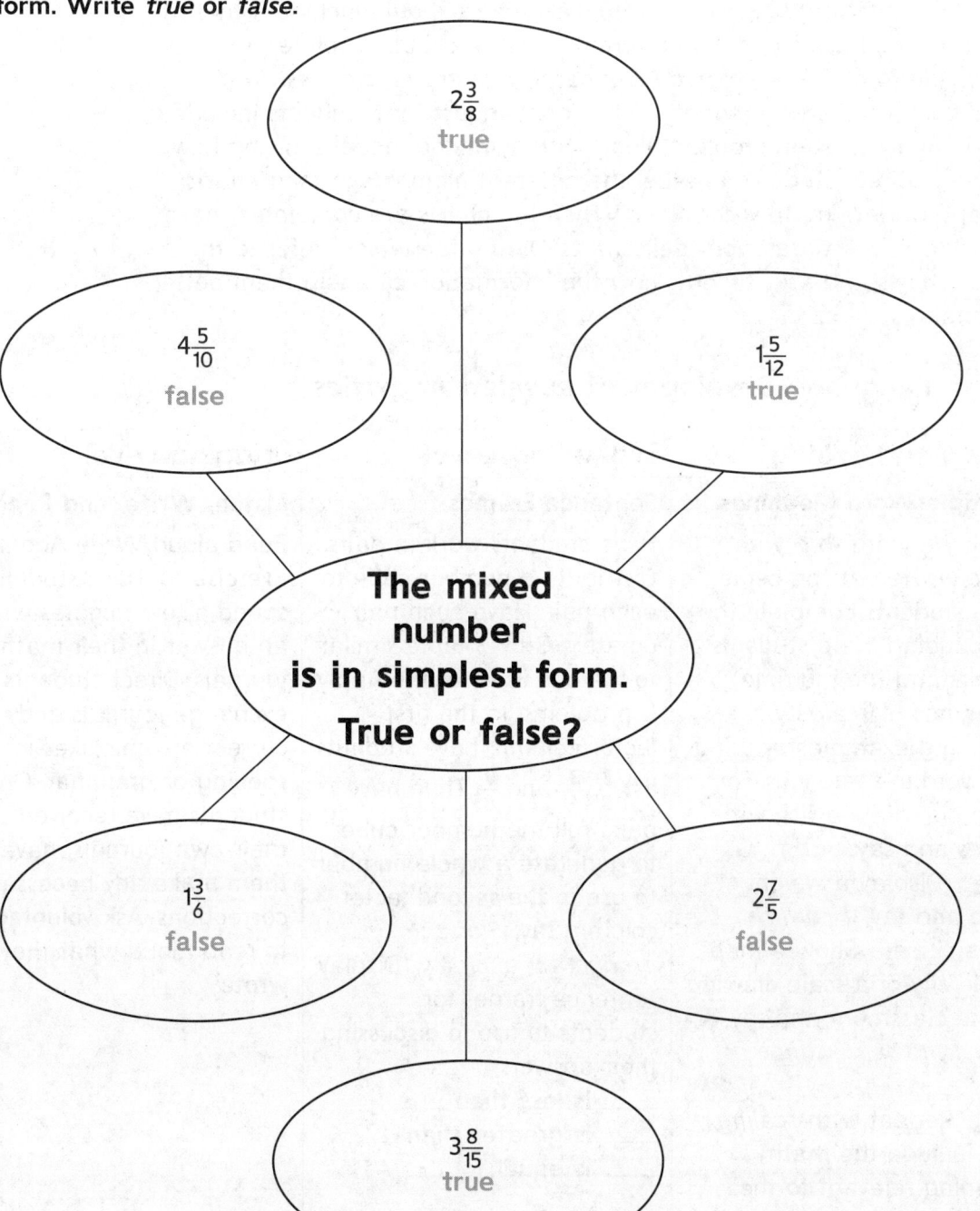

- $2\frac{3}{8}$ — true
- $4\frac{5}{10}$ — false
- $1\frac{5}{12}$ — true
- **The mixed number is in simplest form. True or false?**
- $1\frac{3}{6}$ — false
- $2\frac{7}{5}$ — false
- $3\frac{8}{15}$ — true

Lesson 8 Inquiry/Hands On: Multiplication as Scaling

English Learner Instructional Strategy

Vocabulary Support: Anchor Chart

Divide students into four groups. Say, *Make an anchor chart showing what you know about multiplying fractions.* Each chart should include a title at the top of the poster and definitions for math vocabulary related to multiplying fractions, such as *denominator, numerator, simplest form, mixed number, factor, product,* and so on. Suggest that students include a fraction multiplication problem along with a shaded model showing how to solve it. Direct students to label the different elements in their charts with appropriate math vocabulary. When the charts are completed, have groups display and describe their charts. During the lesson, refer to the anchor charts and ask students how the information can help them better understand scaling.

English Language Development Leveled Activities

Emerging Level	Expanding Level	Bridging Level
Multiple Word Meanings Display a word web with *scale* written at the center. Help students complete the surrounding. Help students understand the multiple meanings of *scale* by providing examples or using the word in sentences. For example, draw a fish with scales and say, *A fish has scales.* Display a weight scale and say, *I can measure using a scale.* Show a map of the city or a scale drawing of the classroom and say, *This is the same shape as the ____, but at a smaller scale.* Repeat with *scaling*, and include the math meaning relevant to the lesson.	**Sentence Frames** Have students work in pairs. Distribute a number cube to each pair. Have each pair create a 3-by-3 table similar to the one used for the Apply It problems. In the first factor column, have students list $\frac{3}{8}$, $\frac{3}{4}$, and $\frac{9}{8}$. Then have pairs roll the number cube to generate a whole number to use in the second factor column. Say, *Find the product for each row.* Display sentence frames for students to use in discussing their answers: ____ is less than ____. ____ is greater than ____. ____ is equal to ____.	**Listen, Write, and Read** Read aloud Write About It Exercise 18. Have students spend a few minutes writing an answer in their math journals. Direct students to exchange journals and correct any mistakes in spelling or grammar. Once students have received back their own journals, have them make any necessary corrections. Ask volunteers to read aloud what they wrote.

Teacher Notes:

NAME _____ DATE _____

Lesson 8 Vocabulary Definition Map

Inquiry/Hands On: Multiplication as Scaling

Use the definition map to write a description and list characteristics about the vocabulary word or phrase. Write or draw math examples. Share your examples with a classmate.

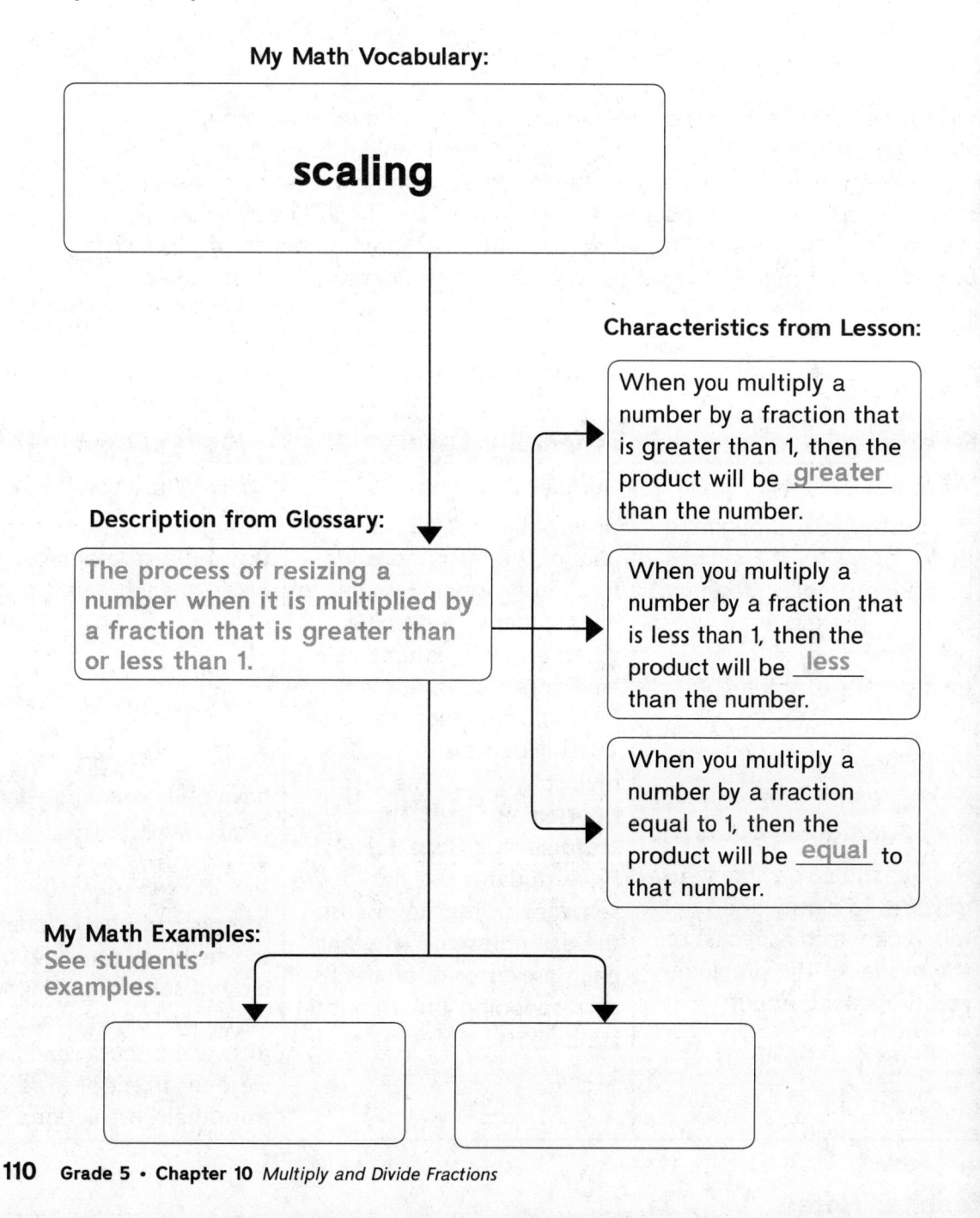

My Math Vocabulary:

scaling

Description from Glossary:
The process of resizing a number when it is multiplied by a fraction that is greater than or less than 1.

Characteristics from Lesson:

When you multiply a number by a fraction that is greater than 1, then the product will be _greater_ than the number.

When you multiply a number by a fraction that is less than 1, then the product will be _less_ than the number.

When you multiply a number by a fraction equal to 1, then the product will be _equal_ to that number.

My Math Examples:
See students' examples.

110 Grade 5 · Chapter 10 *Multiply and Divide Fractions*

Lesson 9 Inquiry/Hands On: Division with Unit Fractions

English Learner Instructional Strategy

Collaborative Support: Numbered Heads Together

Display anchor charts, word webs, KWL charts, or any other graphic organizers from previous lessons related to division and/or fractions. Be sure to include the classroom cognate chart as well. Have students take turns coming up to the graphic organizers and sharing a piece of information about fractions and division with the other students.

During the Practice It part of the lesson, organize students into groups of four and number off students as 1-4. Assign one problem to each group. The students should discuss the problem, agree on how to draw a model to find the solution, and ensure that everyone in the group understands and can give the answer. Afterward, call out a random number from 1 to 4. Have students assigned to that number raise their hands, and when called on, answer for their team.

English Language Development Leveled Activities

Emerging Level	Expanding Level	Bridging Level
Tiered Questions	**Number Recognition**	**Show What You Know**
Guide students through the Build It and Try It Exercises by asking simple questions that can be answered with one-word answers or by a gesture. For example, *Show me which tile to use. How many tiles? Do I use a $\frac{1}{5}$ tile or a $\frac{1}{10}$ tile? Is the answer 10 or $\frac{1}{10}$?* and so on. As you ask the questions, be sure to use gestures to clarify and reinforce which aspect(s) of the model or the problem you are asking about.	During the Talk About It part of the lesson, provide fraction division examples to help students determine whether each statements is true or false. Display the following sentence frames: **The dividend is ____. The divisor is ____. The quotient is ____. The statement is (true/false).** Have students use the sentence frames to discuss the examples and whether each proves or disproves its corresponding Talk About It statement.	Have students work in pairs. Distribute several index cards to each pair. On the board, write $9 \div \frac{1}{3}$. Say, *On your index cards, write each step describing how you would solve this problem using fraction tiles.* Then have pairs exchange cards. Say, *Follow the steps shown on the cards you were given to solve the problem.* Afterward, survey students to see which pairs did or did not solve the problem correctly. For incorrect answers, discuss which steps were incorrect or missing from their instructions.

Teacher Notes:

Student page

NAME _____ DATE _____

Lesson 9 Guided Writing

Inquiry/Hands On: Division with Unit Fractions

How do you divide with unit fractions?

Use the exercises below to help you build on answering the Essential Question. Write the correct word or phrase on the lines provided.

1. Rewrite the question in your own words.
   See students' work.

2. What key words do you see in the question?
   divide, unit fractions

3. A numerator is the __top__ number in a fraction. A numerator is the part of the fraction that tells the number of __parts__ you have.

4. The numerator in the fraction $\frac{1}{5}$ is __1__. A unit fraction is a fraction with __1__ as its numerator. The fraction $\frac{1}{5}$ is a __unit__ fraction.

5. When you are dividing $2 \div \frac{1}{5}$ you are trying to find how many __groups__ of $\frac{1}{5}$ are in __2__.

6. How many $\frac{1}{5}$-fraction tiles represent one whole? __5__

7. How many $\frac{1}{5}$-fraction tiles represent two wholes? __10__

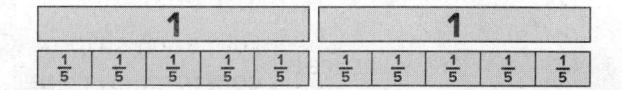

8. $2 \div \frac{1}{5} =$ __10__

9. How do you divide with unit fractions?
   Model the whole number and then model using the unit fraction to determine how many of the unit fraction will be needed to represent the whole number.

Grade 5 • Chapter 10 *Multiply and Divide Fractions* 111

Lesson 10 Divide Whole Numbers by Unit Fractions

English Learner Instructional Strategy

Vocabulary Support: Frontload Academic Vocabulary

Write the word *unit* and the Spanish cognate *unidad* on a classroom cognate chart. Explain that *unit* can mean "part or section," and then provide a model, such as a bar diagram or circle graph, to model the meaning of *unit fraction*. Distribute a number cube to pairs of students. Have pairs roll the number cube to determine a denominator. Direct student pairs to draw a bar diagram and use it to model a unit fraction for the denominator. For example, students would model $\frac{1}{4}$ by dividing the bar into fourths and shading one section. Afterward, have pairs describe their model using the following sentence frames: **Our unit fraction is ___. We shaded ___ part out of ___ parts.**

English Language Development Leveled Activities

Emerging Level	Expanding Level	Bridging Level
Number Sense Write $2 \div \frac{1}{2}$ and draw a long rectangle. Say, *This bar shows the number 2.* Label the bar diagram with a 2. Draw a vertical line dividing the bar in equal halves. Have students count the sections aloud with you: **1, 2.** Then say, *One whole and one whole. Two is made up of two wholes.* Draw vertical lines dividing each section into equal halves. Say, *Two wholes divided by half.* Have the class count aloud the sections with you: **1, 2, 3, 4.** Say, *Four one-half sections. Two divided by one-half equals four.* Have pairs work together to model $2 \div \frac{1}{4}$ and $2 \div \frac{1}{8}$.	**Recognize and Act It Out** Write $3 \div \frac{1}{3}$. Use three equal-sized strips of construction paper to represent 3. Have the class count the sections. Say, *There are three one whole sections in the number 3. We will find how many $\frac{1}{3}$ sections there are in 3.* Fold each strip of construction paper into three equal-sized sections. Explain that each section represents $\frac{1}{3}$. Have the class count the sections. Say, *There are 9 one-third sections in 3.* Write = 9 beside the expression. Give students construction paper to model the expression $5 \div \frac{1}{2}$ independently.	**Partner Work** Create two piles of cards. One pile has a whole number written on each card. The second pile has a unit fraction written on each card. Direct pairs to choose one card from each pile. Say, *Divide the whole number by the unit fraction.* Encourage pairs to model with bar diagrams or fraction tiles. One student in each pair will work to find the quotient. The other student will list the steps taken and will check the answer using multiplication. Then have pairs report back to you. Have pairs switch roles and repeat the activity using new cards.

Teacher Notes:

NAME _____ DATE _____

Lesson 10 Vocabulary Cognates
Divide Whole Numbers by Unit Fractions

Use the Glossary to define the math word in English and in Spanish in the word boxes. Write a sentence using your math word.

numerator	numerador
Definition The top number in a fraction; the part of the fraction that tells the number of parts you have.	**Definición** Número que se encuentra en la parte superior de una fracción; indica cuántas de las partes iguales se usan.

My math word sentence:

Sample answer: The numerator in the fraction $\frac{2}{5}$ is 2.

unit fraction	fracción unitaria
Definition A fraction with 1 as its numerator.	**Definición** Fracción con un numerador 1.

My math word sentence:

Sample answer: The fraction $\frac{1}{8}$ is a unit fraction.

112 Grade 5 • Chapter 10 *Multiply and Divide Fractions*

Lesson 11 Divide Unit Fractions by Whole Numbers

English Learner Instructional Strategy

Language Structure Support: Tiered Questions

Gear questions during the lesson to elicit responses that align with students' levels of English language proficiency. Emerging students may be able to respond only with gestures or single-word answers, so ask questions such as: *Do we divide by 3 or 4? Show me the numerators. Do we multiply or divide these two numbers?*

For expanding students, ask questions that can be answered with short phrases or simple sentences: *What do we do first? Which numbers do we multiply? What do we do next?*

For bridging students, ask questions that require more elaborate answers: *Why do we need to simply? How can we check our answer?*

English Language Development Leveled Activities

Emerging Level	Expanding Level	Bridging Level
Number Sense	**Modeled Talk**	**Report Back**
Write examples of unit fractions. Ask, *What is the same in these fractions?* Give students a chance to answer, and then say, *They all have 1 as the numerator. These are called **unit fractions**.* Have students chorally repeat **unit fractions**. Choose a unit fraction, such as $\frac{1}{5}$, and model it using a bar diagram. Say, *The bar shows one whole. The denominator of the unit fraction is five. I will divide the bar into five equal sections.* Divide the bar into five equal sections. Say, *Each section is $\frac{1}{5}$.* Label one section with $\frac{1}{5}$. Repeat with another unit fraction.	Write $\frac{1}{2} \div 4$. Draw a long rectangle on the board. Say, *This bar represents the number 1.* Divide the rectangle into equal halves. Say, *Each section represents $\frac{1}{2}$.* Label one section with $\frac{1}{2}$. Point to the whole number in the expression and say, *We need to divide $\frac{1}{2}$ by 4. We will divide each section into four equal parts.* Divide each section into four equal parts, resulting in eight sections total. Count the sections aloud with students. Say, *Now each section shows $\frac{1}{8}$. One half divided by four is one eighth.* Write $= \frac{1}{8}$ beside the expression.	Divide students into three groups and assign each group one of the Problem Solving Exercises 8–10. Have students work together to solve their assigned problem, recording the steps necessary to find the answer. Display the following sentence frames to help groups report back after they are finished: **First we** _____. **Then we** _____. **Next we** _____. **Last we** _____. **The answer is** _____.

Teacher Notes:

NAME _____ DATE _____

Lesson 11 Vocabulary Definition Map
Divide Unit Fractions by Whole Numbers

Use the definition map to write a description and list characteristics about the vocabulary word or phrase. Write or draw math examples. Share your examples with a classmate.

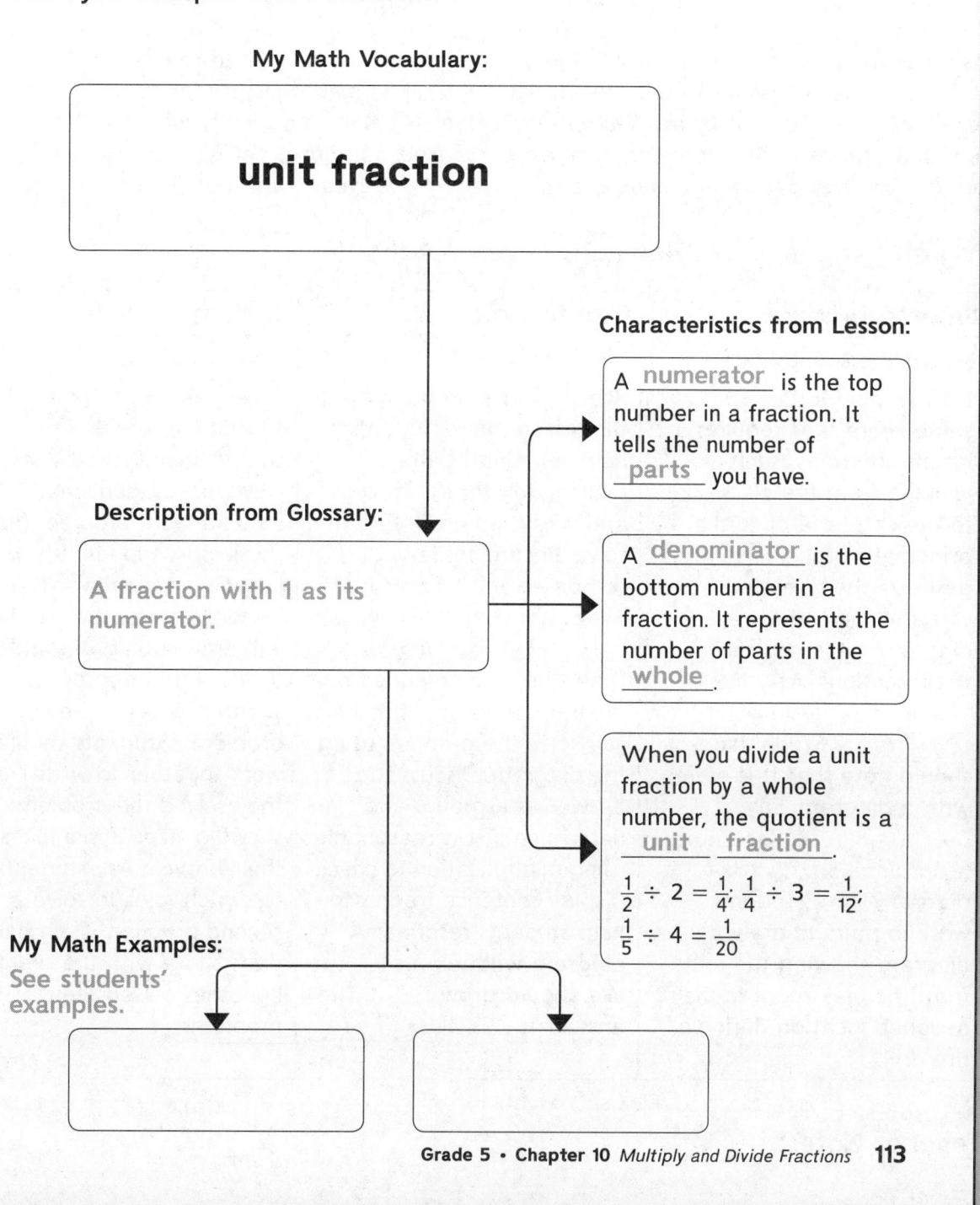

My Math Vocabulary:

unit fraction

Description from Glossary:

A fraction with 1 as its numerator.

Characteristics from Lesson:

A __numerator__ is the top number in a fraction. It tells the number of __parts__ you have.

A __denominator__ is the bottom number in a fraction. It represents the number of parts in the __whole__.

When you divide a unit fraction by a whole number, the quotient is a __unit__ __fraction__.

$\frac{1}{2} \div 2 = \frac{1}{4}$; $\frac{1}{4} \div 3 = \frac{1}{12}$; $\frac{1}{5} \div 4 = \frac{1}{20}$

My Math Examples:
See students' examples.

Lesson 12 Problem-Solving Investigation Strategy: Draw a Diagram

English Learner Instructional Strategy

Vocabulary Support: Utilize Resources

Write the word *diagram* and the Spanish cognate, *diagrama*, on a classroom cognate chart. Provide a concrete model by displaying a diagram from a book or other source, or by drawing a diagram on the board.

As students work through the Apply the Strategy exercises, be sure to remind them that they can refer to the Glossary or the Multilingual eGlossary for help, or direct students to other translation tools if they are having difficulty with non-math terms in the problems, such as: *purchased, sand toys, DVDs, originally, save, decorating, cookie, bird sanctuary, ice cream,* and *batches*.

English Language Development Leveled Activities

Emerging Level	Expanding Level	Bridging Level
Academic Vocabulary	**Modeled Talk**	**Academic Language**
Explain how to get somewhere that requires a lot of directions, such as walking from the classroom to the cafeteria or to the principal's office. First, describe the directions verbally. Next, draw a diagram while you repeat the directions. Ask, *Which way was easier to understand?* Students should note that it is easier with a diagram. Say, *A diagram can help you solve a problem. Say diagram.* **diagram** Have students work in pairs to draw a diagram showing the route from the classroom to their assigned location during a fire drill.	Read aloud a problem from the lesson. Have students help you identify the information that is known and what you are trying to solve. Record student responses on the board. Ask, *What type of diagram will help solve the problem?* Draw the recommended bar diagram. Have students help you solve the problem using the diagram. Discuss the answer as a group, and determine if it is reasonable using multiplication to check. Display sentence frames to help students respond: **A diagram will help us ____. We should draw ____. The answer is ____.**	Have pairs work together solving two problems of their choice from the lesson. Have one student read aloud a word problem from the lesson and identify what is known and what needs to be found. The other student will draw a diagram and divide it into appropriate sections to solve the problems. Students should work together to write how they solved the problem, showing an example to check their answer. Ask students to switch roles to solve a second problem. Then have pairs share with the group how they solved their problems.

Teacher Notes:

NAME _____ DATE _____

Lesson 12 Problem-Solving Investigation

STRATEGY: Draw a Diagram

Draw a diagram to solve each problem.

1. **Mrs. Vallez** purchased sand toys that were **originally** $20. She (Mrs. Vallez) received $\frac{1}{4}$ **off** of the **total** price. How much did she **save**?

Understand	Solve
I know:	
I need to find:	
Plan	**Check**
|⎯⎯⎯⎯⎯⎯20⎯⎯⎯⎯⎯⎯|	

2. **Sue** has **four** DVDs and **Terry** has **six** DVDs. **They** put all their DVDs **together** and sold them for $10 for **two** DVDs. How **much money** will they earn if they sell **all** of their DVDs?

Understand	Solve
I know:	
I need to find:	
Plan	**Check**

114 Grade 5 • Chapter 10 *Multiply and Divide Fractions*

Chapter 11 Measurement

What's the Math in This Chapter?

Mathematical Practice 6: Attend to precision

On the floor of the classroom, tape off a line that is 15 feet long (or shorter if necessary, but make sure it is an increment of 3 for conversion to yards). Ask students to estimate the length of the line using a variety of strategies. Record the estimates on the board, but do not include any of the mentioned units. After the list is compiled, discuss the different estimations and if noted by the students, highlight that the numbers may vary greatly because different units of measurement were used.

Divide students into small groups and have each group measure the line. Allow groups to determine the units they wish to measure. Record the actual measurements, including units, next to the list of estimates. Discuss with students that the measurements that they made in their small groups were precise because they used measurement tools to accurately record the length of the line.

Say, *Now I need to know the length of this line in yards. We are in a hurry though so we can't measure again. How can we figure this out?* **We can convert one of the measurements that are in feet. 3 feet = 1 yard. The line is 3 yards long.**

Display a chart with Mathematical Practice 6. Restate Mathematical Practice 6 and have students assist in rewriting it as an "I can" statement, for example: **I can be clear when sharing measurements by including units.** Post the new "I can" statement.

Inquiry of the Essential Question:

How can I use measurement conversions to solve real-world problems?

Inquiry Activity Target: **Students come to a conclusion that using units is very important in measurement.**

As an introduction to the chapter, present the Essential Question to students. The inquiry graphic organizer will offer opportunities for students to observe, make inferences, and apply prior knowledge of measurement representing the Essential Question. As they investigate, encourage students to draw, write, and collaborate with peers to demonstrate their observations and thinking. Then have students present additional questions they may have to a peer to extend discussions.

Regroup students and restate Mathematical Practice 6 and the Essential Question. Pose questions to reflect on what has been learned to guide students in making connections between the Mathematical Practice and the Essential Question.

NAME _____ DATE _____

Chapter 11 Measurement

Inquiry of the Essential Question:

How can I use measurement conversions to solve real-world problems?

Read the Essential Question. Describe your observations (I see...), inferences (I think...), and prior knowledge (I know...) of each math example. Write additional questions you have below. Then share your ideas and questions with a classmate.

45 ft = _____ yd	I see...
Since feet are smaller units than yards, divide.	
3 feet = 1 yard, so divide by 3.	I think...
45 ÷ 3 = 15	
So, 45 feet = 15 yards.	I know...

6 c = _____ fl oz	I see...
Since cups are larger units than fluid ounces, multiply.	
8 fluid ounces = 1 cup, so multiply by 8.	I think...
6 × 8 = 48	
So, 6 cups = 48 fluid ounces.	I know...

Distance Walked (mi)

(line plot with X's above 1/8, 1/4, and 1/2 on a number line from 0 to 1)

I see...

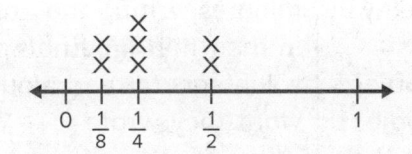

I think...

I know...

Fair Share: $\frac{1}{2} + \frac{1}{2} + \frac{1}{4} + \frac{1}{4} + \frac{1}{4} + \frac{1}{8} + \frac{1}{8}$
= 2 ÷ 7 = $\frac{2}{7}$ mile

Questions I have...

_____
_____
_____

Grade 5 • Chapter 11 *Measurement* **115**

Lesson 1 Inquiry/Hands On: Measure with a Ruler

English Learner Instructional Strategy

Vocabulary Support: Activate Prior Knowledge

Display a KWL chart. Ask, *What are some things we measure? How do we measure them?* Record what students already know about measuring in the first column of the chart. Display a ruler and ask, *What can we measure with a ruler?* Record what students hope to learn during the lesson, including how to measure objects to the nearest half- or quarter-inch, in the second column. After the lesson, have students describe what they learned. Record student responses in the third column of the KWL chart.

Display the following sentence frames to help students during the lesson:
Emerging: ___ inches
Expanding: **The length of ___ is ___ inches.**
Bridging: ___ **is about** ___ **inches long measured to the nearest** ___.

English Language Development Leveled Activities

Emerging Level	Expanding Level	Bridging Level
Basic Vocabulary Write *length* and *width*. Say each word and have students repeat chorally. Underline *leng* in *length*, and above the word write *long*. Say, *Length is how long something is.* Underline *wid* in *width*, and above the word write *wide*. Say, *Width is how wide something is.* Use objects from the classroom to clarifying length versus width, such as a pencil or a book. Run a finger along the length as you say, *This is the length of the ___.* Similarly demostrate width. Using other objects, prompt student responses by asking, *Is this the length or the width?*	**Exploring Language Structures** Divide students into small groups. Give each group several small objects and a ruler. Say, *Measure the length of each object to the nearest quarter-inch. Record the measurements. Then arrange the objects from shortest to longest.* Display the following sentence frames: **The shortest object is ___. The longest object is ___. ___ is longer than ___. ___ is shorter than ___. ___ is ___ inches long.** Have students use the sentence frames to describe how their groups arranged the objects.	**Turn & Talk** Read aloud Apply It Exercise 15 and say, *Turn to the student nearest you and discuss your answer.* Give students a chance to discuss their ideas. Then come together again as a group to continue the discussion. Have students spend a few minutes writing the answer in their math journals. Ask volunteers to read aloud what they wrote.

Multicultural Teacher Tip

As the metric system is the standard throughout most parts of the world, ELs will most likely be more familiar with units of metric measurement than they will be with standard units. Students who have worked only with the metric system in the past will be more familiar with partial amounts written as decimals, not fractions.

NAME _____ DATE _____

Lesson 1 Note Taking
Inquiry/Hands On: Measure with a Ruler

Read the question. Write words you need help with and research each word. Use your lesson to write your Cornell notes. Write or draw math examples to explain your thinking. Share your examples with a classmate.

Building on the Essential Question

How do you use a ruler to measure?

Words I need help with:
See students' words.

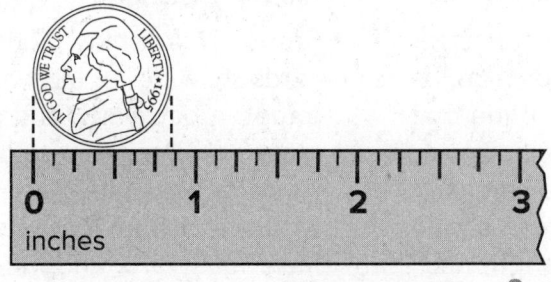

Notes:

A ruler can be used to measure __length__. Length is the measurement of the __distance__ between two points.

You can think of a __ruler__ like a number line.

$0 \quad \frac{1}{8} \quad \frac{1}{4} \quad \frac{3}{8} \quad \frac{1}{2} \quad \frac{5}{8} \quad \frac{3}{4} \quad \frac{7}{8} \quad 1$

When measuring an object using an inch ruler, it is important to line up the __zero__ on the ruler with the object.

The right side of the nickel is between __0__ and __1__. The end of the nickel is closest to __1__ inch.

If you use a smaller unit of measure, you will get a more __accurate__ measurement.

To the nearest fourth inch, the nickel is **about** $\frac{3}{4}$ inches long.

My Math Examples:
See students' examples.

116 Grade 5 • Chapter 11 *Measurement*

Lesson 2 Convert Customary Units of Length

English Learner Instructional Strategy

Vocabulary Support: Make Cultural Connections

Write the new vocabulary terms and their Spanish cognates, on a classroom cognate chart. Provide pictorial or concrete examples to model the meaning of each word.

Then write and explain *customary system*. Underline the word *custom* in *customary* and say, *Something that is customary is done or used as part of a custom.* Explain that a custom is a habit or practice and that many cultures have their own customs, or ways of doing things. Invite students to share examples of customs from their cultures. Then say, *There are even different customs for measuring things. In this lesson we will use systems of measurement used in the United States.*

English Language Development Leveled Activities

Emerging Level	Expanding Level	Bridging Level
Word Meaning Write: *inch, inches, foot, feet, yard,* and *yards*. Use a ruler to draw a line that measures one inch and then three lines that measure 3, 6, and 11 inches. Label each line with its length and then say the lengths aloud as you point to the corresponding words on the board: *one inch; three inches; six inches; 11 inches.* Have students chorally repeat. Be sure they are saying the /ch/ and /chuz/ sounds correctly. Using a yardstick, repeat the procedure to demonstrate *foot/feet* and *yard/yards*.	**Memory Device** Create and display picture cards showing an inchworm, a foot, a backyard, and a racetrack. Use them to help students associate pictures with the customary units: *inch, foot, yard,* and *mile*. Display the following sentence frames: **An inchworm is about one ____ long. A foot is about one ____ long. A yard is measured in ____. Four laps around the track is one ____.** Have students help you order the images from shortest to longest, and then have them use the sentence frames to describe the pictures.	**Building Oral Language** Have students work in pairs. Give each pair two number cubes and a set of My Vocabulary Cards for *inch, foot, yard,* and *mile*. Have each student draw a card at random and roll a number cube. Say, *The number you rolled is the measurement for the unit of length on your card.* Then have pairs use a sentence frame to compare their measurements: **____ is greater than ____.** Sample example: **4 feet is greater than 2 inches.** Have pairs repeat the activity several times.

Teacher Notes:

Lesson 2 Vocabulary Chart
Convert Customary Units of Length

Use the three-column chart to organize the vocabulary in this lesson. Write the word in Spanish. Then write the correct terms to complete each definition.

English	Spanish	Definition
convert	convertir	To change one __unit__ to another.
customary system	sistema usual	The units of __measurement__ most often used in the United States. These include foot, pound, and quart.
foot (ft)	pie	A __customary__ unit for measuring length. Plural is __feet__. 1 foot = 12 __inches__
inch (in.)	pulgada (pulg)	A __customary__ unit for measuring length. Plural is __inches__.
mile (mi)	milla (mi)	A __customary__ unit for measuring length. Plural is __miles__. 1 mile = 5,280 __feet__
yard (yd)	yarda (yd)	A __customary__ unit of length equal to 3 __feet__ or 36 __inches__.

Lesson 3 Problem-Solving Investigation Strategy: Use Logical Reasoning

English Learner Instructional Strategy

Collaborative Support: Think-Pair-Share

List and review the following strategies: *Use Logical Reasoning, Draw a Diagram, Look for a Pattern, Solve a Simpler Problem.* Have students work in pairs. Assign each pair one problem from Review the Strategies. Say, *Read your problem. Then plan a strategy that will help you solve the problem.* Give students several minutes to choose a plan. Then ask, *What is your plan?* Have pairs share their plans using the following sentence frame: **Our plan is to _____.**

Have each pair use their chosen strategy to solve the problem. Afterward, ask pairs to share their answers with the group. Discuss each plan and why it did or did not lead to a correct solution.

English Language Development Leveled Activities

Emerging Level	Expanding Level	Bridging Level
Listen and Write	**Synthesis**	**Academic Language**
Use a 5-column chart to model solving the following problem: Four cars are parked in a line: a red car, blue car, yellow car, and green car. The green car is directly in front of the blue car and directly behind the yellow car. The blue car is neither first nor last in line. What is the order of the cars? Label the columns: First, Second, Third, Last and label the rows: Red, Blue, Yellow, Green. Have students draw similar charts and follow along as you mark *Xs* in the chart to eliminate situations that are not logical. Write *Yes* when a car's position is clear.	Write and then read aloud the following problem: *TJ, Abby, Max and Akil ran a race. Max could not catch up with Abby or Akil. TJ did not come in second or third. Abby pulled ahead of Akil right at the finish line to take first place. What is the order of the runners at the finish line?* Have students work in pairs. Guide pairs in creating a table to use to organize the information from the problem, and then have them use logical reasoning to solve. Display the following sentence frame to help students share their answers: _____ came in _____ place.	Have pairs work together on two My Homework problems. Have one student read aloud the word problem and identify what is known and what they are trying to solve. The other student will decided on a method of logical reasoning to use to solve the problem. Have the students work together to solve the problem and then check the answer. Have pairs switch roles and work together on another problem. Afterward, have students explain how logical reasoning helped them find the answers.

Teacher Notes:

NAME _____ DATE _____

Lesson 3 Problem-Solving Investigation

STRATEGY: Use Logical Reasoning

Use logical reasoning to solve each problem.

1. An after-school club is building a clubhouse that has a **rectangular** floor that is **8 feet** by **6 feet**.
   What is the **total** floor **area** in **square inches** of the club?

Understand	Solve
I know the measurements in feet are: I need to find:	The total area in square __inches__ is:
Plan The measurements in inches are:	**Check**

2. There is a **red**, a **green**, and a **yellow** bulletin board hanging in the hallway.
   All of the bulletin boards are **rectangular** with a **height** of **4** feet.
   Their **lengths** are **6** feet, **5** feet, and **3** feet.
   The **red** bulletin board has the **largest** area and the **yellow** one has the **smallest** area.
   What is the **area** of the **green** bulletin board?

Understand	Solve
I know : Largest area = __red__ board Smallest area = __yellow__ board I need to find:	The area of the __green__ board is:
Plan The three areas will be: 4 feet by __6__ feet 4 feet by __5__ feet 4 feet by __3__ feet	**Check**

118 Grade 5 • Chapter 11 *Measurement*

Lesson 4 Inquiry/Hands On: Estimate and Measure Weight

English Learner Instructional Strategy

Language Structure Support: Modeled Talk

Display the following sentence frames to help students describe the objects and their weights in the Measure It and Try It parts of lesson: ____ weighs ____ ounces. ____ weighs more than ____. ____ weighs less than ____. ____ pound(s) weigh(s) ____ ounces.

As you work through the lesson, model using each sentence frame so students can hear standard English pronunciation. Be sure to enunciate each work clearly. After saying each sentence, ask a question that will allow a volunteer to respond using the same sentence frame. For example, *How much does the ____ weigh? How many ounces in ____ pounds?* and so on.

As students discuss the answers to the Talk About It Exercises, monitor their English usage and pronunciation and offer gentle corrections as needed, including a model of correct pronunciation.

English Language Development Leveled Activities

Emerging Level	Expanding Level	Bridging Level
Look, Listen, and Identify	**Pairs Check**	**Show What You Know**
Write *light* and *heavy*. Toss around a light object, such as wad of paper, and say, *This is light.* Then exaggerate the heaviness of a large textbook and say, *This is heavy.* Display photos or drawings of light or heavy objects, such as a feather, leaf, car, bowling ball, and so on. For each image, ask *Heavy or light?* Have students answer chorally. Tape the image next to *light* or *heavy* as appropriate. Then display sentence frames for students to use in comparing the objects: ____ **is heavier than** ____. ____ **is lighter than** ____.	Have students work in pairs to solve Apply It Exercises 13 and 14. Have one student complete Exercise 13 as the other student provides support and suggestions. Then have students switch roles to complete Exercise 14. Have each pair meet with another pair to compare answers. Once students have agreed on the correct answers, have them share with class. Provide sentence frames to help students share their answers: **The greater measurement is** ____.	Say, *Find an object in the classroom that you think weighs 12 ounces.* Give students time to locate objects. Place 12 ounces on one side of a balance scale. Have students take turns placing their object on the other side of the scale to determine if it weighs more or less than 12 ounces. Each student should use the following sentence frame to describe the object: ____ **weighs [more/less] than 12 ounces.** Record the results in a tally chart. Provide a small reward of some kind to any student with an object that is exactly 12 ounces in weight.

Teacher Notes:

Grade 5 • Chapter 11 *Measurement*

NAME _____ DATE _____

Lesson 4 Guided Writing

Inquiry/Hands On: Estimate and Measure Weight

How do you estimate and measure weight?

Use the exercises below to help you build on answering the Essential Question. Write the correct word or phrase on the lines provided.

1. Rewrite the question in your own words.
   See students' work.

2. What key words do you see in the question?
   estimate, measure, weight

3. Weight is a measurement that tells how __heavy__ an object is.

4. You can use a __balance__ to find the weight of objects. When objects on one side of the balance and objects on the other side are __level__, then they are equal in weight.

5. Some of the __units__ used to measure weight are ounces (oz) and pounds (lb).

6. Which unit of weight is greater, an ounce or a pound?
   a pound

7. There are __16__ ounces in one pound.
   There are __32__ ounces in two pounds.
   There are __48__ ounces in three pounds.

8. A slice of bread weighs about 1 __ounce__ and a loaf of bread weighs about 1 __pound__.

9. When measuring flour to make one loaf of bread, __ounces__ would give you a more precise measurement.

10. How do you estimate and measure weight?
    Determine if the object being weighed would be measured in ounces or pounds. Then place the object on one side of the scale and add weights to the other side until both sides of the scale are level.

Grade 5 • Chapter 11 *Measurement* 119

Lesson 5 Convert Customary Units of Weight

English Learner Instructional Strategy

Vocabulary Support: Build Background Knowledge

Before the lesson, write: *ounce (oz), pound (lb)*. Help students remember the atypical abbreviations by explaining their origins. Discuss that *oz* is based on the old Italian word *onza*, which means "ounce." Write *onza* and underline its *o* and *z*. Discuss that *onza* is also the Spanish cognate for *ounce*. Then discuss that *lb* is based on the ancient Roman term *libra pondo*, which means "pound weight." Write *libra pondo* on the board and underline its *l* and *b*. Have students write this information in their math journals.

English Language Development Leveled Activities

Emerging Level	Expanding Level	Bridging Level
Word Knowledge Write *wait*. Ask a student to stand at the front of the class. Say, *Wait until I say go, then return to your seat.* Stress the word *wait*. After a moment, say *go* and have the student sit down. Write *weights*. Show students images of weightlifting equipment. Say, *You can lift weights to develop muscles.* Stress *weights*. Point to the word *weights* on the board. Erase the *s* from *weights*. Place an object from the classroom on a scale. Say, *I will find the weight of the _____.* Stress the word *weight*. Then say, *The weight is _____.* Have students chorally repeat.	**Memory Device** Create picture cards showing the following: five quarters, four sticks of butter, a giraffe. Use them to associate a picture with each customary unit. As you display the pictures, say, *Five quarters weigh about one ounce. Four sticks of butter weigh about one pound. A giraffe weighs about a ton.* Have students help you order the images from lightest to heaviest and write the associated unit under each card. Display the following sentence frames to help students describe the units and pictures: *_____ is heavier than _____. _____ is lighter than _____.*	**Building Oral Language** Have students work in pairs. Give each pair a 5-part spinner numbered 1–5. Say, *The number you spin is your measurement of weight in tons.* Have pairs spin to find a weight in tons and then have them convert the weight to pounds. Display a sentence frame to help them answer: *_____ **tons is equal to** _____ **pounds**.* Then have pairs spin again to find a weight in pounds and convert it to ounces. Display another sentence frame to help them answer: *_____ **pounds are equal to** _____ **ounces**.*

Teacher Notes:

T120 Grade 5 • Chapter 11 *Measurement*

NAME _____ DATE _____

Lesson 5 Vocabulary Definition Map
Convert Customary Units of Weight

Use the definition map to write a description and list characteristics about the vocabulary word or phrase. Write or draw math examples. Share your examples with a classmate.

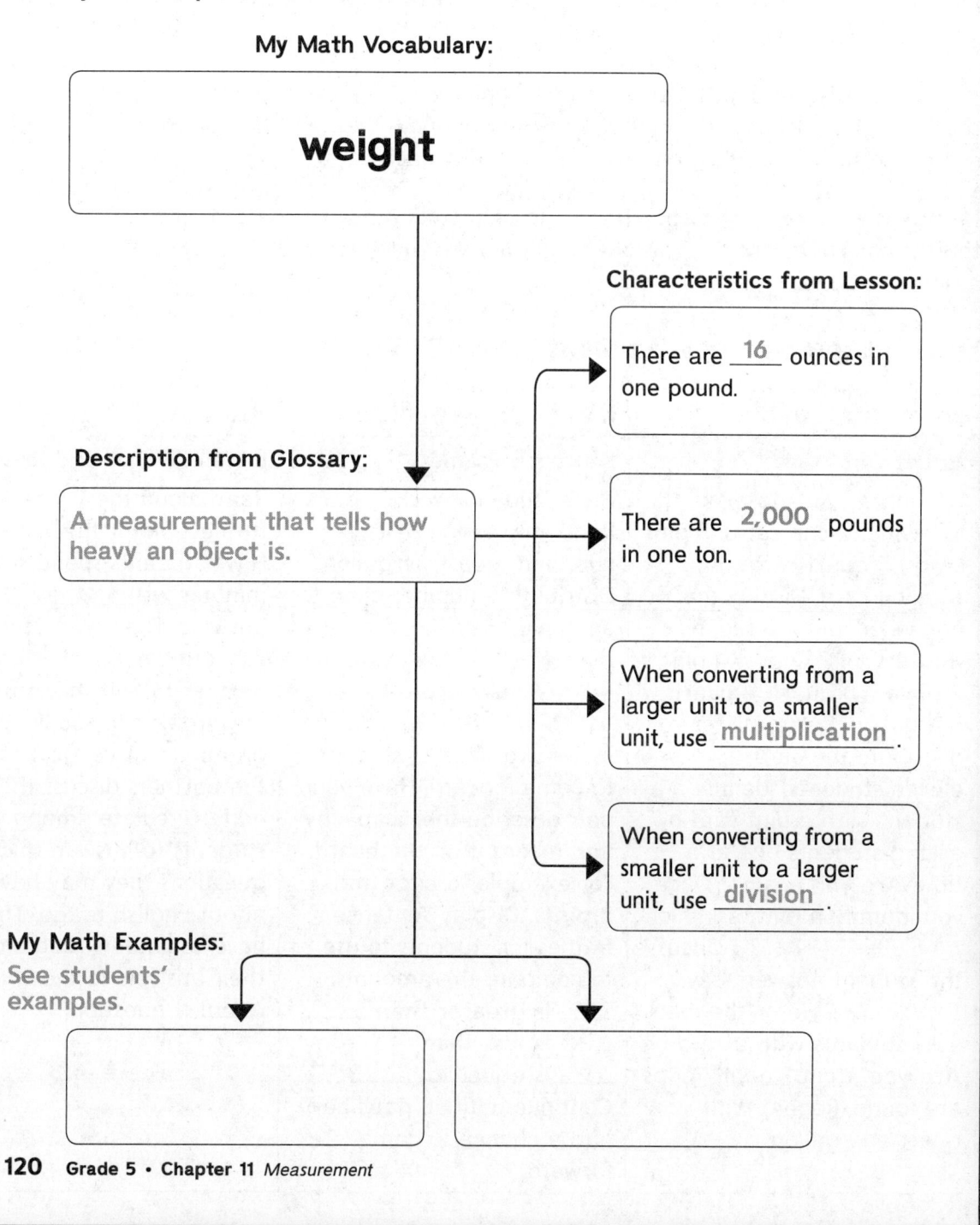

My Math Vocabulary:

weight

Description from Glossary:

A measurement that tells how heavy an object is.

Characteristics from Lesson:

There are __16__ ounces in one pound.

There are __2,000__ pounds in one ton.

When converting from a larger unit to a smaller unit, use __multiplication__.

When converting from a smaller unit to a larger unit, use __division__.

My Math Examples:
See students' examples.

120   Grade 5 • Chapter 11 *Measurement*

Lesson 6 Inquiry/Hands On: Estimate and Measure Capacity

English Learner Instructional Strategy

Graphic Support: Word Web

Divide students into four small groups. Be sure each group contains students of varying levels of English proficiency. Assign each group as cup, pint, quart, or gallon. Distribute a word web to each group. Say, *Write your unit of capacity in the center. Fill in the other parts of the web with math or nonmath words, phrases, descriptions, or examples related to your word.* After groups have completed their webs, call on students to share one item from their groups' webs. Discuss how the item does or does not relate to the word's meaning as a unit of capacity.

Display the following sentence frames to help students participate during the lesson: **There are _____ in one _____. My estimate is _____. The actual amount is _____.**

English Language Development Leveled Activities

Emerging Level	Expanding Level	Bridging Level
Act It Out	**Sentence Frames**	**Listen, Write, and Read**
Use index cards to make the following: 2 cup cards, 3 pint cards, 5 quart cards, and 1 gallon card. Display the equivalencies covered in this lesson: 2 cups = 1 pint, 2 pints = 1 quart, 4 quarts = 1 gallon. Randomly distribute the cards to eleven students. Identify a student with a pint card by asking students one at a time **Are you a pint?** Once you identify a pint, ask, *How many cups in 1 pint?* Once the student answers, say, *Find 2 cups.* He or she then asks students with a card **Are you a cup?** until 2 cups are found. Repeat with quart and gallon.	Have students work in pairs. Randomly assign pairs as cup, pint, quart, or gallon. Distribute a number cube to each pair. Say, *Roll your number cube. That is how many units you are of your assigned unit of capacity.* Invite two pairs of students to come forward. Have each pair describe their capacity and record it on the board. For example, **6 cups** and **2 pints.** Display sentence frames for students to use to compare the amounts: _____ **is greater than** _____. _____ **is less than** _____. _____ **is equal to** _____. Continue until all pairs have had a chance to come forward.	Read aloud the Write About It question for the lesson. Have students spend several minutes writing their answers. Then have students exchange papers with a partner to edit the writing for mistakes in spelling, grammar, or usage. Have the partners discuss the edits. Circulate among students to answer any questions they may have about English usage. Then have volunteers read aloud their answers to the Write About It question.

Teacher Notes:

NAME _____ DATE _____

Lesson 6 Note Taking

Inquiry/Hands On: Estimate and Measure Capacity

Read the question. Write words you need help with and research each word. Use your lesson to write your Cornell notes. Write or draw math examples to explain your thinking. Share your examples with a classmate.

Building on the Essential Question	Notes:
How do you estimate and measure capacity?	Capacity is a measurement that tells the __amount__ a container can hold. Some of the __units__ used to measure capacity are cups (c), pints (pt) quarts (qt), and gallons (gal). There are __2__ cups in one pint. There are __2__ pints in one quart. There are __4__ quarts in one gallon. A pint container can hold __2__ cups. A container twice that size will be 2 pints in capacity and can hold __4__ cups. A quart container can hold __2__ pints. A container four times that size will be 4 quarts in capacity and can hold __8__ pints.
Words I need help with: See students' words.	

Container	Cups	Pints	Quarts
2 pints	4	2	1
1 gallon	16	8	4

My Math Examples:
See students' examples.

Grade 5 • Chapter 11 *Measurement* 121

Lesson 7 Convert Customary Units of Capacity

English Learner Instructional Strategy

Sensory Support: Pictures/Realia

Write new vocabulary with Spanish cognates on a classroom cognate chart. For all new vocabulary, provide visual references to reinforce meanings. Display labeled containers and photographs of objects that would hold or be measured in the various units of capacity discussed in this lesson. For example: a thimble, a milk jug, a cup, quart container, a fish tank, a pool, a bucket, a spoon, bottle of water, and so on.

English Language Development Leveled Activities

Emerging Level	Expanding Level	Bridging Level
Word Knowledge Write *capacity* and display a small clear container. Say, *I am going to fill this container with water.* Pour dyed water into the container until it is full. Say, *No more water can fit in the container. I have filled it to capacity.* Stress *capacity* as you say it again and have students chorally repeat. Have students help you fill containers with water. Use containers with exact capacities of 1 fluid ounce, 1 cup, 1 pint, 1 quart, and 1 gallon. Ask either/or questions to help students identify units of capacity, for example, *Is the capacity one cup or one gallon?*	**Recognize and Act It Out** Gather five labeled containers with the following capacities: one fluid ounce, one cup, one pint, one quart, one gallon. Hold up each container to display its labeled capacity as you say, *This container has a capacity of ___.* Compare units of capacity by filling two unit capacity containers with dyed water, and then empty the water from each container into two large, identical containers. Have students use the following sentence frame to compare capacities: **One is larger than one ___.** Continue until all units have been compared.	**Building Oral Language** Write *128 fluid ounces*, and ask, *Do you multiply or divide to find cups?* **divide** Discuss that when changing a smaller unit to a larger unit you divide. Then have pairs work to find the capacity in cups. Display a sentence frame to help students answer: ___ **fluid ounces is equal to ___ cups.** Then have pairs find the number of pints, quarts, and gallons in 128 fluid ounces using the following sentence frame: ___ **fluid ounces is equal to ___ (pints/quarts/gallons).** Repeat the exercise with a new measurement of capacity.

Teacher Notes:

NAME _____ DATE _____

Lesson 7 Vocabulary Chart
Convert Customary Units of Capacity

Use the three-column chart to organize the vocabulary in this lesson. Write the word in Spanish. Then write the correct terms to complete each definition.

English	Spanish	Definition
capacity	capacidad	The amount a container can _hold_.
fluid ounce (fl oz)	onza liquida	A _customary_ unit of capacity.
gallon (gal)	galón (gal)	A customary unit for measuring capacity for _liquids_. 1 gallon = 4 _quarts_
cup (c)	taza	A customary _unit_ of capacity equal to 8 fluid _ounces_.
pint (pt)	pinta (pt)	A customary unit for _measuring_ capacity. 1 pint = 2 _cups_
quart (qt)	cuarto (ct)	A customary unit for measuring _capacity_. 1 quart = 4 _cups_

Lesson 8 Display Measurement Data on a Line Plot
English Learner Instructional Strategy

Graphic Support: KWL Chart

Display a KWL chart. In the first column, record what students already know about data and line plotting from previous lessons or grades. In the second column, record what students hope to learn during the lesson, including how to plot fractions and find the fair share. After the lesson, display the following sentence frame. Have students use it to describe what they learned: **I learned that** ____. Record student responses in the third column of the KWL chart.

During the Talk Math section of the lesson, pair emerging students with more proficient English speakers who share the same native language. Allow the emerging student to answer in his or her native language, and then have the bilingual student translate to report back to you.

English Language Development Leveled Activities

Emerging Level	Expanding Level	Bridging Level
Word Recognition Use red circles to represent three apples. Cut one apple in half and two apples into quarters. Unevenly distribute some of the pieces to four student volunteers. Ask each student to identify their fraction using the sentence frame: **I have** ____ **of an apple.** Say, *Some students have more than others. No one has an equal amount. No one has a fair share.* Have students chorally repeat, **fair share.** Collect the apple pieces and redistribute them so each student has the same amount. Say, *Now each student has the same amount. Everyone has a fair share.*	**Recognize and Act It Out** Write the following list of numbers: $\frac{1}{4}, \frac{1}{2}, \frac{1}{2}, \frac{1}{2}, \frac{1}{2}, \frac{3}{4}, 1, 1$. Say, *This list is data showing how many fractions of a mile someone ran on different days.* Then create a number line titled: Distance Ran (in miles) with $\frac{1}{4}, \frac{1}{2}, \frac{3}{4}$ and 1 marked. Say, *The title tells us what data will be plotted on the number line.* Have students guide you in plotting the data using the following sentence frames: **Put** ____ **Xs above** ____. Afterward, discuss the data. Prompt student responses with questions such as, *Which distance was run the most/least?*	**Academic Language** Write the following list of numbers: $\frac{1}{4}, \frac{1}{2}, \frac{1}{2}, \frac{1}{2}, \frac{1}{2}, \frac{3}{4}, 1, 1$. Say, *This data is the distance in miles that someone ran on different days.* Have student pairs create line plots using the data. Then ask pairs to trade line plots and find the fair share of the line plot they have received. Display a sentence frame for students to use to share their answer: **The fair share is** ____. Determine the correct answer, and then for any pairs with a different answer, discuss as a group where a mistake was made, either in the plotting of data or in determining the fair share.

Teacher Notes:

Student page

NAME _____ DATE _____

Lesson 8 Vocabulary Definition Map
Display Measurement Data on a Line Plot

Use the definition map to write a description and list characteristics about the vocabulary word or phrase. Write or draw math examples. Share your examples with a classmate.

My Math Vocabulary:

> **fair share**

Characteristics from Lesson:

To find a fair share, first __add__ the separate parts to find the whole. Then __divide__ the whole by the number of parts.

Use the __measurements__ in a table to make a __line__ plot.

When using a line __plot__ to find fair shares, add the fractions with __like__ denominators first.

Description from Glossary:

An amount divided equally.

My Math Examples:
See students' examples.

Grade 5 • Chapter 11 *Measurement* 123

Lesson 9 Inquiry/Hands On: Metric Rulers

English Learner Instructional Strategy

Language Structure Support: Tiered Questions

During the lesson, be sure to ask questions according to ELs' level of English comprehension. For example, ask emerging students simple questions that elicit one-word answers or can be answered with a gesture: *Meters or centimeters? What fraction is it, $\frac{1}{4}$ or $\frac{1}{10}$? How many centimeters? Do I put the ruler here, or here? Show me a centimeter.*

For expanding students, ask more complex questions that elicit simple phrases or short sentences: *How long is the toy car? What is the difference between a metric ruler and a customary ruler? What could be measured in kilometers?*

For bridging students, ask questions that require more complex answers: *Why would we round to millimeters rather than centimeters? Why is it better to measure the _____ in meters rather than centimeters?*

English Language Development Leveled Activities

Emerging Level	Expanding Level	Bridging Level
Show What You Know Have each student measure his or her thumb to the nearest centimeter using a metric ruler. Have students use a sentence frame to describe the measurement: **My thumb is _____ centimeters long.** Model pronunciation as needed. Then have students measure again, this time to the nearest millimeter. Display another sentence frame: **My thumb is _____ millimeters long.**	**Word Web** Divide students into four groups. Assign the groups as millimeter, centimeter, meter, and kilometer. Distribute a word web to each group. Say, *Write your assigned metric unit of length in the center. Fill in the web with examples of what you would measure with that unit.* Afterward, have groups share their examples. Discuss each group's examples, including whether or not another unit of measurement might have been a better choice and why.	**Listen, Write, and Read** Display the following sentence frames: **Future: I will measure _____ using _____. Present: I am measuring _____. Past: I measured _____, and it was _____ long.** Invite students to come forward and measure objects from the classroom using either centimeters or millimeters. Have them use the sentence frames appropriately as they describe what they will do, what they are doing, and what they did.

Teacher Notes:

Student page

NAME _____ DATE _____

Lesson 9 Guided Writing

Inquiry/Hands On: Metric Rulers

How do you use metric rulers to measure?

Use the exercises below to help you build on answering the Essential Question. Write the correct word or phrase on the lines provided.

1. Rewrite the question in your own words.
   See students' work.

2. What key words do you see in the question?
   metric, rulers, measure

3. The metric system is a __decimal__ system of measurement.

4. __Metric__ units of length include millimeters, centimeters, meters, and kilometers. Length is the measurement of the __distance__ between two points.

5. A metric ruler can be used to measure __length__ in centimeters and millimeters. 1 centimeter = __10__ millimeters

6. When measuring an object using a metric ruler, it is important to line up the __zero__ on the ruler with the left side of the object.

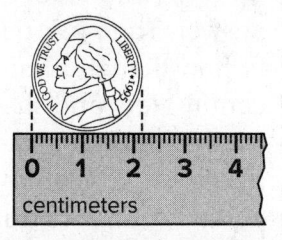

7. The right side of the nickel is between __2__ and __3__. The end of the nickel is closest to __2__ centimeters.

8. If you use a smaller unit of measure, you will get a more __accurate__ measurement.

9. To the nearest millimeter, the nickel is about __22__ millimeters long.

10. How do you use metric rulers to measure?
    Line up the object you are measuring with the zero mark on the metric ruler. Find the length of the object by locating the nearest centimeter/millimeter to the end of the object.

124 Grade 5 • Chapter 11 *Measurement*

Lesson 10 Convert Metric Units of Length

English Learner Instructional Strategy

Sensory Support: Photographs/Illustrations

Write new vocabulary with Spanish cognates on the classroom cognate chart. To introduce and reinforce meaning, display photographs and illustrations of objects that would be measured using the various metric units of length. For example, a paper clip, a ladybug, a chair, a highway, a CD, a tree, and so on. Discuss the appropriate metric unit to measure each object, then label each with that unit. Display in the classroom for students to reference during the lesson.

During Talk Math, provide the following sentence frame to help students answer: **I move the decimal _____ places to the _____. So, 7.38 kilometers equals _____ meters.**

English Language Development Leveled Activities

Emerging Level	Expanding Level	Bridging Level
Word Recognition	**Listen and Write**	**Building Oral Language**
Write *cent*. Display a penny and a dollar bill. Say, *A penny is one cent. There are 100 cents in a dollar.* A cent is one hundredth of a dollar. Write $\frac{1}{100}$. Write *centipede* and underline *cent-*. Display an image of a centipede. Point to its legs and say, *A centipede has 100 legs. One leg is one hundredth of the total number of legs.* Write $\frac{1}{100}$. Write *centimeter* and underline *cent-*. Point to a centimeter on a meter stick and say, *This is a centimeter. There are 100 centimeters in a meter. Each centimeter is one hundredth of a meter.* Write $\frac{1}{100}$.	Create a three-column chart labeled: *Milli-*, *Centi-*, and *Kilo-*. Have students work in pairs to create their own versions of the chart as you model. In the first row, write the value for each prefix: $\frac{1}{1,000}$, $\frac{1}{100}$, 1,000. In the second row, write the corresponding metric unit of length for each prefix: *millimeter, centimeter, kilometer*. For the third row, have students first write names of objects that correspond approximately to each length or would be measured in that length. Then discuss students' examples and record the most reasonable in your chart.	Write: 1.5 kilometers. Ask, *If I want to find this length in centimeters, do I multiply or divide?* Have students answer. Then ask pairs to find the length in centimeters and share the answer using a sentence frame: **_____ kilometers is equal to _____ centimeters.** Provide additional measurements in millimeters, centimeters, meters, or kilometers, and have pairs work together to convert them to another specified metric unit of length. Provide the following sentence frame to have students share their answers: **_____ multiplied/divided by _____ equals _____.**

Teacher Notes:

T125 Grade 5 • Chapter 11 *Measurement*

NAME _____ DATE _____

Lesson 10 Concept Web
Convert Metric Units of Length

Use the concept web to identify the unit of metric measurement that would be best to measure each item.

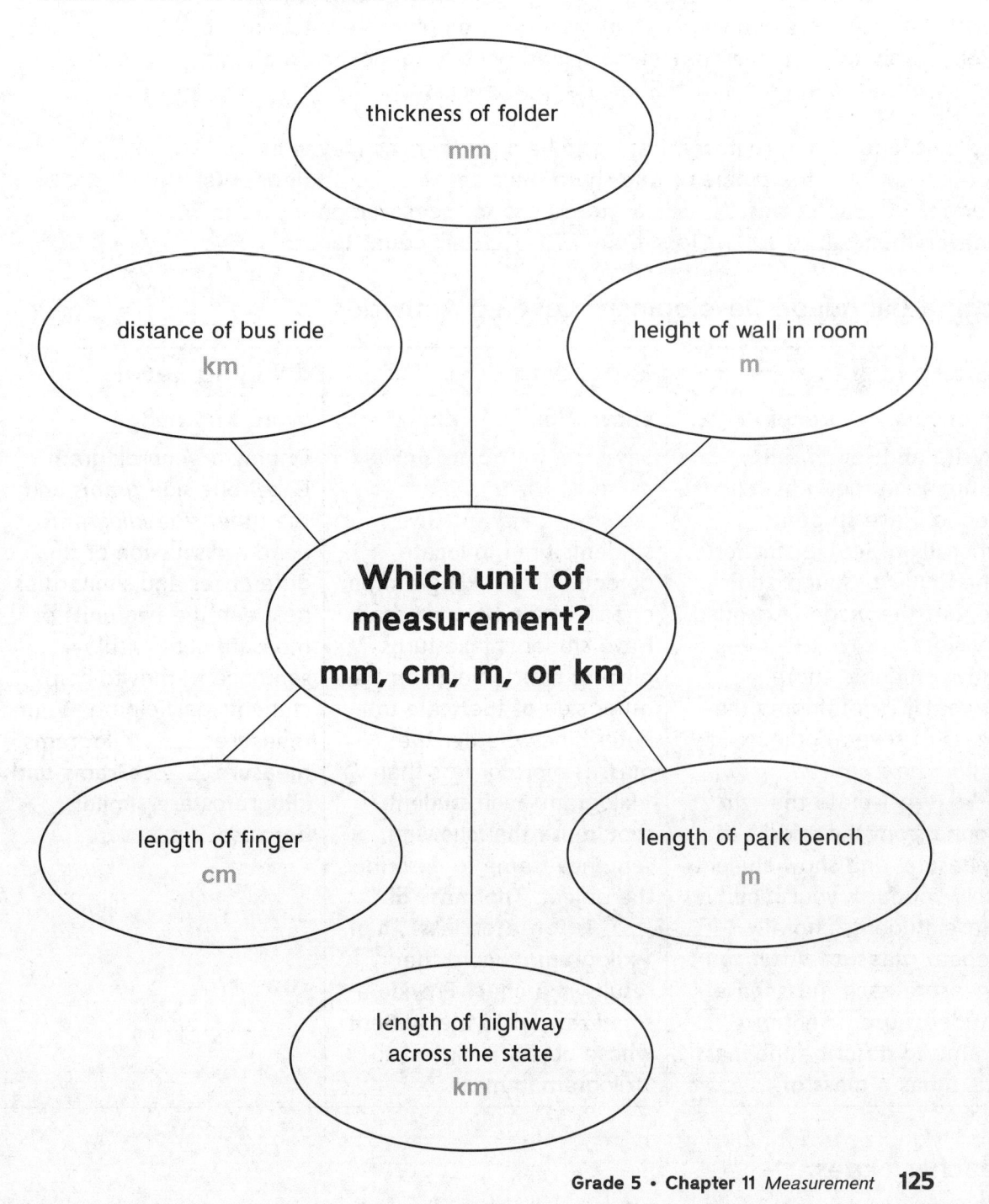

- thickness of folder
  mm

- distance of bus ride
  km

- height of wall in room
  m

**Which unit of measurement?
mm, cm, m, or km**

- length of finger
  cm

- length of park bench
  m

- length of highway across the state
  km

Grade 5 • Chapter 11 *Measurement*   **125**

Lesson 11 Inquiry/Hands On: Estimate and Measure Metric Mass

English Learner Instructional Strategy

Collaborative Support: Partner Reading

Have students work in pairs to solve the Apply It Exercises. Try to pair emerging students with expanding or bridging students. Have partners take turns reading aloud information from each problem. Tell students to listen closely to what their partners say and politely suggest corrections for any mistakes in pronunciation or usage as necessary.

Display the following sentence frames to help partners as they work together to solve the problems and share their answers: ____ kilogram(s) is equal to ____ grams. ____ is a more precise measurement. ____ is greater then ____. ____ is less than ____. ____ is equal to ____.

English Language Development Leveled Activities

Emerging Level	Expanding Level	Bridging Level
Phonemic Awareness Write and say, *grams*. Emphasize the final /z/ sound. Have students chorally repeat. Listen for the final /z/ sound and repeat the model if needed. Write and say, *kilograms*. Have students chorally repeat. Model finding the mass of several objects. Say, *I will measure the mass. Measure.* Isolate the /zh/ sound from the middle of *measure*, and show students how you form your mouth. Have students chorally repeat **measure**. After you measure each mass, have students use a sentence frame to describe the mass: ____ **has a mass of** ____.	**Show What You Know** Say, *Find an object in the classroom with a mass of exactly 1 kilogram.* Give students time to locate objects. Place 1 kilogram on one side of a balance scale. Have students take turns placing their objects on the other side of the scale to determine whether the mass is more or less than 1 kilogram. Each student should use the following sentence frame to describe the object: **The mass of ____ is [greater/less] than 1 kilogram.** Record the results in a chart. Provide a small reward to the student whose object is closest to 1 kilogram in mass.	**Word Knowledge** Display a Venn diagram. Label one side *grams* and the other side *kilograms*. Lead a discussion of the differences and similarities between the two units of measurement. Display sentence frames to help students participate: **Grams measure ____. Kilograms measure ____. Grams and kilograms are similar because ____.**

Teacher Notes:

NAME _____ DATE _____

Lesson 11 Vocabulary Chart

Inquiry/Hands On: Estimate and Measure Metric Mass

Use the three-column chart to organize the vocabulary in this lesson. Write the word in Spanish. Then write the correct terms to complete each definition.

English	Spanish	Definition
mass	masa	Measure of the amount of _matter_ in an object.
gram	gramo (g)	A _metric_ unit for measuring mass.
kilogram	kilogramo (kg)	A metric unit for measuring _mass_. 1 kilogram = 1,000 _grams_.
metric system	sistema métrico (SI)	The _decimal_ system of measurement. Includes units such as meter, _gram_, and liter.

Lesson 12 Convert Metric Units of Mass

English Learner Instructional Strategy

Collaborative Support: Small Groups

Write *mass, gram, milligram,* and *kilogram* and the Spanish cognates, *mass, gramo, miligramo,* and *kilogramo,* on a classroom cognate chart. Provide concrete examples by displaying classroom objects whose masses would be best measured in each unit or represent an equivalency to each unit.

Divide students into two groups and assign each as Milligrams or Kilograms. Write a mass in grams. Then say, *Convert this mass to your unit of mass.* Display a sentence frame to help groups answer: ____ **grams is equal to ____ milligrams/kilograms.** The first group to answer correctly scores a point. Continue until one group has five points. Afterward, discuss how mental math can be used to make the conversion and how the process differs between milligrams and kilograms.

English Language Development Leveled Activities

Emerging Level	Expanding Level	Bridging Level
Word Recognition	**Listen and Write**	**Building Oral Language**
Gather a wide variety of images of objects with varying masses, such as a grain of rice, a car, an apple, a dog, a sheet of paper, and so on. Display each image to a volunteer as you pose an either/or statement about the metric unit of mass most likely used to measure the object. For example, display the image of an apple and say, *Pick either milligrams or grams to measure with.* After the student chooses, hand over the image. Once all images have been identified, have students gather according to their object's unit of mass. Have groups chorally say their unit of mass.	Divide students into three groups, and assign each group as one of the following: Milligrams, Grams, Kilograms. Say, *Think of objects that would be measured in your unit of metric mass.* Give groups several minutes to brainstorm a list. Then provide the following sentence frame for students to use as they share items from their group's list: **A ____ would be measured in ____.** Discuss groups' examples for reasonableness.	Gather nutritional fact labels from several different food products. Having students work in pairs, give each pair one of the food labels. Say, *Find two metric units of mass on your label and compare them.* Provide the following sentence frame to help students share their comparisons: **____ is greater in mass than ____.** Then have pairs choose one of the measurements and convert it to a different metric unit of mass. Provide the following sentence frame for students to use in sharing the conversion: **____ is equal in mass to ____.**

Teacher Notes:

NAME _____ DATE _____

Lesson 12 Multiple Meaning Word
Convert Metric Units of Mass

Complete the four-square chart to review the multiple meaning word.

Everyday Use	Math Use in a Sentence
Sample answer: A mass can be when something relates to or is done by a large group of people. For example, the event had mass appeal.	Sample sentence: Metric units of mass are gram, milligram, and kilogram.
Math Use	**Example From This Lesson**
Measure of the amount of matter in an object.	Sample answer: Converting units of mass in the metric system will require division or multiplication by 1,000.

(center: **mass**)

Write the correct term on each line to complete the sentence.

Use <u>multiplication</u> to convert from a larger unit of mass to a smaller unit.
Use <u>division</u> to convert from a smaller unit of mass to a larger unit.

Grade 5 • Chapter 11 *Measurement*

Lesson 13 Convert Metric Units of Capacity
English Learner Instructional Strategy

Vocabulary Support: Utilize Resources

Write *liter* and *milliliter* and the Spanish cognates, *litro* and *mililitro*, on a classroom cognate chart. Provide concrete examples of the meanings by displaying an eyedropper of water and a liter container of water.

For Independent Practice Exercises 5–12, have students work in pairs. Afterward, ask volunteers to share their answers using the following sentence frame: ____ **is equal to** ____.

As students work through the Problem Solving exercises, be sure to remind them that they can refer to the Glossary or the Multilingual eGlossary for help, or direct students to other translation tools if they are having difficulty with non-math language in the problem.

English Language Development Leveled Activities

Emerging Level	Expanding Level	Bridging Level
Word Knowledge	**Academic Vocabulary**	**Building Oral Language**
Display pairs of images or objects from the classroom for students to compare for capacity. For example, display a jar of baby food and a jar of pickles. Ask, *Which has greater capacity?* Have students answer by pointing or using a simple sentence frame such as: ____ **has greater capacity than** ____. Continue in this manner until all students have had a turn. Then use the same sentence frame to compare something measured in milliliters to something measured in liters.	Gather containers that show milliliter measurements and liter measurements. Hold up each container and identify its capacity in either milliliters or liters. Then hold up images of items that would be measured using either milliliters or liters. For example, a spoonful of milk would have a capacity in milliliters and a bucket of water would have a capacity in liters. Display the following sentence frame for students to use as they identify the reasonable metric unit of capacity to use: **Use** ____ **to measure the capacity of** ____.	Write: 0.18 liters. Ask, *Do we multiply or divide to find milliliters?* **multiply** Then have students work in pairs to find the capacity in milliliters. Ask a volunteer to share the answer using the sentence frame: ____ **liters is equal to** ____ **milliliters**. Then have students find a capacity in liters when given the capacity in milliliters. Next, provide a pair of measurements, one in liters and one in milliliters. Ask, *Which capacity is greater?* Have pairs convert to determine the answer and report back using the sentence frame: ____ **is greater than** ____.

Teacher Notes:

NAME _____ DATE _____

Lesson 13 Vocabulary Cognates

Convert Metric Units of Capacity

Use the Glossary to define the math word in English and in Spanish in the word boxes. Write a sentence using your math word.

liter	litro (L)
Definition A metric unit for measuring volume or capacity.	**Definición** Unidad métrica para medir el volume o la capacidad.

My math word sentence:
Sample answer: 1 liter = 1,000 milliliters

milliliter	mililitro (mL)
Definition A metric unit used for measuring capacity.	**Definición** Unidad métrica para medir la capacidad.

My math word sentence:
Sample answer: 1,000 milliliters = 1 liter

Chapter 12 Geometry

What's the Math in This Chapter?

Mathematical Practice 7: Look for and make use of structure

Distribute one geoboard and one rubber band to each student. Say, *Create your own geometric shape on the board.* Give students time to create shapes. Ask students to form groups with other students based on similar shape attributes, such as angle measurement or number of sides, etc. Have groups share with the class how their shapes are similar. Continue to have students regroup themselves based on other attributes of their shapes.

After students have had time to classify their shape several times, have students return to their seats. Ask students to discuss the different attributes that were represented in the shapes and create a list on the board. Discuss with students that these attributes helped them find similarities and differences in all of the polygons that were created. Emphasize the *structure*, such as the number of angles and parallel sides helped them sort and organize the geometric shapes.

Display a chart with Mathematical Practice 7. Restate Mathematical Practice 7 and have students assist in rewriting it as an "I can" statement, for example: **I can organize geometric figures by their structure.** Post the new "I can" statement.

Inquiry of the Essential Question:

How does geometry help me solve problems in everyday life?

Inquiry Activity Target: **Students come to a conclusion that geometric figures have similarities and differences that can be used in problem-solving.**

As an introduction to the chapter, present the Essential Question to students. The inquiry graphic organizer will offer opportunities for students to observe, make inferences, and apply prior knowledge of geometric figures representing the Essential Question. As they investigate, encourage students to draw, write, and collaborate with peers to demonstrate their observations and thinking. Then have students present additional questions they may have to a peer to extend discussions.

Regroup students and restate Mathematical Practice 7 and the Essential Question. Pose questions to reflect on what has been learned to guide students in making connections between the Mathematical Practice and the Essential Question.

NAME _____ DATE _____

Chapter 12 Geometry

Inquiry of the Essential Question:

How does geometry help me solve problems in everyday life?

Read the Essential Question. Describe your observations (I see..), inferences (I think...), and prior knowledge (I know...) of each math example. Write additional questions you have below. Then share your ideas and questions with a classmate.

The figure has 6 sides. The sides are not congruent and the angles are not congruent. So, the polygon is a hexagon that is *not* regular.

I see ...

I think...

I know...

The triangle has no congruent sides and one right angle, So, it is a scalene right triangle.

I see ...

I think...

I know...

The figure has all right angles. Opposite sides are congruent and parallel. So, it is a rectangle.

I see ...

I think...

I know...

Questions I have...

_____

_____

_____

Grade 5 • Chapter 12 *Geometry* 129

Lesson 1 Polygons
English Learner Instructional Strategy

Language Structure Support: Tiered Sentence Frames

Write the New Vocabulary with Spanish cognates for this lesson on a classroom cognate chart. Provide concrete examples. After students have completed Example 3, display the following sentence frames to help them describe the shapes they've drawn:

Emerging: **This is a _____. It has _____ sides.**

Expanding: **I drew a _____. It is not regular because _____.**

Bridging: **I drew a _____. It is similar to a regular _____ because _____, but it is not a regular _____ because _____.**

English Language Development Leveled Activities

Emerging Level	Expanding Level	Bridging Level
Memory Device	**Academic Vocabulary**	**Academic Language**
Display a picture of an octopus. Say, *This is an octopus. It has eight tentacles.* Point to each tentacle as you count them. Write the word octopus and underline the prefix *oct-*. Draw an octagon. Point to each side as you count them. Say, *This polygon has eight sides. It is called an octagon.* Write the word *octagon* and underline the prefix *oct-*. Say *octagon* again and have students chorally repeat. Draw other examples and non-examples of octagons. Have students give a thumbs-up if the shape is an octagon and a thumbs-down if it is not an octagon.	Display a square and a rectangle. Cut strings the length of each side for the square and the rectangle. Display the strings from the square. Say, *All sides are the same length. When all sides of a polygon are the same length, it is a regular polygon.* Label the square *regular polygon*. Display the strings from the rectangle and say, *These sides are not all the same length, so this is not a regular polygon.* Label the rectangle *not regular*. Display regular and irregular polygons and have students identify them using the following sentence frame: **This polygon is/is not a regular polygon.**	Draw polygons on index cards. Make each polygon different, but create matching pairs of polygon types, such as regular pentagons; not regular pentagons; regular hexagons; not regular hexagons; and so on. Label each card with a different number. Distribute one or more cards to each student. Have students work together to find their match or matches by comparing the number of sides and side lengths. Have students describe their pairs using the following sentence frame: **Numbers _____ and _____ are a match because _____.**

Teacher Notes:

T130 Grade 5 • Chapter 12 *Geometry*

NAME _____ DATE _____

Lesson 1 Vocabulary Chart
Polygons

Use the three-column chart to organize the vocabulary in this lesson. Write the word in Spanish. Then write the correct terms to complete each definition.

English	Spanish	Definition
congruent angles	ángulo recto	Angles of a figure that are __equal__ in measure. This triangle has congruent angles.
congruent sides	lados congruentes	Sides of a figure that are __equal__ in length.
hexagon	hexágono	A polygon with __six__ sides and __six__ angles.
octagon	octágono	A polygon with __eight__ sides.
pentagon	pentágono	A polygon with __five__ sides.
polygon	poligono	A __closed__ figure made up of line segments that do not __cross__ each other.
regular polygon	poligono regular	A polygon in which all **sides** are __congruent__ and all __angles__ are congruent.

Lesson 2 Inquiry/Hands On: Sides and Angles of Triangles
English Learner Instructional Strategy

Vocabulary Support: Anchor Chart

Divide students into four groups. Say, *Make an anchor chart showing what you know about polygons, including what you have learned in previous grades and the first lesson of Chapter 12.* Discuss that each chart should include a title at the top of the poster and definitions for math vocabulary related to polygons, such as *sides, angles, regular polygon, congruent,* and so on. Suggest that students include examples of several polygons, including a triangle. Direct students to label different elements in their charts with appropriate math vocabulary. When the charts are completed, have groups display and describe their charts. Afterward, discuss how the anchor charts can help students better understand how to analyze and discuss the attributes of triangles.

English Language Development Leveled Activities

Emerging Level	Expanding Level	Bridging Level
Report Back	**Listen and Identify**	**Sentence Frames**
Review the meaning of *congruent*. Display the following sentence frames: ____ **congruent sides.** ____ **congruent angles.** As you model measuring the attributes of similar triangles, pause to ask, *How many congruent angles/sides?* Allow students to report back with a single word answer, but then model using the sentence frames and have students chorally repeat.	Draw three triangles: one with three congruent sides, one with two congruent sides, and one with no congruent sides. As you draw each triangle, say, *This triangle has ____ congruent sides.* Once you have finished drawing, ask, *Which triangle has ____ congruent sides?* Have volunteers identify the correct triangle. Say, *Which sides are congruent?* Have the student identify the congruent sides or say, **There are no congruent sides.** Draw three new triangles with two, three, and no congruent angles. Ask new volunteers to identify them.	Have students work in small groups to solve Apply It Exercise 12. Display the follow sentence frames: **We know ____. We need to find out ____. First we ____. Then we ____. The answer is ____. We can check the answer by ____.** Have groups complete each sentence frame as they write a description of how they solved the problem. Ask volunteers to read aloud their descriptions.

Teacher Notes:

Student page

NAME _____ DATE _____

Lesson 2 Note Taking

Inquiry/Hands On: Sides and Angles of Triangles

Read the question. Write words you need help with and research each word. Use your lesson to write your Cornell notes. Write or draw math examples to explain your thinking. Share your examples with a classmate.

Building on the Essential Question	**Notes:**
How do you describe triangles using sides and angles?	A _polygon_ is a closed figure made up of line segments that do not cross each other.
	A polygon with three sides and three angles is a _triangle_.
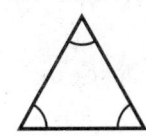	Each angle in this triangle is an _acute_ angle, which means it measures less than _90_ degrees.
	An angle that measures exactly _90_ degrees is a right angle.
Words I need help with: See students' words.	An angle that measures greater than _90_ degrees is an obtuse angle.
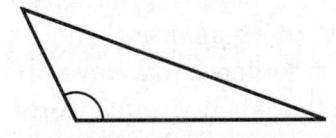	When **angles** have the **same** measure, they are called _congruent_ angles.
	When **sides** of a triangle measure the **same** length, they are called _congruent_ sides.
	The _perimeter_ of a triangle can be found by adding the side lengths.

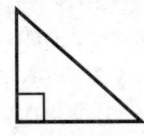

My Math Examples:

See students' examples.

Grade 5 • Chapter 12 Geometry   **131**

Lesson 3 Classify Triangles
English Learner Instructional Strategy

Collaborative Support: Show What You Know

Divide students into three groups. Randomly distribute the following My Vocabulary Cards to groups: *isosceles triangle, equilateral triangle, scalene triangle*. Have one student in the group read aloud term and definition. Have a second student, guided by the others in the group, draw an example of the triangle. Then have a third student stand and explain why the triangle is an example of the type on their card. Repeat the activity with the cards for *acute, obtuse,* and *right triangles*. Be sure any students who did not participate the first time play a role this time.

English Language Development Leveled Activities

Emerging Level	Expanding Level	Bridging Level
Word Knowledge	**Memory Device**	**Academic Language**
Display a picture of a tricycle. Say, *This is a tricycle. It has three wheels.* Point to each wheel as you count them. Write *tricycle* and underline the prefix *tri-*. Draw a triangle. Point to each angle as you count them. Say, *This polygon has three sides and three angles. It is called a triangle.* Write *triangle* and underline the prefix *tri-*. Say *triangle* again and have students chorally repeat. Say, *The prefix tri- means "three."* Draw other examples and nonexamples of triangles. Have students give a thumbs-up if the polygon is a triangle and a thumbs-down if it is not.	Display an equilateral, isosceles, and scalene triangle. Say, *These are all triangles, but how are they different?* Cut strings for the lengths of each side of the triangles. Display the strings for the equilateral triangle and say, *All three sides are **equal** in length. This is an **equilateral** triangle.* Write *equilateral* and underline *equil*. Have students say, **equilateral**. Display the strings for the isosceles and scalene triangles, describe their sides, and label them. Underline the *n* in *scale<u>n</u>e* and stress *no* as you say, *Scalene triangles have **no** sides that are the same length.*	Create pairs of similar but not congruent triangles out of card stock. Each pair should be different. Label each triangle with a different number. Distribute one or more triangles to each student. Have students work together to find their match or matches. Once a match has been made, ask them to compare the side lengths and angles of their pairs using the following sentence frame: **Numbers ___ and ___ are a match because ___.** Repeat the activity so that students can experience matching different types of triangles.

Teacher Notes:

T132 Grade 5 • Chapter 12 *Geometry*

NAME _____ DATE _____

Lesson 3 Concept Web

Classify Triangles

Use the concept web to identify each triangle as an isosceles, an equilateral, or a scalene triangle.

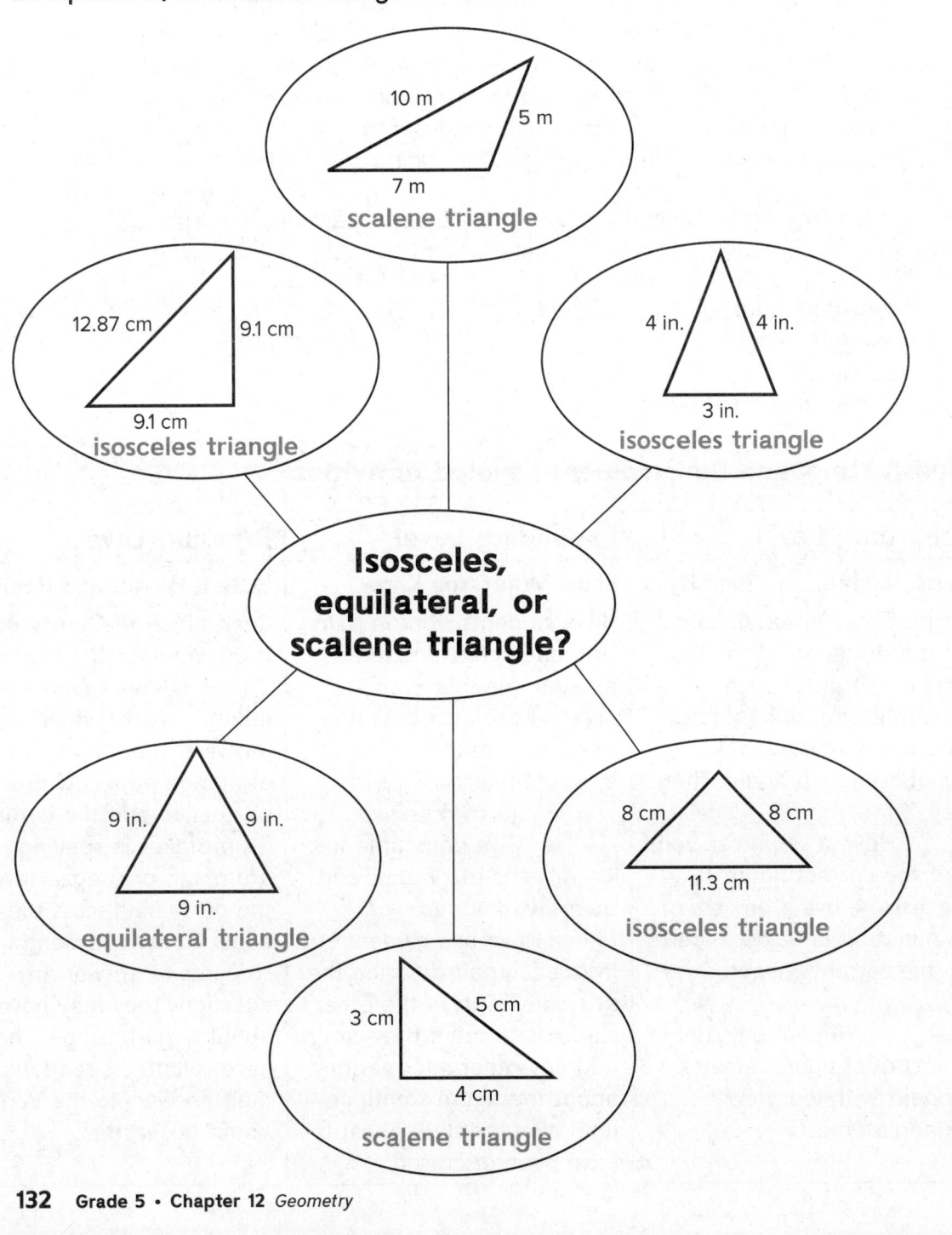

- 10 m, 5 m, 7 m — scalene triangle
- 12.87 cm, 9.1 cm, 9.1 cm — isosceles triangle
- 4 in., 4 in., 3 in. — isosceles triangle

Isosceles, equilateral, or scalene triangle?

- 9 in., 9 in., 9 in. — equilateral triangle
- 8 cm, 8 cm, 11.3 cm — isosceles triangle
- 3 cm, 5 cm, 4 cm — scalene triangle

Lesson 4 Inquiry/Hands On: Sides and Angles of Quadrilaterals

English Learner Instructional Strategy

Vocabulary Support: Activate Prior Knowledge

Display anchor charts, word webs, KWL charts, or any other graphic organizers from previous lessons related to polygons, including examples of how to classify triangles by their attributes. Be sure to include the classroom cognate chart as well. Have students take turns coming up to the graphic organizers and sharing a piece of information about polygons with the other students. Ask students how their knowledge of classifying triangles might be applied to classifying other kinds of polygons, including quadrilaterals.

Display the following sentence frames to help students participate during the lesson:

____ congruent sides
____ congruent angles
____ parallel sides
Figures ____ have congruent ____.

English Language Development Leveled Activities

Emerging Level	Expanding Level	Bridging Level
Look, Listen, and Identify	**Show What You Know**	**Listen, Write, and Read**
Display four different kinds of quadrilaterals. Describe one attribute of each quadrilateral, but try to choose attributes that are unique to each figure. Then say, *Show me the figure with ____.* Have a volunteer point out the correct figure. If necessary, use a process of elimination to guide students to the correct answer: *Does this figure have ____? No, it has ____.* Repeat until only the correct figure remains. Repeat with four new quadrilaterals.	Have students work in pairs. Use index cards to create enough quadrilateral examples to distribute one to each student pair. Say, *Discuss the attributes of your figure with your partner.* Give pairs time to identify the attributes, and then say, *Describe your figure.* Have one student from each pair describe the attributes. Survey the other students to see if they can add any other information about the figure. Continue until all quadrilateral cards have been discussed.	Read aloud the Write About It question for the lesson. Have students spend several minutes writing their answers. Then have students exchange papers with a partner to edit the writing for mistakes in spelling, grammar, or usage. Have the partners discuss the edits. Circulate among students to answer any questions they may have about English usage. Then have volunteers read aloud their answers to the Write About It question.

Teacher Notes:

NAME _____ DATE _____

Lesson 4 Guided Writing

Inquiry/Hands On: Sides and Angles of Quadrilaterals

How do you describe quadrilaterals using sides and angles?

Use the exercises below to help you build on answering the Essential Question. Write the correct word or phrase on the lines provided.

1. Rewrite the question in your own words.
   See students' work.

2. What key words do you see in the question?
   quadrilateral, sides, angles

3. A __polygon__ is a closed figure made up of line segments that do not cross each other.

4. A polygon that has four sides and four angles is a __quadrilateral__.

5. On each of these quadrilaterals, the arrows point at angles that are across from each other, not next to each other. These angles are called __opposite__ angles.

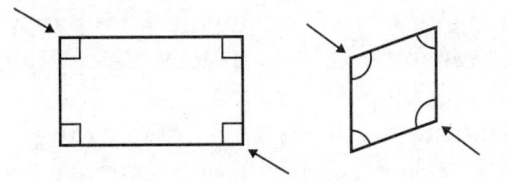

6. Angles that have the same measure are __congruent__ angles.

7. Sides that are the same __length__ are congruent sides.

8. Parallel lines are lines that are the __same__ distance apart.

9. Opposite sides are sides that are across from each other. The sides do not meet. Opposite sides that are the same distance apart are __parallel__.

10. How do you describe quadrilaterals using sides and angles?
    Quadrilaterals are polygons that have 4 sides and 4 angles. Measure the angles and measure the length of each side. Determine if any sides are parallel or congruent. Determine if any angles are congruent.

Grade 5 • Chapter 12 *Geometry* 133

Lesson 5 Classify Quadrilaterals
English Learner Instructional Strategy

Graphic Support: Venn Diagram

Write the New Vocabulary with Spanish cognates for this lesson on a classroom cognate chart. Provide concrete examples for the vocabulary by displaying models for each quadrilateral.

Display a Venn diagram, and use it to compare and contrast quadrilateral pairs. For example, write *Square* above the left side and draw an example of a square below. Do the same for *Rectangle* on the right side. Then display the following sentence frames for students to use in comparing and contrasting the shapes: A ____ is similar to a ____ because ____. A ____ is different from a ____ because ____. Continue until all quadrilateral types have been compared.

English Language Development Leveled Activities

Emerging Level	Expanding Level	Bridging Level
Word Recognition Draw several pairs of lines on the board. Some pairs should be parallel, some should intersect, and others should be unparallel but not intersecting. Write *parallel*, then point to each pair and identify it as parallel or not parallel. Have students chorally repeat **parallel** or **not parallel**. Extend the *ll*s in *parallel* to make them look like a pair of parallel lines. Say, *The two Ls are parallel.* Display images of other parallel lines using real world examples (stripes, stair rails, road lines), and have students identify any parallel lines in the images.	**Academic Vocabulary** Draw a set of parallel lines and a parallelogram, and label them as *parallel* and *parallelogram*. Extend the *ll*s in each word to make them look like pairs of parallel lines, and then identify them as such to reinforce meaning. Draw a rectangle on the board. Point to each set of opposite, parallel sides and identify them as parallel. Have students identify the number of pairs of parallel sides in the shape using the frame: **There are ____ pairs of parallel sides in this shape.** Repeat with a square, trapezoid, rhombus, and parallelogram.	**Building Oral Language** Have student pairs create a two-column chart labeled *Rectangle* and *Parallelogram*. Then ask each pair to cut out a parallelogram and a rectangle from paper or card stock. Say, *Place each shape in the appropriate column.* Display the following sentence frames: **A rectangle and a parallelogram are similar because ____. A rectangle and a parallelogram are different because ____.** Have students use the sentence frames to write sentences comparing and contrasting the two shapes. Ask students to read aloud their sentences.

Teacher Notes:

NAME _____ DATE _____

Lesson 5 Vocabulary Chart
Classify Quadrilaterals

Use the three-column chart to organize the vocabulary in this lesson. Write the word in Spanish. Then write the correct terms to complete each definition.

English	Spanish	Definition
parallelogram	paralelogramo	A quadrilateral with __four__ sides in which each pair of opposite sides are __parallel__ and __congruent__.
rectangle	rectángulo	A quadrilateral with __four__ right angles; __opposite__ sides are equal and __parallel__.
rhombus	rombo	A __parallelogram__ with __four__ congruent sides.
trapezoid	trapecio	A quadrilateral with exactly __one__ pair of __parallel__ sides.
square	cuadrado	A __rectangle__ with __four__ congruent sides.

Lesson 6 Inquiry/Hands On: Build Three-Dimensional Figures

English Learner Instructional Strategy

Vocabulary Support: Cognates

Write *three-dimensional figure* and the Spanish cognate, *figura tridimensional,* on a classroom cognate chart. Provide a concrete example of meaning by placing the net for a three-dimensional figure onto a flat surface. Say, *This is a two-dimensional figure. It has two dimensions: length and width.* Run your hand along the length and width as you identify them. Have students say: **two-dimensional figure**. Fold the net to form the three-dimensional figure. Say, *This is a three-dimensional figure. It has three dimensions: length, width, and height.* Run your hand along the length, width, and height as you identify them. Have students say: **three-dimensional figure**.

Draw several common polygons on the board. Describe each as a two-dimensional figure. Ask students to find three-dimensional figures in the classroom that have at least one face matching the shape of a polygon on the board.

English Language Development Leveled Activities

Emerging Level	Expanding Level	Bridging Level
Act It Out Display the following sentence frames: ____ **squares.** ____ **rectangles.** Place a rectangular prism onto a sheet of graph paper. Trace each side of the prism to create a net for the three-dimensional figure. Ask, *How many faces are rectangles? How many are squares?* Have students answer using the sentence frames. Model cutting out the net and folding it to form a rectangular prism. Ask the same questions as before and have students answer. Repeat the activity with a cube, but have volunteers do the tracing, cutting out, and folding.	**Think-Pair-Share** Read aloud both parts of Apply It Exercise 10. Say, *Think about the answer, and then find a partner and discuss the answer with him or her.* After students have had a chance to talk with a partner, display the following sentence frames: **The sides of the building are ____ . There are ____ faces.** Ask volunteers to share their answers using the sentence frames. Display a photograph of another building and repeat the activity. Ask students to compare the two buildings' shapes.	**Numbered Heads Together** Have students get into groups of four. Ask the students in each group to number off as 1–4. Display the following sentence frames: **First ____ . Then ____ . Next ____ . Finally ____ .** Have the students in each group work together to write an answer to Write About It exercise 13 using the sentence frames to describe the steps needed to draw a net and create a three-dimensional figure. Afterward, choose numbers from 1–4 to designate which student in each group will use the sentence frames to describe their group's answer.

Teacher Notes:

NAME _____   DATE _____

Lesson 6 Note Taking

Inquiry/Hands On: Build Three-Dimensional Figures

Read the question. Write words you need help with and research each word. Use your lesson to write your Cornell notes. Write or draw math examples to explain your thinking. Share your examples with a classmate.

Building on the Essential Question	Notes:
How do I build three-dimensional figures?	A __three__ - __dimensional__ figure is a figure that has length, width, and height. A __prism__ that has rectangular bases is a rectangular prism. A flat surface is called a __face__. A __cube__ is a rectangular prism with faces that are congruent squares. Two figures having the same size and the same shape are called __congruent__ figures. A two-dimensional pattern of a three-dimensional figure is called a __net__. The net below is for a rectangular __prism__, it is made up of __6__ rectangles.
Words I need help with: See students' words.	
My Math Examples: See students' examples.	

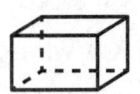

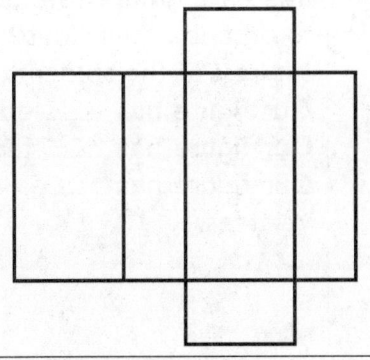

Grade 5 • Chapter 12 *Geometry* **135**

Lesson 7 Three-Dimensional Figures
English Learner Instructional Strategy

Sensory Support: Realia

Ahead of time, be sure there are at least a few examples of triangular prisms around the classroom, along with cubes and rectangular prisms. Then divide students into three groups and say, *We are going on a shape hunt.* Randomly assign each group as either Cube, Rectangular prism, or Triangular prism. Say, *Find examples of your shape in the classroom.* After a few minutes, have groups describe the objects they found using this sentence frame: ____ is a ____ because ____.

English Language Development Leveled Activities

Emerging Level	Expanding Level	Bridging Level
Word Knowledge Make labels on sentence strips for *Two-dimensional* and *Three-dimensional*. Gather a wide variety of objects to use as examples for two-dimensional and three-dimensional shapes. Display a circle drawn on paper. Say, *This shape is flat. It is two-dimensional.* Place it by the Two-dimensional label. Then set a ball or other sphere shaped object onto the paper circle. Say, *This shape sticks out. It is three-dimensional.* Set it by the other label. Repeat, with the other examples. Have students chorally identify each as **flat and two-dimensional** or **sticks out and three-dimensional**.	**Word Recognition** Write and say *edge* as you point to and count the edges of a three-dimensional figure. Repeat the process for *face* and *vertex*. Divide students into small groups and distribute one random three-dimensional figure to each group. Say, *Count the edges, faces, and vertices for your figure.* Display the following sentence frames and have groups use them to describe their figure to the rest of the students: **Our figure has ____ edges. Our figure has ____ faces. Our figure has ____ vertices.**	**Academic Language** Have students work in pairs. Confidentially distribute random three-dimensional figures to students. Ensure that students do not see their partners' figures. Say, *Describe your figure and have your partner guess what it is.* Display the following sentence frames for students to use in their descriptions: **The figure has ____ edges. The figure has ____ faces. The figure has ____ vertices.** If the figure is a prism, **The figure has a ____ base.** Once pairs have guessed correctly, have students secretly exchange shapes, and have pairs repeat the activity.

Teacher Notes:

NAME _____ DATE _____

Lesson 7 Vocabulary Chart

Three-Dimensional Figures

Use the three-column chart to organize the vocabulary in this lesson. Write the word in Spanish. Then write the correct terms to complete each definition.

English	Spanish	Definition
base	base	One of the two parallel _congruent_ faces in a prism.
cube	cubo	A rectangular prism with _six_ faces that are congruent _squares_.
prism	prisma	A three-dimensional figure with _two_ parallel, congruent faces, called _bases_. At least three faces are _rectangles_.
rectangular prism	prisma rectangular	A prism that has _rectangular_ bases.
three-dimensional figure	figura tridimensional	A figure that has length, width, and _height_.
triangular prism	prisma triangular	A prism that has _triangular_ bases.
vertex	vértice	The point where two _rays_ meet in an angle or where three or more _faces_ meet on a three-dimensional figure.
edge	arista	The line segment where two _faces_ of a three-dimensional figure meet.
face	cara	A _flat_ surface.

136 Grade 5 • Chapter 12 *Geometry*

Lesson 8 Inquiry/Hands On: Use Models to Find Volume

English Learner Instructional Strategy

Sensory Support: Manipulatives

Write *cubic unit* and *unit cube* and the Spanish cognates, *unidad cúbica* and *cubo unitario,* on a classroom cognate chart. As you introduce and use the terms during the lesson, refer back to the cognate chart to assist native Spanish speakers with understanding. Place a spinner and unit cubes on a table. Invite a student to come to the table, spin to generate a number, and then use that number of unit cubes to construct a three-dimensional figure. Ask, *How many unit cubes did you use?* Display a sentence frame for the student to use: ____ **unit cubes.** Then ask the other students, *What is the volume of the figure?* Display a sentence frame to help them answer: ____ **cubic units.** As time allows, have students take turns coming to the table and creating new figures based on the numbers they spin.

English Language Development Leveled Activities

Emerging Level	Expanding Level	Bridging Level
Tiered Questions	**Academic Vocabulary**	**Partners Work**
As you work through the Model the Math, Build It, and Try It parts of the lesson, monitor students' comprehension by asking questions appropriate to their level of English proficiency. Ask questions that can be answered with a gesture or a single word. For example, *Is this an edge or the base? Show me Layer 2. Is the volume 12 or 48? How many cubes in this row?* and so on. List vocabulary on the board that students will need during the lesson: *layer, volume, cubic centimeters, cube.*	Display a Venn diagram. Label one side *area* and the other side *volume*. Lead a discussion of the differences and similarities between the two ways of measuring. Display sentence frames to help students participate: **Area is used to measure ____. Volume is used to measure ____. Area and volume are similar because ____.**	Have students work in pairs. Say, *Write a short paragraph explaining to a younger student how to find the volume of a rectangular prism.* After pairs have completed the task, ask volunteers to read aloud their explanations. Gently correct any mistakes in English usage or grammar by providing a model. Discuss any steps that were missing in students' explanations.

Multicultural Teacher Tip

Most ELs have had math education in their native countries and are familiar with basic math concepts. However, mathematical discourse in an unfamiliar language can be intimidating and confusing, and students may struggle with even seemingly simple steps leading to a solution. Manipulatives are a helpful option for ELs. By utilizing concrete objects to model familiar concepts or to learn new ones, students can work around language barriers that might make verbal or written explanations too difficult. Keep in mind that manipulatives are not always used in other cultures, and the student may need time and encouragement to become comfortable using them instead of solving on paper.

NAME _____ DATE _____

Lesson 8 Vocabulary Cognates

Inquiry/Hands On: Use Models to Find Volume

Use the Glossary to define the math word in English and in Spanish in the word boxes. Write a sentence using your math word.

volume	volumen
Definition The amount of space inside a three-dimensional figure.	**Definición** Cantidad de espacio dentro de una figura tridimensional.
My math word sentence: Sample answer: The volume of a prism can be found by counting centimeter cubes used to build the prism.	

cubic unit	unidad cúbica
Definition A unit for measuring volume, such as a cubic inch or a cubic centimeter.	**Definición** Unidad para medir el volume, como una pulgada cúbica o un centímetro cúbico.
My math word sentence: Sample answer: The volume of a prism built using centimeter cubes has the cubic unit of cm³.	

unit cube	cubo unitario
Definition A cube with a side length of one unit.	**Definición** Cubo con lados de una unidad de longitud.
My math word sentence: Sample answer: A centimeter cube has the side length of 1 centimeter. It is a unit cube.	

Lesson 9 Volume of Prisms
English Learner Instructional Strategy

Collaborative Support: Round the Table

Write *volume* and the Spanish cognate *volumen*, on a classroom cognate chart. Model with realia to demonstrate meaning.

Place students into multilingual groups of 4 or 5. Assign Independent Practice Exercises 3 and 7 to each group. Have one student draw Exercise 3 on a large piece of paper. Have another student label the prism. Then have the students continue to work jointly to solve the problem by passing the paper around the table. Each student will perform one step in multiplying to find the volume. Direct each student in a group to write with a different color to ensure all students participate in solving the problem. Afterward, choose one student to report back to you. Instruct students to repeat with Exercise 7.

English Language Development Leveled Activities

Emerging Level	Expanding Level	Bridging Level
Word Knowledge Use objects or images to demonstrate different meanings of *volume*, such as a single volume from a set of books, a radio for volume as sound level, and an empty container to model volume as capacity. Afterward, fill the container with rice. Say, **Volume** *is the amount of space inside.* Say *volume* again and have students chorally repeat. Display several clear rectangular prism containers, and then work with students to order them from least to greatest volume. Ask comparative questions, such as, *Does this one hold more or less than that one?*	**Act It Out** Display a small clear rectangular prism container. Hold up a centimeter cube and say, *Each cube is one cubic centimeter.* Use centimeter cubes to fill the container. Afterward, count the cubes and state the volume: *The volume is ____ cubic centimeters.* Distribute a variety of clear rectangular prism containers, one to each student pair. Direct pairs to find the volumes of their containers in the same manner you modeled using centimeter cubes. Have pairs share their answers using a sentence frame: **The volume is ____ cubic centimeters.**	**Academic Language** Create cards with images of rectangular swimming pools. Label each pool's dimensions in feet. Distribute cards to student pairs and ask them to describe the pool on their card using the sentence frame: **Our pool is ____ feet wide, ____ feet long, and ____ feet deep.** Record the dimensions and ask students to guess which pool has the greatest volume. Have pairs calculate their pool's volume and share it using the sentence frame: **The volume of our pool is ____ cubic feet.** Discuss which pool has the most volume and compare it to the guesses.

Teacher Notes:

NAME _____ DATE _____

Lesson 9 Multiple Meaning Word
Volume of Prisms

Complete the four-square chart to review the multiple meaning word.

Everyday Use	**Math Use in a Sentence**
Sample answer: Quantity of sound. For example, you can turn up the volume on a song.	Sample sentence: The volume of a cube is found by multiplying the side length by itself three times.
Math Use	**Example From This Lesson**
The amount of space inside a three-dimensional figure.	Sample answer: Volume = base × height The base can be found by multiplying length by width.

(center: volume)

Write the correct term on each line to complete the sentence.

The volume of a container can be found by __multiplying__ the length by the __width__ by the height.

Lesson 10 Inquiry/Hands On: Build Composite Figures

English Learner Instructional Strategy

Language Structure Support: Sentence Frames

Write *composite figure* and the Spanish cognate, *figura compuesta,* on a classroom cognate chart. Underline *compos* and say, *When you compose, you build something or put it together. When you put together two or more three-dimensional figures, you build a composite figure.* Provide a concrete example by creating a composite figure from two figures.

Display the following sentence frames to help students participate in the lesson:

____ centimeter cubes
The layer has ____ centimeter cubes.
I added ____ and ____.
The volume is ____ cubic centimeters.

English Language Development Leveled Activities

Emerging Level	Expanding Level	Bridging Level
Activate Prior Knowledge Write *volume = length × width × height*. Create a simple rectangular prism using centimeter cubes. Display the following sentence frames: **The length is ____. The width is ____. The height is ____. The volume is ____.** Model finding the volume by first counting the number of centimeter cubes in each dimension. Pause after counting each dimension and ask, *What is the length/width/height/volume?* Allow students to answer with a number, but then model using the sentence frame and have students chorally repeat.	**Academic Word Knowledge** Distribute centimeter cubes to each student. Say, *Use the cubes to make a rectangular prism that is one or two layers high.* Pair up students, and say, *Use your two prisms to build a composite figure. Find the volume of the composite figure.* Give students time to complete the task. Have pairs share the volume of their composite figures and explain how they arrived at the answer. Display a sentence frame for students to use in sharing their answers: **The volume of the composite figure is ____.**	**Pass the Pen** Divide students into small groups, and distribute centimeter cubes to each group. Say, *Use the cubes to make two different rectangular prisms. Combine the prisms to build a composite figure. Find the volume of the composite figure.* Have students work jointly on the problem, with students taking turns to complete each step. Direct each member of the group to write a sentence or phrase on a strip of paper describing the step he or she completed. Afterward, have groups share their work, and have each student read the sentence he or she wrote.

Teacher Notes:

T139 Grade 5 • Chapter 12 *Geometry*

NAME _____ DATE _____

Lesson 10 Guided Writing

Inquiry/Hands On: Build Composite Figures

How do you build composite figures using centimeter cubes?

Use the exercises below to help you build on answering the Essential Question. Write the correct word or phrase on the lines provided.

1. Rewrite the question in your own words.
   See students' work.

2. What key words do you see in the question?
   composite figure, centimeter cube

3. <u>Volume</u> is the amount of space inside a three-dimensional figure.

4. A <u>composite</u> figure is made up of two or more three-dimensional figures.

5. The <u>volume</u> of a composite figure made with centimeter cubes can be found by <u>counting</u> the total number of centimeter cubes used to build the composite figure.

6. The volume of the rectangular prism below is <u>12</u> cubic centimeters.

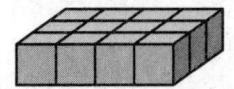

7. The volume of the rectangular prism below is <u>6</u> cubic centimeters.

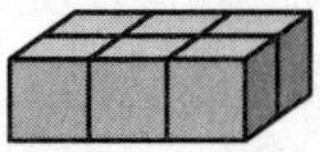

8. The volume of the composite figure below is equal to the <u>sum</u> of the volume of each rectangular prism above. The volume of this composite figure is <u>18</u> cubic centimeters.

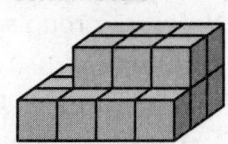

9. How do you build composite figures?
   Build rectangular prisms using centimeter cubes. Then combine the rectangular prisms to create a composite figure. The volume of the composite figure is the total number of cubes used to create the figure, which is equal to the sum of the volumes of the prisms used to create the composite figure.

Lesson 11 Volume of Composite Figures

English Learner Instructional Strategy

Language Structure Support: KWL Chart

Display a KWL chart. In the first column, record what students already know about three-dimensional shapes and finding volume. In the second column, record what students hope to learn during the lesson, including how to find the volume of composite figures. After the lesson, display the following sentence frame. Have students use it to describe what they learned: **I learned that** ____. Record student responses in the third column of the KWL chart.

Pair emerging students with more proficient English speaking mentors and assign Exercise 11. Have them work together to describe the steps in finding the volume of a composite figure, and then have the emerging student report back their answer to you, guided by their mentor.

English Language Development Leveled Activities

Emerging Level	Expanding Level	Bridging Level
Word Knowledge Display an empty box. Say, *Here is a figure. What is the volume?* Measure the sides, calculate the volume, and record it. Display another, smaller box and say, *Here is another figure. What is the volume?* Measure the sides, calculate the volume, and record it. Next, stack the boxes and say, *We had two figures. Now they are put together as one. This is a composite figure.* Say *composite* again and have students chorally repeat. Say, *We add the volumes of the two boxes. This is the volume of the composite figure.* Model the addition.	**Recognize and Act It Out** Give each student three number cubes, and then have students get into pairs. Say, *Roll your cubes. Use the numbers you roll as the width, length, and height of a figure. Build your figure with centimeter cubes.* Have partners combine their figures to create a single composite figure. Say, *Find the volume of your composite figure.* Display the following sentence frame and have pairs use it to describe the volume of their composite figure: **The volume of our composite figure is ____ cubic centimeters.**	**Academic Language** Divide students into an even number of groups. Randomly distribute 24 or more centimeter cubes to groups so that two groups have 24, two groups have 28, two groups have 32, and so on. Say, *Use your cubes to make a composite figure.* After the task is complete, have each group move to sit by a different group's figure. Say, *Find the volume of the composite figure.* Then direct groups to work together to determine which composite figures match in volume. Discuss how different composite figures can have the same volume.

Teacher Notes:

Student page

NAME _____ DATE _____

Lesson 11 Vocabulary Definition Map

Volume of Composite Figures

Use the definition map to write a description and list characteristics about the vocabulary word or phrase. Write or draw math examples. Share your examples with a classmate.

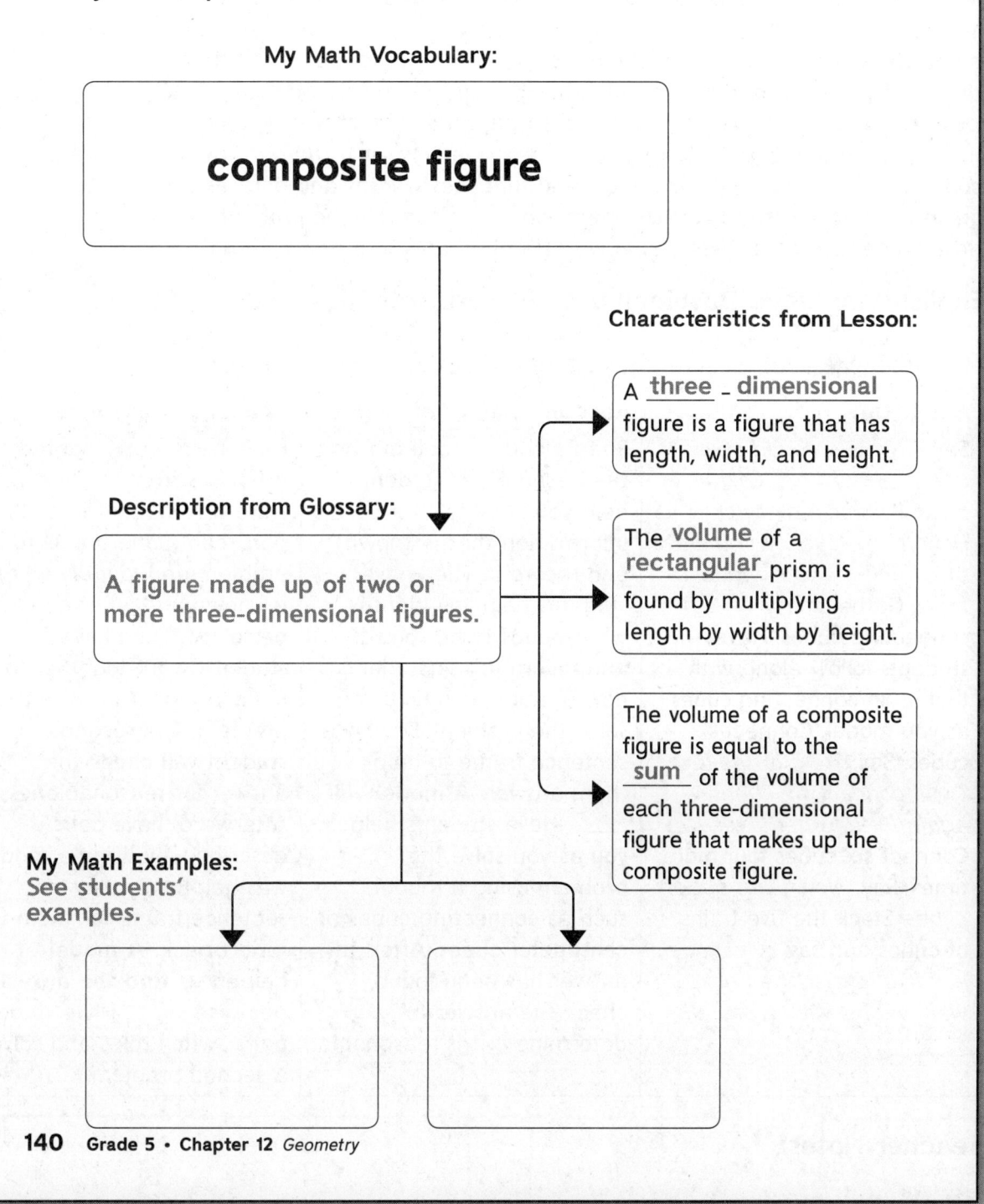

My Math Vocabulary:

composite figure

Description from Glossary:

A figure made up of two or more three-dimensional figures.

Characteristics from Lesson:

A <u>three</u> - dimensional figure is a figure that has length, width, and height.

The <u>volume</u> of a <u>rectangular</u> prism is found by multiplying length by width by height.

The volume of a composite figure is equal to the <u>sum</u> of the volume of each three-dimensional figure that makes up the composite figure.

My Math Examples:
See students' examples.

140 Grade 5 • Chapter 12 *Geometry*

Lesson 12 Problem-Solving Investigation Strategy: Make a Model

English Learner Instructional Strategy

Language Structure Support: Utilize Resources

As students work through the lesson exercises, be sure to remind them that they can refer to the Glossary or the Multilingual eGlossary for help with math terms, or direct students to other translation tools if they are having difficulty with non-math language in the problems, such as: *mail, package, assembly line, work station, store, cans of food, display, center ring, circus, clown, ball of yarn, practicing,* or *trumpet*. Remind students to look closely at signal words and phrases that appear in story problems, such as: *How many in each, only once, describe how,* or *if the pattern continues,* and help students understand that these words and phrases will guide them towards the correct operation to use to solve the problem. Have students record these signal words and phrases in a math journal.

English Language Development Leveled Activities

Emerging Level	Expanding Level	Bridging Level
Act It Out	**Recognize and Act It Out**	**Academic Language**
Say, *I want to build a wall using thirty bricks. Each* **layer** *will have six bricks. How many layers will there be? I will* **make a model** *to solve.* Gather thirty connecting cubes. Have students follow along with their own connecting cubes as you model. Connect six cubes. Say, *Here is one* **layer**. *I will connect six cubes again for another layer.* Connect six cubes four more times. Say, *Now I am out of cubes.* Stack the five trains of cubes and say, *My model has five layers. The brick wall will be five layers tall.*	Read aloud a problem from the lesson. Have students help you identify information that is known and record it. Then have students help you identify what needs to be solved and record it. Ask students how a model can help you solve the problem. Display a sentence frame to help them answer: **A model will ____.** Have students help you as you solve the problem using a model, such as connecting cubes or centimeter cubes. After an answer has been found, check the answer to determine if it is reasonable.	Have pairs work together on problems from the lesson. Have one student in each pair read aloud the word problem and identify what is known and what needs to be solved. Then have students work together to make a model to solve the problem. The second student will check the answer for reasonableness. Afterward, have pairs describe why using a model was helpful. Display a sentence frame for them to report back: **A model helped us find the answer because ____.** Have student pairs switch roles and solve a second problem.

Teacher Notes:

Student page

NAME _____ DATE _____

Lesson 12 Problem-Solving Investigation

STRATEGY: Make a Model

Solve each problem by making a model.

1. On an assembly line that is **150** feet **long**, there is a work station **every** **15** feet. The **first station** is at the **beginning** of the line. How many work **stations** are there?

Understand	Solve
I know:	
I need to find:	
Plan	**Check**
Label this line to represent the assembly line.	

2. A store is stacking cans of food into a **rectangular prism** display. The **bottom** layer has **8** cans **by** 5 cans. There are **5 layers**. How **many** cans are in the **display**?

can

Understand	Solve
I know:	
I need to find:	
Plan Model the bottom layer. How many cans are in the bottom layer?	**Check**

Grade 5 • Chapter 12 *Geometry* **141**

An Interview with Dinah Zike Explaining Visual Kinesthetic Vocabulary®, or VKVs®

What are VKVs and who needs them?

> VKVs are flashcards that animate words by kinesthetically focusing on their structure, use, and meaning. VKVs are beneficial not only to students learning the specialized vocabulary of a content area, but also to students learning the vocabulary of a second language.

Dinah Zike | Educational Consultant
Dinah-Might Activities, Inc. – San Antonio, Texas

Why did you invent VKVs?

> Twenty years ago, I began designing flashcards that would accomplish the same thing with academic vocabulary and cognates that Foldables® do with general information, concepts, and ideas—make them a visual, kinesthetic, and memorable experience.

I had three goals in mind:

- **Making two-dimensional flashcards three-dimensional**

- **Designing flashcards that allow one or more parts of a word or phrase to be manipulated and changed to form numerous terms based upon a commonality**

- **Using one sheet or strip of paper to make purposefully shaped flashcards that were neither glued nor stapled, but could be folded to the same height, making them easy to stack and store**

Why are VKVs important in today's classroom?

> At the beginning of this century, research and reports indicated the importance of vocabulary to overall academic achievement. This research resulted in a more comprehensive teaching of academic vocabulary and a focus on the use of cognates to help students learn a second language. Teachers know the importance of using a variety of strategies to teach vocabulary to a diverse population of students. VKVs function as one of those strategies.

An Interview with Dinah Zike Explaining Visual Kinesthetic Vocabulary®, or VKVs®

How are VKVs used to teach content vocabulary to EL students?

❝ VKVs can be used to show the similarities between cognates in Spanish and English. For example, by folding and unfolding specially designed VKVs, students can experience English terms in one color and Spanish in a second color on the same flashcard while noting the similarities in their roots. ❞

What organization and usage hints would you give teachers using VKVs?

❝ Cut off the flap of a 6" x 9" envelope and slightly widen the envelope's opening by cutting away a shallow V or half circle on one side only. Glue the non-cut side of the envelope into the front or back of student workbooks or journals. VKVs can be stored in the pocket.

Encourage students to individualize their flashcards by writing notes, sketching diagrams, recording examples, forming plurals (radius: radii or radiuses), and noting when the math terms presented are homophones (sine/sign) or contain root words or combining forms (kilo-, milli-, tri-).

As students make and use the flashcards included in this text, they will learn how to design their own VKVs. Provide time for students to design, create, and share their flashcards with classmates. ❞

Dinah Zike's book Foldables, Notebook Foldables, & VKVs for Spelling and Vocabulary 4th-12th won a Teachers' Choice Award in 2011 for "instructional value, ease of use, quality, and innovation"; it has become a popular methods resource for teaching and learning vocabulary.

Dinah Zike's **Visual Kinesthetic Vocabulary**®

Chapter 1

✂ cut on all dashed lines ▭ fold on all solid lines

punto

decimal

How is a decimal different from a whole number? (¿En qué se diferencia un decimal de un número entero?)

equivalent decimals

Write an equivalent decimal for 2.54. (Escribe un decimal equivalente a 2.54.)

What is another word for *equivalent*? (¿Cuál es otra palabra para decir *equivalente*?)

Chapter 1 Visual Kinesthetic Learning **VKV3**

Chapter 1

cut on all dashed lines   fold on all solid lines

point

decimal

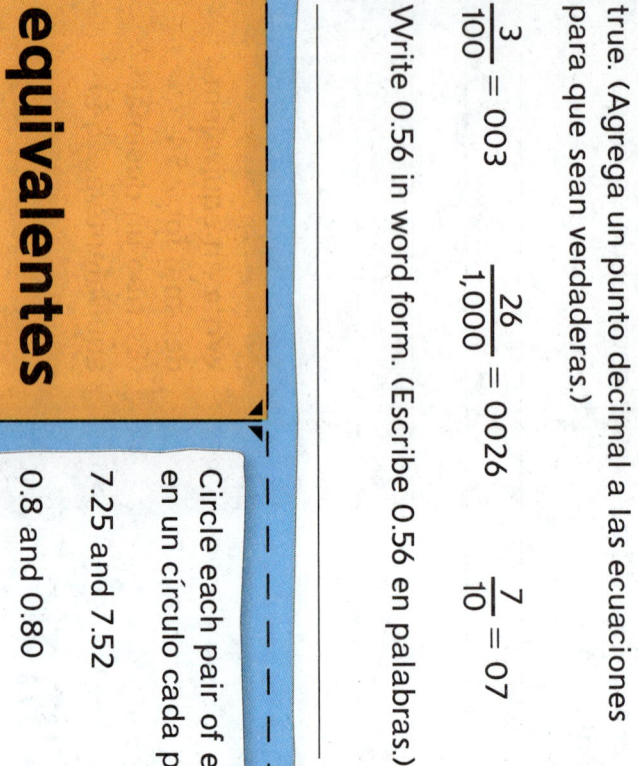

equivalentes

decimales

Add a decimal point to each equation to make it true. (Agrega un punto decimal a las ecuaciones para que sean verdaderas.)

$\dfrac{3}{100} = 003$ $\dfrac{26}{1,000} = 0026$ $\dfrac{7}{10} = 07$

Write 0.56 in word form. (Escribe 0.56 en palabras.)

Circle each pair of equivalent decimals. (Encierra en un círculo cada par de decimales equivalentes.)

7.25 and 7.52   0.005 and 0.050
0.8 and 0.80   05.2 and 5.20
00.77 and 0.077   3.40 and 3.4

VKV4   Chapter 1 Visual Kinesthetic Learning

Chapter 2

✂ cut on all dashed lines   fold on all solid lines

Which shows the prime factorization of 12? (¿Cuál muestra la descomposición en factores primos de 12?)

1 × 3 × 4
2 × 2 × 3
1 × 2 × 6

When a number is cubed, it is raised to the _____ power. (Cuando un número es elevado al cubo, es elevado a la _____ potencia.)

Define *exponent*. (Define *exponente*.)

prime factorization

cubed

exponent

Circle the prime numbers. (Encierra en un círculo los números primos.)

2   9
5   8

Chapter 2 Visual Kinesthetic Learning   VKV5

Chapter 2

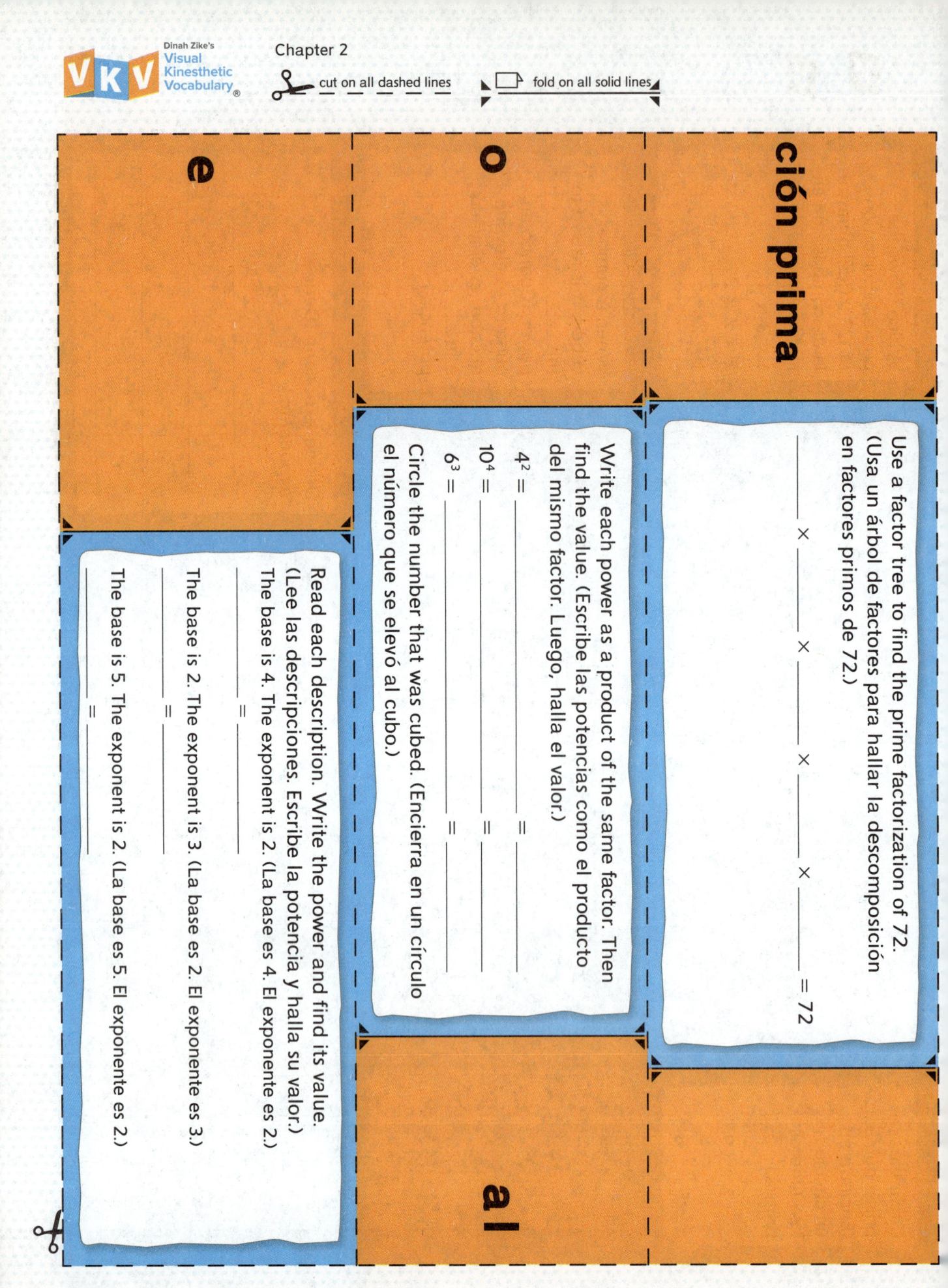

ción prima

e   o   al

Use a factor tree to find the prime factorization of 72. (Usa un árbol de factores para hallar la descomposición en factores primos de 72.)

___ × ___ × ___ × ___ × ___ = 72

Write each power as a product of the same factor. Then find the value. (Escribe las potencias como el producto del mismo factor. Luego, halla el valor.)

4^2 = ___
10^4 = ___
6^3 = ___

Circle the number that was cubed. (Encierra en un círculo el número que se elevó al cubo.)

Read each description. Write the power and find its value. (Lee las descripciones. Escribe la potencia y halla su valor.)

The base is 4. The exponent is 2. (La base es 4. El exponente es 2.)
___ = ___

The base is 2. The exponent is 3. (La base es 2. El exponente es 3.)
___ = ___

The base is 5. The exponent is 2. (La base es 5. El exponente es 2.)
___ = ___

Chapter 2

compatible numbers

Compatible numbers are (Los números compatibles son) _____

product

Circle each product. (Encierra en un círculo los productos.)

$2 \times 3 \times 3 = 18$

$5^3 = 75$

$250 \times 4 = 1,000$

estimate

Name two ways you can estimate. (Nombra dos maneras como se puede estimar.) _____

Chapter 2

r o s

números

Use compatible numbers and mental math to estimate. (Usa números compatibles y cálculo mental para estimar.)

418 × 32 is about (es aproximadamente) _____.

97 × 24 is about (es aproximadamente) _____.

763 × 47 is about (es aproximadamente) _____.

490 × 21 is about (es aproximadamente) _____.

Find each product. (Halla los productos.)

3 × 3 × 5 = _____ 4^3 = _____

2^4 = _____   45 × 3 = _____

74 × 5 = _____ 12^2 = _____

Estimate to find each product. Show how you estimated. (Estima para hallar los productos. Muestra cómo estimaste.)

74 × 59 = _____   488 × 32 = _____

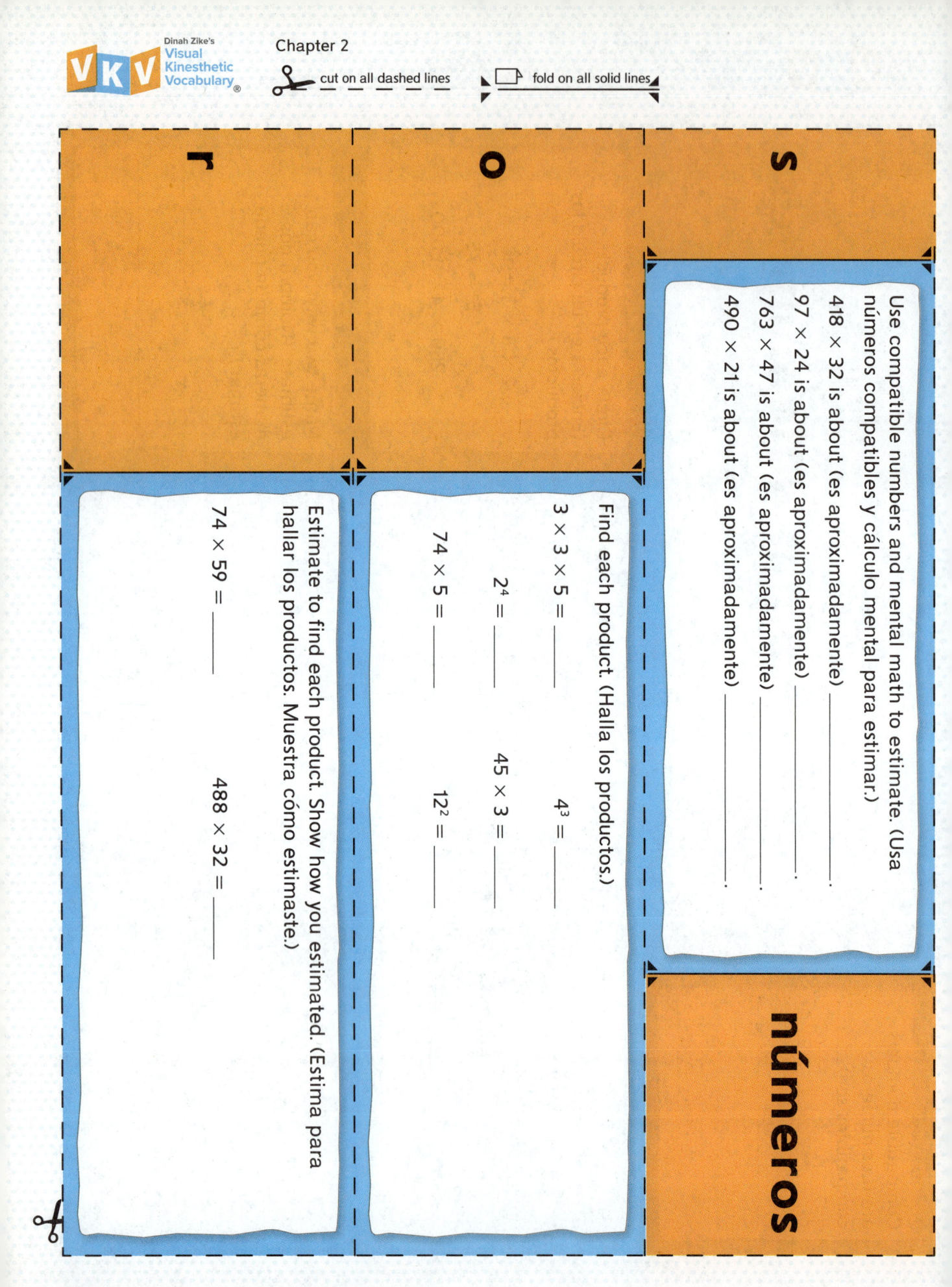

VKV8 Chapter 2 Visual Kinesthetic Learning

Chapter 3

cut on all dashed lines   fold on all solid lines

The dividend is (El dividendo es) _____

Circle the quotient. (Encierra en un círculo el cociente.)
$12\overline{)144}$

Name three multiples of 10. (Nombra tres múltiplos de 10.)

dividend

quotient

multiple

Define *quotient*. (Define *cociente*.) _____

Name three multiples of 7. (Nombra tres múltiplos de 7.)

Chapter 3 Visual Kinesthetic Learning   VKV9

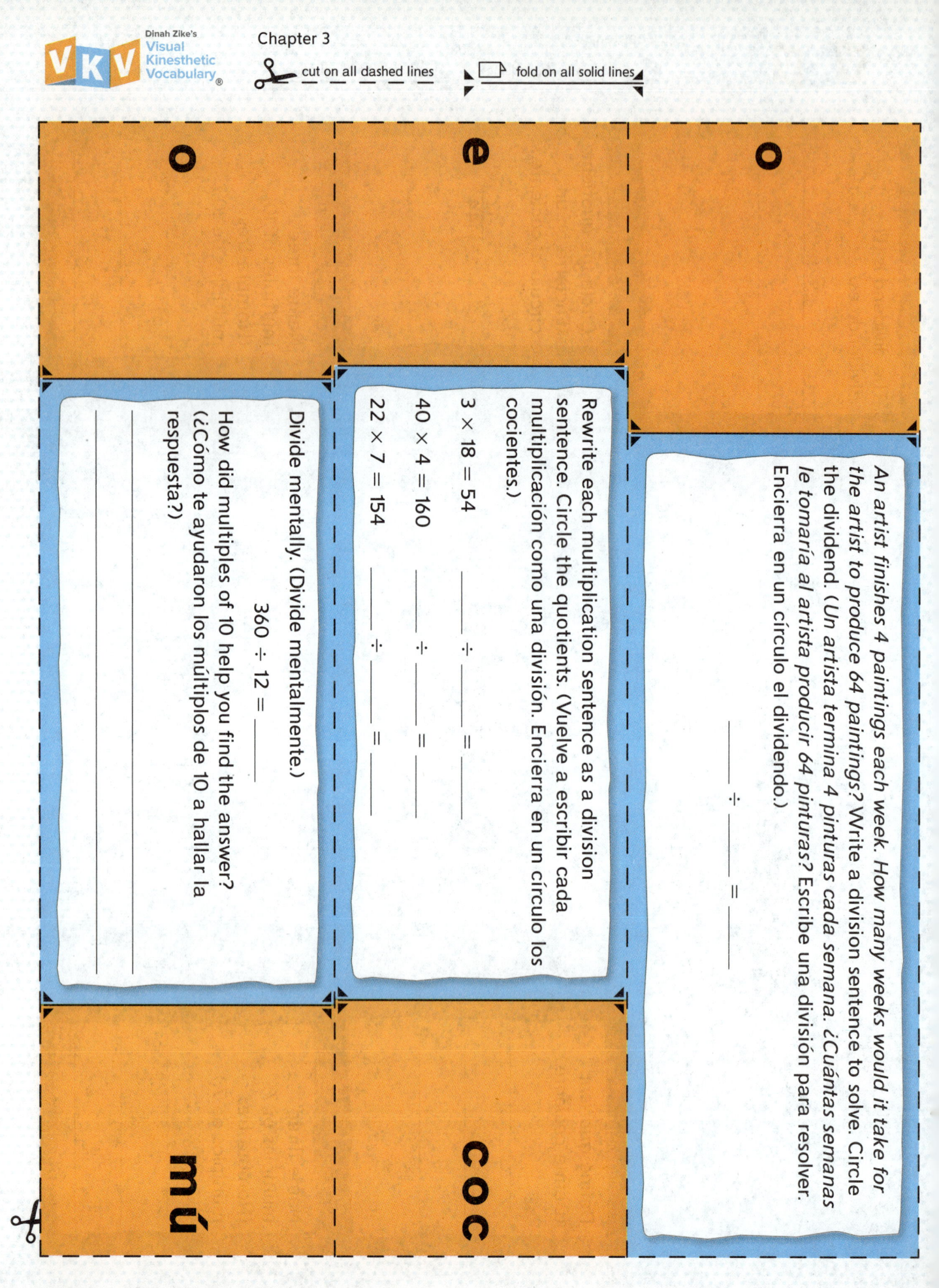

Chapter 3

cut on all dashed lines — fold on all solid lines

o | e | o

An artist finishes 4 paintings each week. How many weeks would it take for the artist to produce 64 paintings? Write a division sentence to solve. Circle the dividend. (Un artista termina 4 pinturas cada semana. ¿Cuántas semanas le tomaría al artista producir 64 pinturas? Escribe una división para resolver. Encierra en un círculo el dividendo.)

____ ÷ ____ = ____

Rewrite each multiplication sentence as a division sentence. Circle the quotients. (Vuelve a escribir cada multiplicación como una división. Encierra en un círculo los cocientes.)

3 × 18 = 54   ____ ÷ ____ = ____
40 × 4 = 160   ____ ÷ ____ = ____
22 × 7 = 154   ____ ÷ ____ = ____

Divide mentally. (Divide mentalmente.)

360 ÷ 12 = ____

How did multiples of 10 help you find the answer? (¿Cómo te ayudaron los múltiplos de 10 a hallar la respuesta?)

_____

mú | coc

VKV10 Chapter 3 Visual Kinesthetic Learning

Chapter 5

✂ cut on all dashed lines   ⬜ fold on all solid lines

Use the Associative Property of Addition to find the sum mentally. Show your steps. (Usa la propiedad asociativa de la suma para hallar mentalmente la suma. Muestra los pasos que tomaste.)

0.7 + 3 + 5.3 = _____

Associative Property of Addition

inverse operations

_____ is the inverse operation of addition. (_____ es la operación inversa de la suma.)

Inverse operations are operations that _____ each other. (Las operaciones inversas son operaciones que _____ entre sí.)

Chapter 5 Visual Kinesthetic Learning   VKV11

Chapter 5

inversas

operaciones

propiedad asociativa de la suma

Which example shows the Associative Property of Addition? Circle the answer. (¿Cuál ejemplo muestra la propiedad asociativa de la suma? Encierra en un círculo la respuesta.)

1. 3.3 + (5.7 + 4.1) = (3.3 + 5.7) + 4.1
2. 1.8 + 4 = 5.8 and 4 + 1.8 = 5.8
3. 9.6 + 0 = 9.6

Solve. Use an inverse operation to check your answer. (Resuelve. Usa una operación inversa para comprobar tu respuesta.)

   52.14   25
 − 2.7   − 12.6

Write the correct terms in the sentence. (Escribe las palabras correctas en la oración.)

The Associative Property of Addition states that the way in which numbers are _____ does not change the _____. (La propiedad asociativa en la suma establece que la manera en la que los números están _____ no cambia la _____.)

VKV12   Chapter 5 Visual Kinesthetic Learning

Chapter 6

propiedad conmutativa

Associative Property of Multiplication

Use the Associative Property of Multiplication to solve mentally. Show your steps. (Usa la propiedad asociativa de la multiplicación para resolver mentalmente. Muestra los pasos que tomaste.)

1.5 × 9 × 4 = ___

Chapter 6

cut on all dashed lines   fold on all solid lines

propiedad asociativa de la multiplicación

Commutative Property

Which property would you use to find the unknown in the equation below? (¿Cuál propiedad usarías para hallar la incógnita en la siguiente ecuación?)

____ × 2.4 = 2.4 × 9

____ Property of Multiplication

(propiedad ____ de la multiplicación)

VKV14   Chapter 6 Visual Kinesthetic Learning

Chapter 7

order of operations

Evaluate the expression. (Evalúa la expresión.)

$5^3 + (3 \times 5) - 42 = $ _____

numerical expression

Is 2^3 a numerical expression? Explain your answer. (¿Es 2^3 una expresión numérica? Explica tu respuesta.) _____

Write an example of a numerical expression. (Escribe un ejemplo de una expresión numérica.) _____

coordinate plane

Circle the origin in the coordinate plane. (Encierra en un círculo el origen en el plano de coordenadas.)

Chapter 7

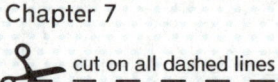

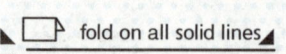

orden de las operaciones

Write 1–4 to show the correct order of operations. (Escribe los números del 1 al 4 para mostrar el orden correcto de las operaciones.)

___ Add and subtract in order from left to right. (Sumar y restar en orden de izquierda a derecha.)

___ Perform operations in parentheses. (Efectuar las operaciones entre paréntesis.)

___ Find value of exponents. (Hallar el valor de los exponentes.)

___ Multiply and divide in order from left to right. (Multiplicar y dividir en orden de izquierda a derecha.)

expresión numérica

Rewrite each numerical expression in a different way. (Vuelve a escribir las expresiones numéricas de otra manera.)

$(3 \times 2) + 5$
_____

$4 + 6 + 4 + 6$
_____

$(2 \times 7) - (5 \times 2)$
_____

plano de coordenadas

Locate and name the ordered pairs in the coordinate plane. (Localiza y representa los pares ordenados en el plano de coordenadas.)

A (___ , ___)
B (___ , ___)
C (___ , ___)
D (___ , ___)

VKV16 Chapter 7 Visual Kinesthetic Learning

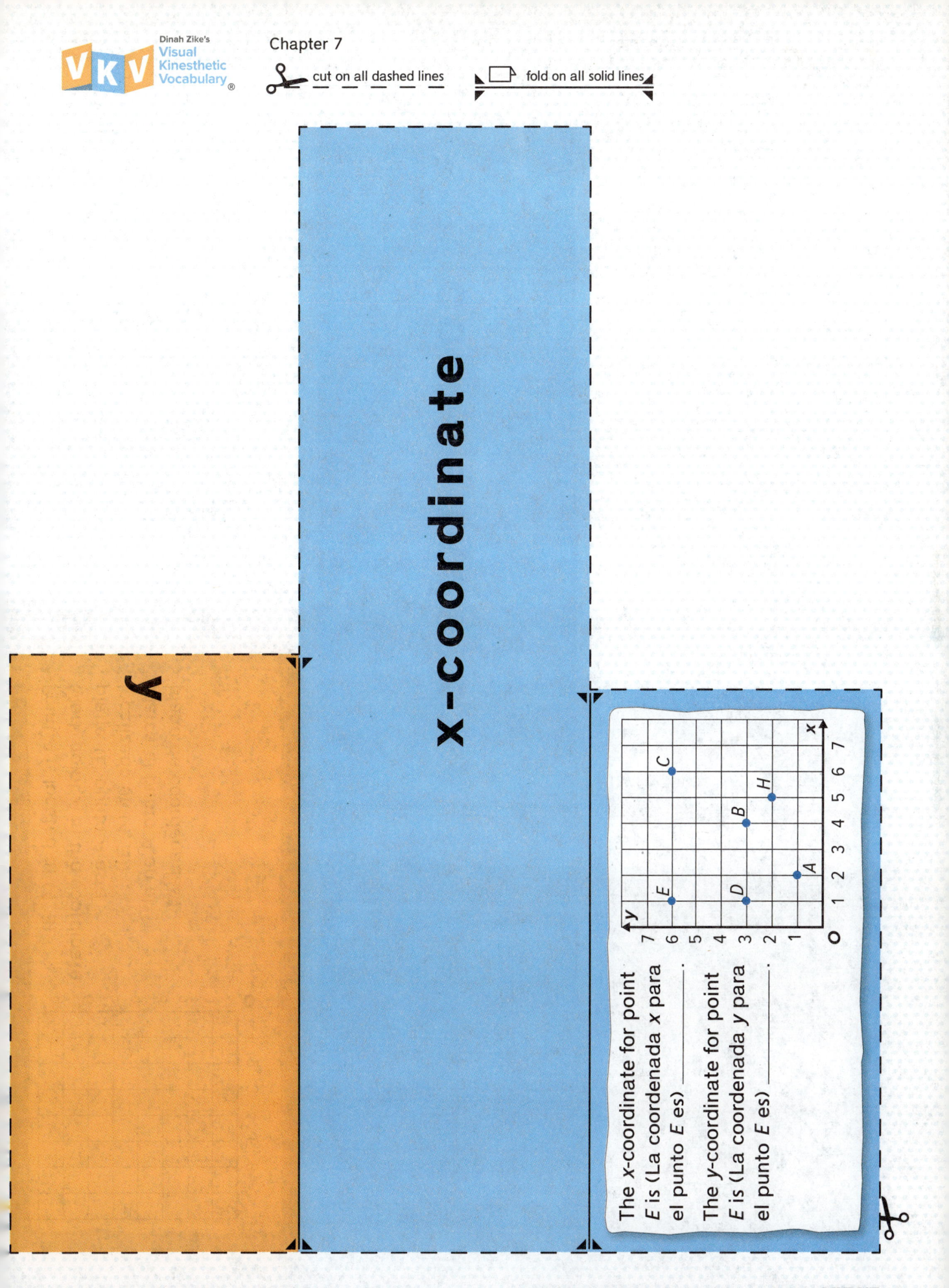

Chapter 7

cut on all dashed lines   fold on all solid lines

x-coordinate

y

The x-coordinate for point E is (La coordenada x para el punto E es) _____.

The y-coordinate for point E is (La coordenada y para el punto E es) _____.

Chapter 7 Visual Kinesthetic Learning VKV17

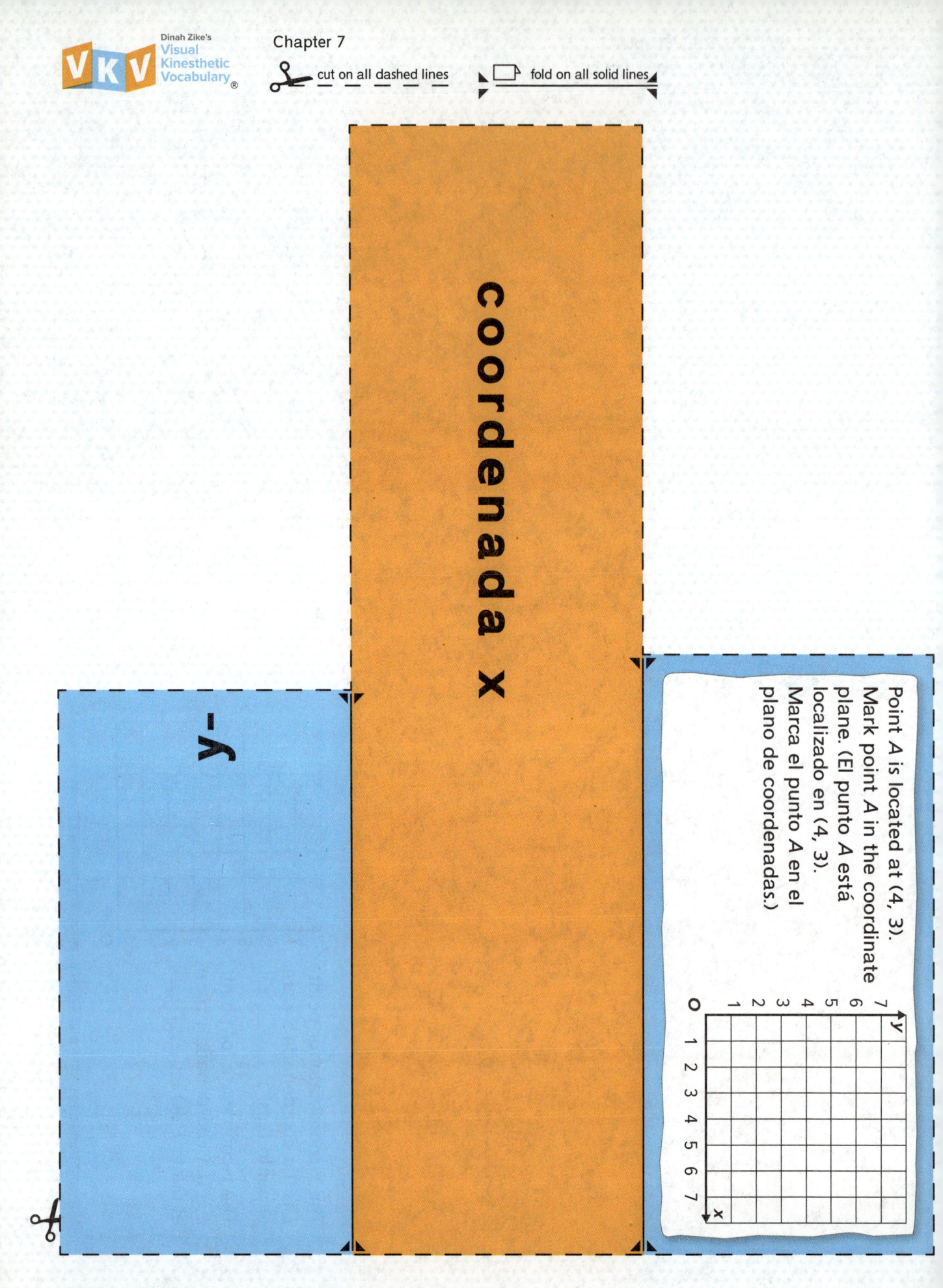

Chapter 7

cut on all dashed lines   fold on all solid lines

coordenada x

y

Point A is located at (4, 3). Mark point A in the coordinate plane. (El punto A está localizado en (4, 3). Marca el punto A en el plano de coordenadas.)

VKV18   Chapter 7 Visual Kinesthetic Learning

Chapter 8

greatest common factor (GCF)

What is the first step in finding the GCF for a set of numbers? (¿Cuál es el primer paso para hallar el máximo común divisor de un conjunto de números?)

equivalent fractions

Circle the fraction that is equivalent to $\frac{7}{21}$. (Encierra en un círculo la fracción que es equivalente a $\frac{7}{21}$.)

$\frac{1}{3}$ $\frac{3}{7}$ $\frac{1}{4}$

Circle the fraction that is equivalent to $\frac{4}{5}$. (Encierra en un círculo la fracción que es equivalente a $\frac{4}{5}$.)

$\frac{8}{15}$ $\frac{2}{3}$ $\frac{20}{25}$

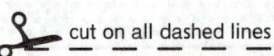

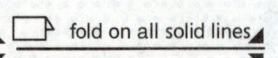

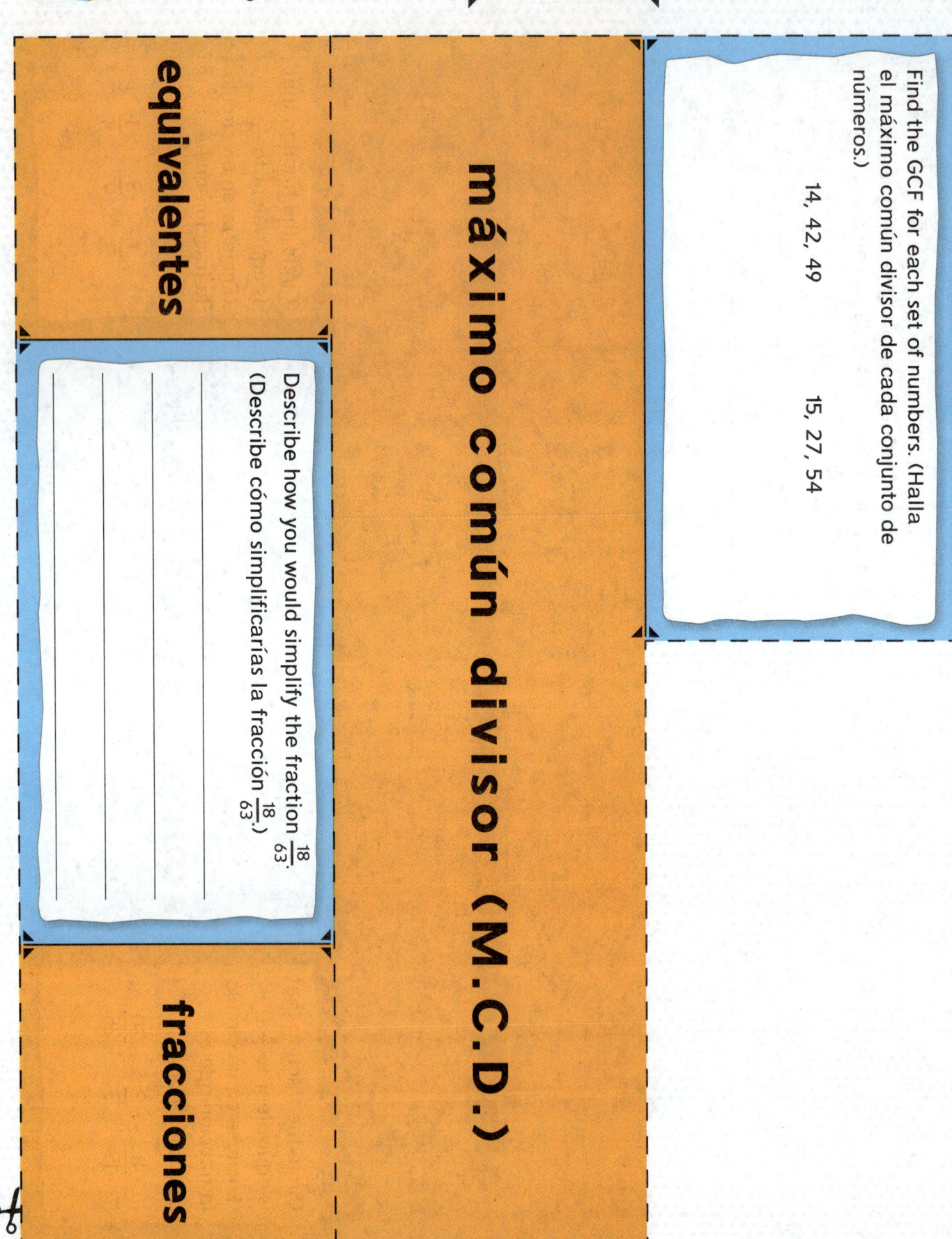

Chapter 8

máximo común divisor (M.C.D.)

equivalentes

fracciones

Find the GCF for each set of numbers. (Halla el máximo común divisor de cada conjunto de números.)

14, 42, 49

15, 27, 54

Describe how you would simplify the fraction $\frac{18}{63}$. (Describe cómo simplificarías la fracción $\frac{18}{63}$.)

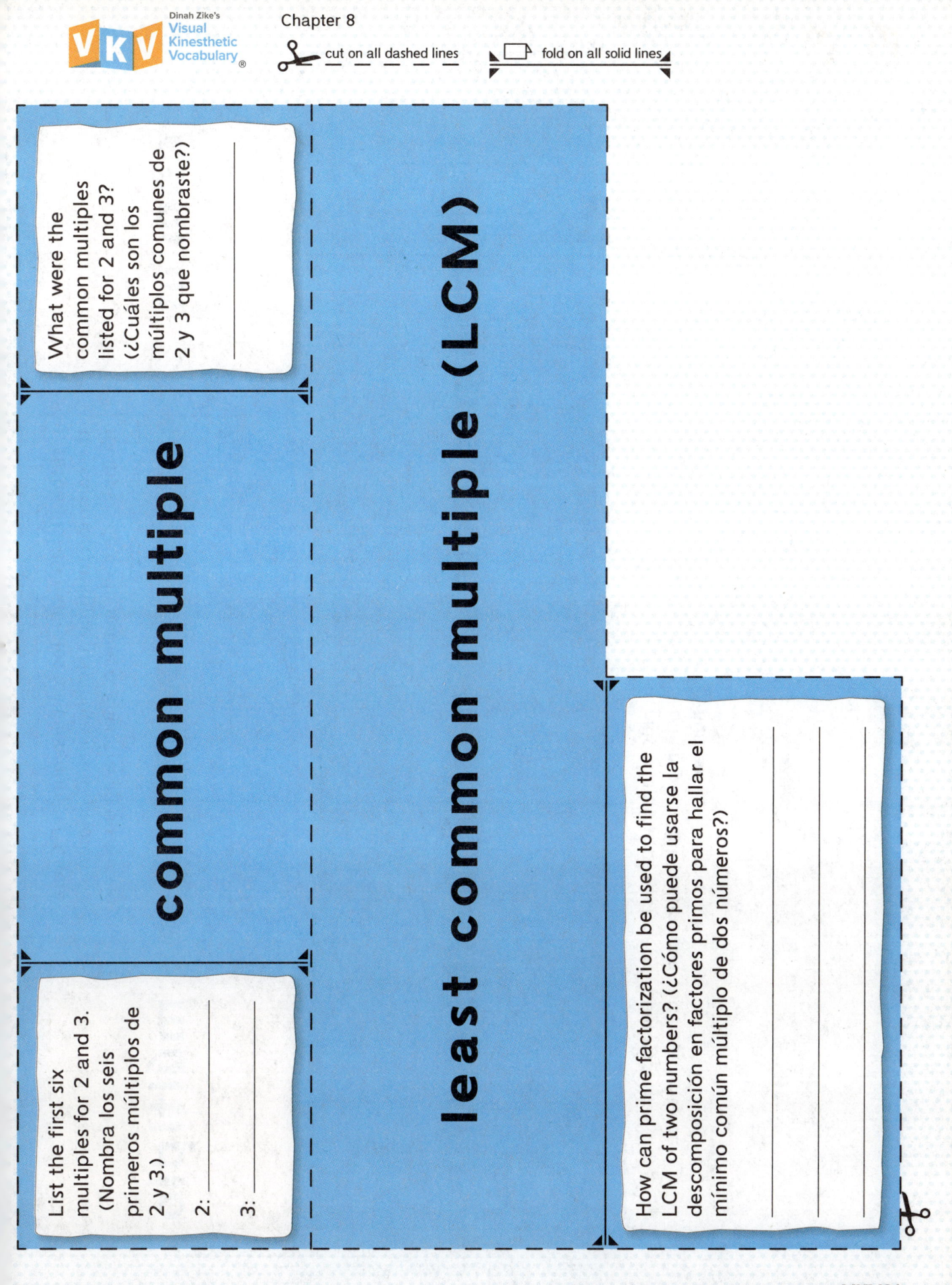

Chapter 8

cut on all dashed lines fold on all solid lines

What were the common multiples listed for 2 and 3? (¿Cuáles son los múltiplos comunes de 2 y 3 que nombraste?) _____

common multiple

least common multiple (LCM)

List the first six multiples for 2 and 3. (Nombra los seis primeros múltiplos de 2 y 3.)
2: _____
3: _____

How can prime factorization be used to find the LCM of two numbers? (¿Cómo puede usarse la descomposición en factores primos para hallar el mínimo común múltiplo de dos números?)
_____
_____
_____
_____

Chapter 8 Visual Kinesthetic Learning VKV21

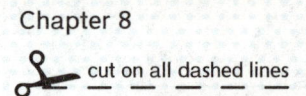

Chapter 8

común

múltiplo

mínimo común múltiplo (M.C.M.)

List the first 10 multiples for each number. Circle the common multiple. (Nombra los 10 primeros múltiplos de cada número. Encierra en un círculo el múltiplo común.)

5: _____

9: _____

Find the LCM of each set of numbers. (Halla el mínimo común múltiplo de cada conjunto de números.)

4, 5, 8   8, 9, 10
_____   _____

VKV22 Chapter 8 Visual Kinesthetic Learning

Chapter 11

cut on all dashed lines — *fold on all solid lines*

Circle the word with the same meaning as *convert*. (Encierra en un círculo la palabra que significa lo mismo que *expresar de otra manera*.)

- measure (medir)
- multiply (multiplicar)
- divide (dividir)
- change (convertir)

Define *capacity*. (Define *capacidad*.)

_____
_____
_____
_____

Name something that is measured in fluid ounces. (Nombra algo que se mida en onzas líquidas.)

_____
_____
_____
_____

---

convert

capacity

fluid ounce

How is a fluid ounce different from an ounce? (¿En qué se diferencia una onza líquida de una onza?)

_____
_____
_____
_____

Chapter 11 Visual Kinesthetic Learning VKV23

Chapter 11

ir

dad

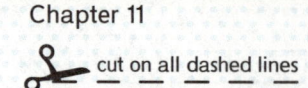

líquida

onza

Customary Units of Length
1 ft = 12 in.
1 yd = 3 ft or 36 in.
1 mi = 5,280 ft or 1,760 yd

Complete the equations to convert each measurement. (Completa las ecuaciones para convertir cada medida.)

24 feet (pies) = _____ yards (yardas)

3,520 yards (yardas) = _____ miles (millas)

3 yards (yardas) = _____ inches (pulgadas)

1.5 feet (pies) = _____ inches (pulgadas)

Customary Units of Capacity
1 cup (taza)(c) = 8 fluid ounces (fl oz)
1 pint (pinta)(pt) = 2 c = 16 fl oz
1 quart (cuarto de galón) (qt) = 2 pt = 32 fl oz
1 gallon (galón) (gal) = 4 qt = 128 fl oz

Use <, >, or = to make each statement true. (Usa <, > o = para hacer que cada enunciado sea verdadero.)

3 pints (pintas) ◯ 1.5 gallons (galones)

1 quart (cuarto de galón) ◯ 3 cups (tazas)

4 cups (tazas) ◯ 1 quart (cuarto de galón)

Draw a line from each unit of capacity to its equivalent measurement in fluid ounces. (Traza una línea de cada unidad de capacidad a su medida equivalente en onzas líquidas.)

1 gallon (galón)   8 fl oz (onzas líquidas)
1 cup (taza)   16 fl oz (onzas líquidas)
1 quart (cuarto de galón)   32 fl oz (onzas líquidas)
1 pint (pinta)   128 fl oz (onzas líquidas)

Chapter 11

Name three things you might measure in kilometers. (Nombra tres cosas que podrías medir en kilómetros.)

There are ____ millimeters in 1 centimeter and ____ centimeters in 1 meter. (Hay ____ milímetros en 1 centímetro y ____ centímetros en 1 metro.)

The metric system is a ____ system of measurement. (El sistema métrico es un sistema de medición ____.)

kilometer

centimeter

metric system

Name two metric units of measurement. (Nombra dos unidades métricas de medida.)

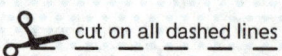

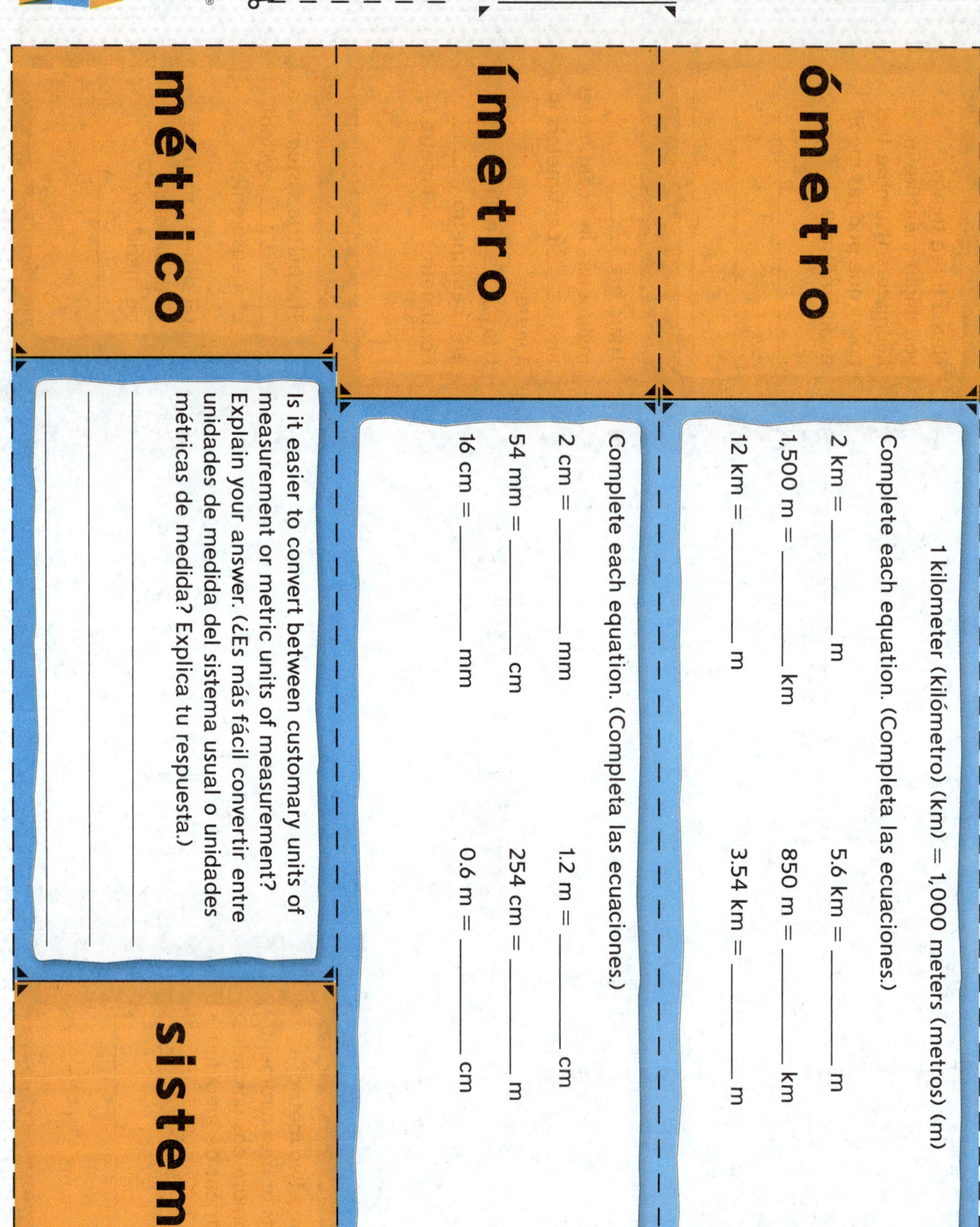

Chapter 11

métrico

ímetro

ómetro

sistema

1 kilometer (kilómetro) (km) = 1,000 meters (metros) (m)

Complete each equation. (Completa las ecuaciones.)

2 km = _____ m   5.6 km = _____ m
1,500 m = _____ km   850 m = _____ km
12 km = _____ m   3.54 km = _____ m

Complete each equation. (Completa las ecuaciones.)

2 cm = _____ mm   1.2 m = _____ cm
54 mm = _____ cm   254 cm = _____ m
16 cm = _____ mm   0.6 m = _____ cm

Is it easier to convert between customary units of measurement or metric units of measurement? Explain your answer. (¿Es más fácil convertir entre unidades de medida del sistema usual o unidades métricas de medida? Explica tu respuesta.)

_____
_____
_____
_____

VKV26 Chapter 11 Visual Kinesthetic Learning

Chapter 11

cut on all dashed lines   fold on all solid lines

gram

A _____ has a mass of about 1 gram.
(_____ tiene una masa de aproximadamente 1 gramo.)

A _____ has a mass of about 1 kilogram.
(_____ tiene una masa de aproximadamente 1 kilogramo.)

A liter is a unit of capacity. There are _____ milliliters in 1 liter.
(Un litro es una unidad _____ de capacidad. Hay _____ mililitros en 1 litro.)

liter

Grams and kilograms are metric units of _____ .
(Los gramos y los kilogramos son unidades métricas de _____ .)

Chapter 11 Visual Kinesthetic Learning   VKV27

Chapter 11

cut on all dashed lines   fold on all solid lines

ro

gramo

kilo

1 kilogram (kilogramo) (kg) = 1,000 grams (gramos) (g)

Use <, >, or = to make each statement true. (Usa <, > o = para hacer que cada enunciado sea verdadero.)

1.75 kg ◯ 17,500 g   350 g ◯ 0.35 kg   12 kg ◯ 1,200 g

Complete each equation. (Completa las ecuaciones.)

2 L = _____ mL   3.2 L = _____ mL

450 mL = _____ L   5,800 mL = _____ L

6.75 L = _____ mL   0.9 L = _____ mL

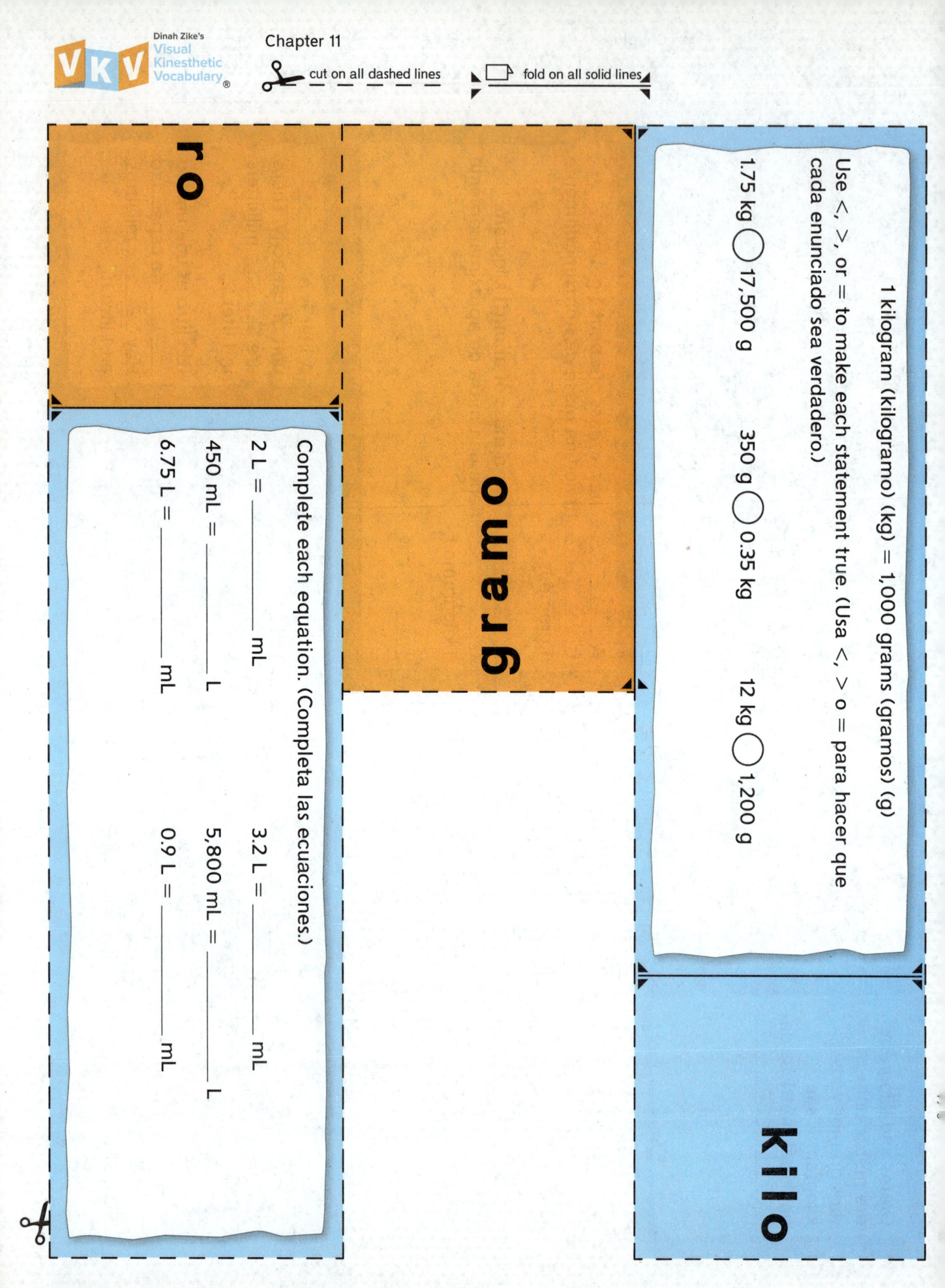

VKV28 Chapter 11 Visual Kinesthetic Learning

Chapter 12

cut on all dashed lines

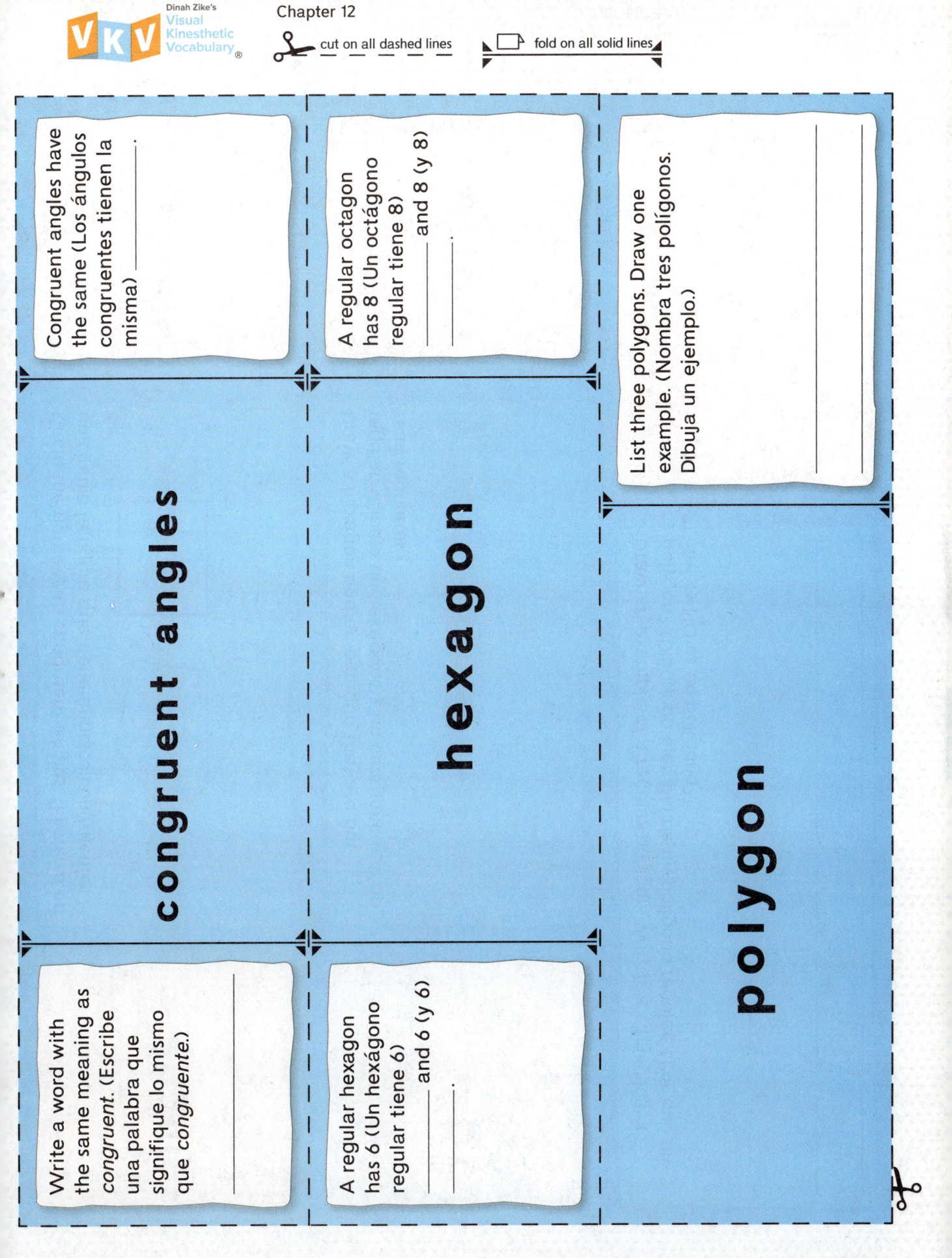

fold on all solid lines

Congruent angles have the same _____. (Los ángulos congruentes tienen la misma _____.)

A regular octagon has 8 _____ and 8 (y 8) _____.

List three polygons. Draw one example. (Nombra tres polígonos. Dibuja un ejemplo.)

congruent angles

hexagon

polygon

Write a word with the same meaning as *congruent*. (Escribe una palabra que signifique lo mismo que *congruente*.)

A regular hexagon has 6 _____ and 6 (y 6) _____.

Chapter 12 Visual Kinesthetic Learning VKV29

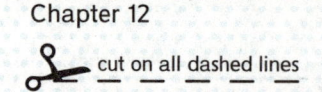

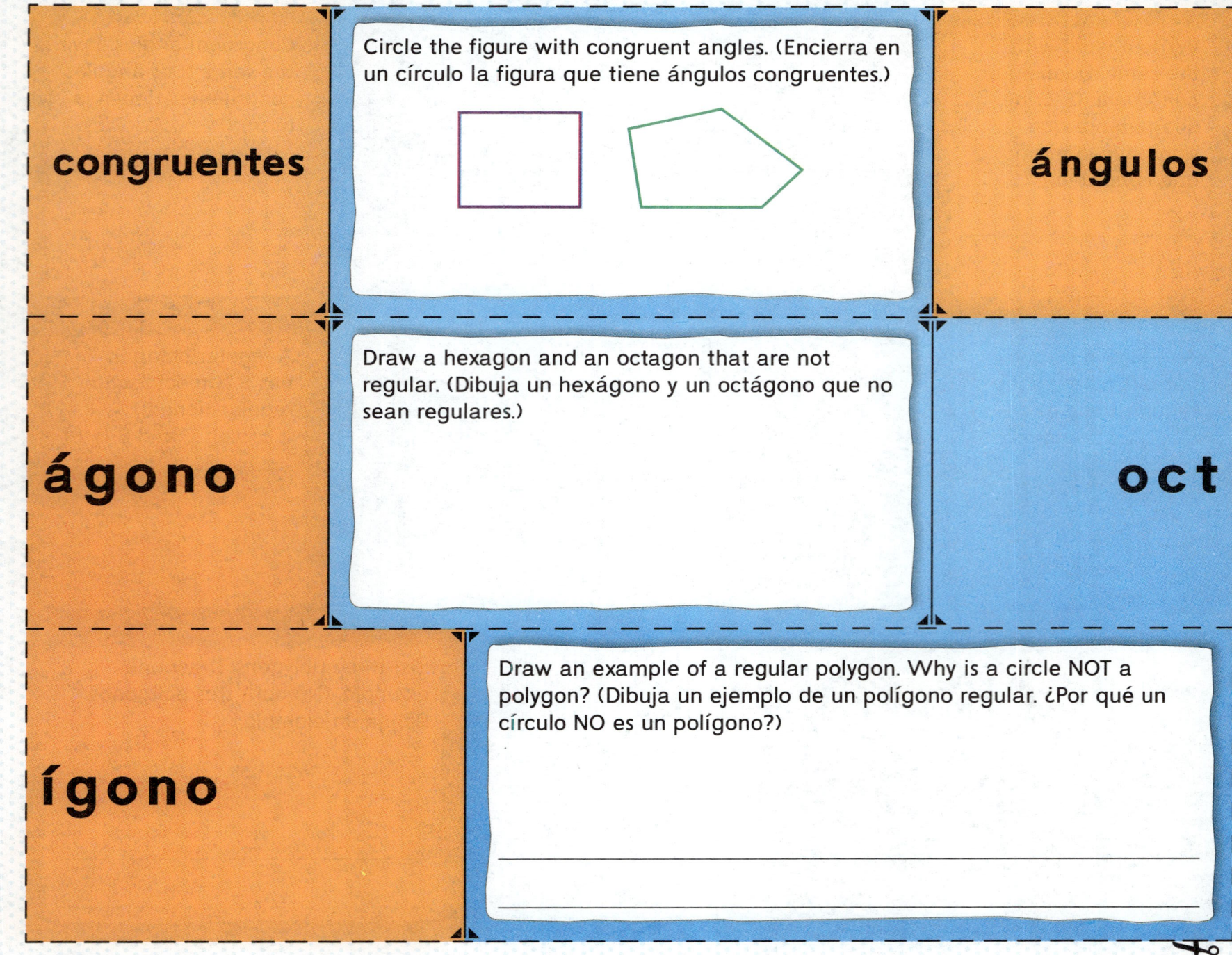

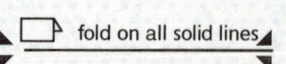

Chapter 12

cut on all dashed lines
fold on all solid lines

ígono

ágono

congruentes

oct

ángulos

Circle the figure with congruent angles. (Encierra en un círculo la figura que tiene ángulos congruentes.)

Draw a hexagon and an octagon that are not regular. (Dibuja un hexágono y un octágono que no sean regulares.)

Draw an example of a regular polygon. Why is a circle NOT a polygon? (Dibuja un ejemplo de un polígono regular. ¿Por qué un círculo NO es un polígono?)

VKV30 Chapter 12 Visual Kinesthetic Learning

Dinah Zike's **Visual Kinesthetic Vocabulary**®

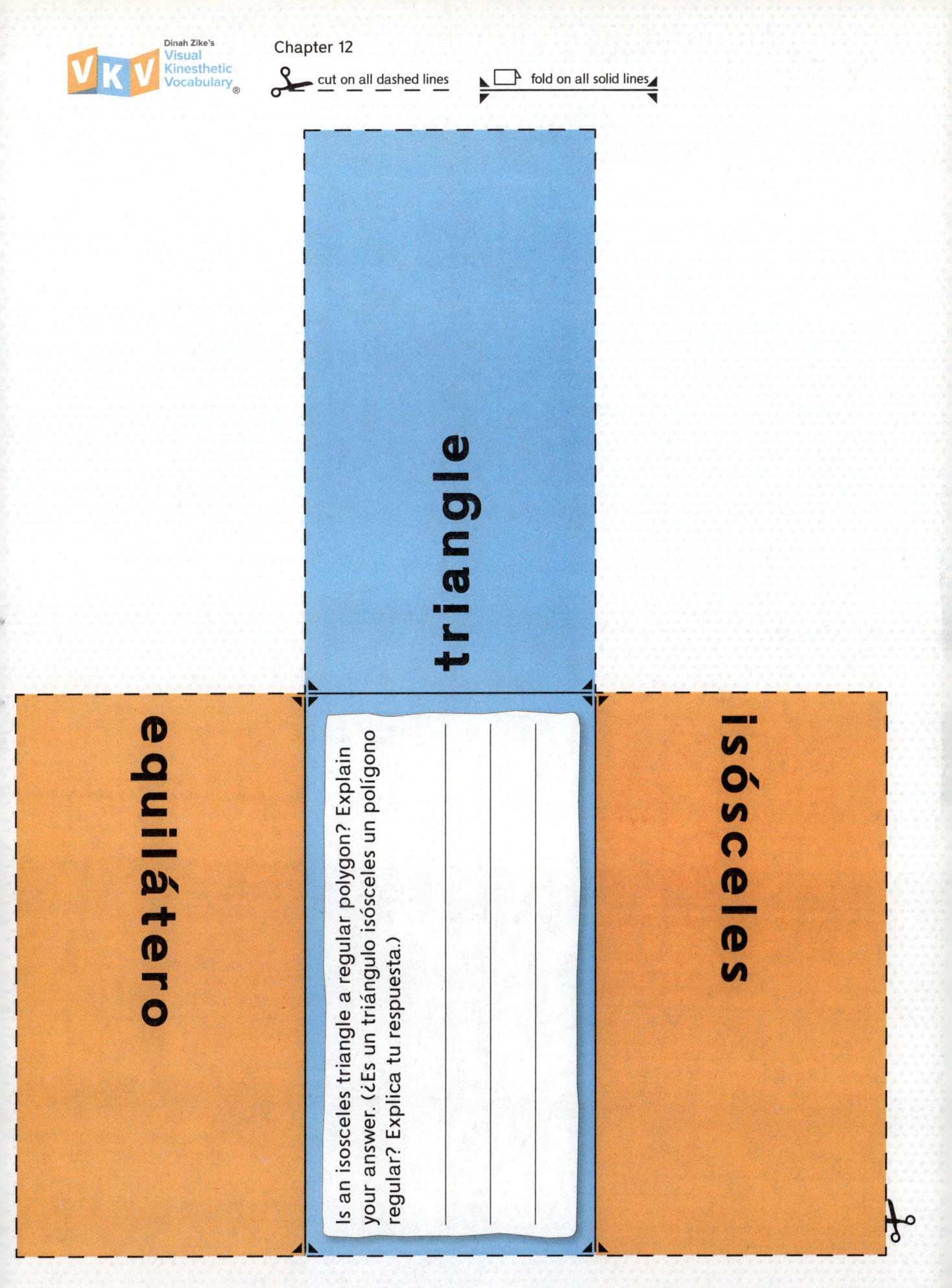

Chapter 12

✂ cut on all dashed lines   📄 fold on all solid lines

triangle

equilátero

Is an isosceles triangle a regular polygon? Explain your answer. (¿Es un triángulo isósceles un polígono regular? Explica tu respuesta.)

isósceles

Chapter 12 Visual Kinesthetic Learning   **VKV31**

Dinah Zike's Visual Kinesthetic Vocabulary®

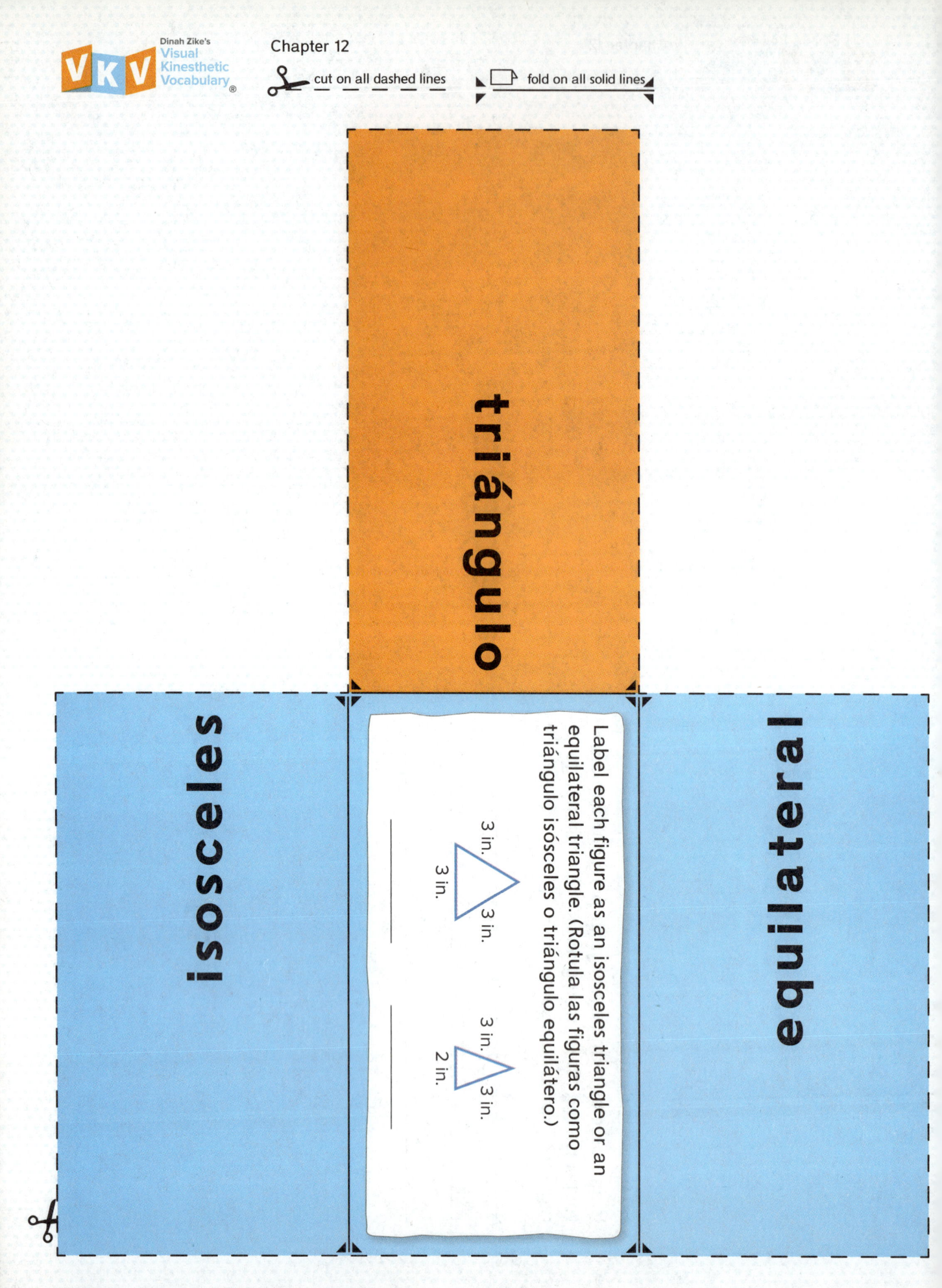

Chapter 12

✂ cut on all dashed lines   fold on all solid lines

triángulo

isosceles

equilateral

Label each figure as an isosceles triangle or an equilateral triangle. (Rotula las figuras como triángulo isósceles o triángulo equilátero.)

3 in.
3 in.
3 in.

3 in.
2 in.
3 in.

VKV32   Chapter 12 Visual Kinesthetic Learning

Chapter 12

✂ cut on all dashed lines   📄 fold on all solid lines

Draw an example of two congruent figures. (Dibuja un ejemplo de dos figuras congruentes.)

At least three faces of a prism are (Al menos tres caras de un prisma son) _____

What is the volume of a prism 2 units long, 3 units wide, and 2 units high? (¿Cuál es el volumen de un prisma de 2 unidades de largo, 3 unidades de ancho y 2 unidades de altura?) _____ cubic units (unidades cúbicas)

congruent figures

prism

cubic unit

Define congruent figures. (Define figuras congruentes.)

A cubic unit is used for (Una unidad cúbica se usa para) _____

Chapter 12 Visual Kinesthetic Learning   VKV33

Chapter 12

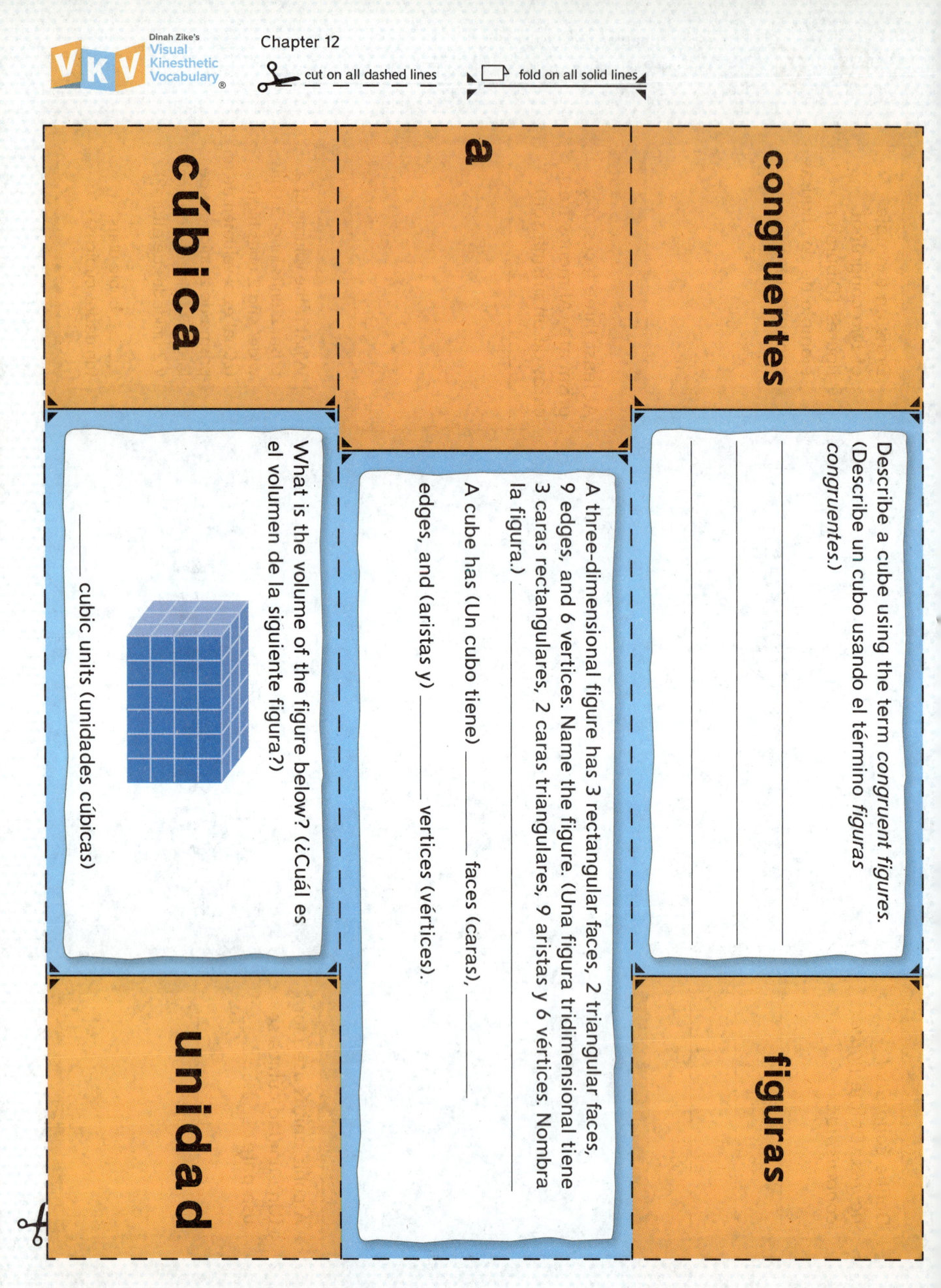

cúbica

congruentes

Describe a cube using the term *congruent figures*. (Describe un cubo usando el término *figuras congruentes*.)

A three-dimensional figure has 3 rectangular faces, 2 triangular faces, 9 edges, and 6 vertices. Name the figure. (Una figura tridimensional tiene 3 caras rectangulares, 2 caras triangulares, 9 aristas y 6 vértices. Nombra la figura.) _____

A cube has (Un cubo tiene) ____ faces (caras), ____ edges, and (aristas y) ____ vertices (vértices).

What is the volume of the figure below? (¿Cuál es el volumen de la siguiente figura?)

____ cubic units (unidades cúbicas)

unidad

figuras

VKV34 Chapter 12 Visual Kinesthetic Learning

VKV Answer Appendix

Chapter 1
VKV3
decimal/decimal point: Sample answer: A decimal has numbers to the right of a decimal point, in the tenths, hundredths, and thousandths places.
equivalent decimals: equal; 2.540

VKV4
demical: 0.03; 0.026; 0.7; 0.45; fifty-six hundredths
decimales equivalentes: 0.8 and 0.80; 05.2 and 5.20; 3.40 and 3.4

Chapter 2
VKV5
prime factorization: 2, 5; $2 \times 2 \times 3$
cubed: third
exponent: Sample answer: It shows how many times the base will be used as a factor.

VKV6
factorización prima: See students' work; $2 \times 2 \times 2 \times 3 \times 3 = 72$
al cubo: $4 \times 4 = 16$; $10 \times 10 \times 10 \times 10 = 10,000$; $6 \times 6 \times 6 = 216$; 6^3 should be circled
exponente: $4^2 = 16$; $2^3 = 8$; $5^2 = 25$

VKV7
compatible numbers: easy to compute mentally.
product: 18, 75, 1,000
estimate: rounding or compatible numbers

VKV8
números compatibles: 12,000; 2,000; 40,000; 10,000
producto: 45, 64, 16, 135, 370, 144
estimar: See students' work.

Chapter 3
VKV9
dividend: the number that is being divided.
quotient: The quotient is the answer to a division problem. See students' work.
multiple: Sample answers: 14, 21, 28; 20, 30, 40

VKV10
dividendo: $64 \div 4 = 16$; 64 should be circled.
cociente: $54 \div 3 = 18$; $160 \div 40 = 4$; $154 \div 7 = 22$. See students' work.
multiplo: 36; Sample answer: 360 is a multiple of 10, so I was able to divide 36 by 12 mentally, and then multiply the result by 10.

Chapter 5
VKV11
Associative Property of Addition: 9; See students' work.
inverse operations: undo; subtraction

VKV12
propiedad asociativa de la suma: grouped, sum (asociados, suma); 1.
operaciones inveresas: 49.44; 14.4; See students' work.

Chapter 6
VKV13
Associative Property of Multiplication: 54; See students' work.

VKV14
propiedad asociativa de la multiplicacion: Commutative

Chapter 7
VKV15
order of operations: 48
numerical expressions: See students' work; Yes, because the exponent implies an operation.
coordinate plane: See students' work.

VKV16
orden de las operaciones: 4, 1, 2, 3
expresíon numérica: Sample answers: $3 + 3 + 5$; $(4 \times 2) + (6 \times 3)$; $(7 + 7) - (5 + 5)$
plano de coordenadas: A(6, 4); B(5, 1); C(4, 5); D(2, 3)

VKV17
x-coordinate: 1; 6

VKV18
coordenada x: See students' work.

Chapter 8
VKV19
Greatest Common Factor: prime factorization of each number
equivalent fractions: $\frac{20}{25}$; $\frac{1}{3}$

VKV20
maximo comun divisor: 7; 3
fracciones equivalentes: Sample answer: Find the GFC of 18 and 63; then divide the numerator and denominator by the GFC.

VKV21
common multiple: 2, 4, 6, 8, 10, 12; 3, 6, 9, 12, 15, 18; 6 and 12
least common multiple: Sample answer: Use prime factorization to find the prime factors; then multiply all the factors together, using any common factors only once.

VKV22
comun multiplo: 5, 10, 15, 20, 25, 30, 35, 40, 45, 50; 9, 18, 27, 36, 45, 54, 63, 72, 81, 90; 45
minimo comun multiplo: 40; 360

Chapter 11
VKV23
convert: change
capacity: Capacity is the measure of how much a container can hold.
fluid ounce: A fluid ounce is a measure of capacity; an ounce is a measure of weight; the amount of liquid in a bottle.

VKV24
convertir: 8 yards; 108 inches; 2 miles; 18 inches
capacidad: <; >; =
onza líquida: 1 gallon = 128 fl oz; 1 cup = 8 fl oz; 1 quart = 32 fl oz; 1 pint = 16 fl oz

VKV25
kilometer: Sample answers: distance between towns; height of mountain; depth of ocean
centimeter: 10; 100
metric system: See students' work; decimal

VKV26
kilómetro: 2,000 m; 5,600 m; 1.5 km; 0.85 km; 12,000 m; 3,540 m
centímetro: 20 mm; 120 cm; 5.4 cm; 2.54 m; 160 mm; 60 cm
sistema métrico: Sample answer: Metric units of measurement; since it is a decimal system, the conversion can often be done mentally.

VKV27
gram: mass; Sample answers: paperclip; loaf of bread
lliter: metric; 1,000

VKV28
gramo: <; =; >
litro: 2,000 mL; 3,200 mL; 0.45 L; 5.8 L; 6,750 mL; 900 mL

Chapter 12
VKV29
congruent angles: Sample answer: same; degree measurement
hexagon: congruent angles; congruent sides
polygon: Sample answers: triangle, square, pentagon; See students' work.

VKV30
ángulos congruentes: See students' work.
octagono: See students' work.
poligono: See students' work; A circle is a closed figure, but it does not have straight sides.

VKV31
triangle: No; all sides of a regular polygon are congruent, but an isosceles triangle has only two congruent sides.

VKV32
traingulo: See students' work.

VKV33
congruent figures: Figures that have the same size and shape; See students' work.
prism: congruent rectangles
cubic unit: measuring volume; 12 cubic units

VKV34
figuras congruentes: Sample answer: A cube is a three-dimensional figure with 6 sides that are all congruent two-dimensional figures.
prisma: triangular prism; 6 faces, 12 edges, 8 vertices
unidad cúbica: 72 cubic units